KT-453-836

Neuroendocrine Regulation of Behavior

JAY SCHULKIN

1501987

CHESTER COLLEGE

ACC. No.	DEPT.
0111096	

CLASS No.

612.8 SCH

LIBRARY

CAMBRIDGE
UNIVERSITY PRESS

PUBLISHED BY THE PRESS SYNDICATE OF THE UNIVERSITY OF CAMBRIDGE
The Pitt Building, Trumpington Street, Cambridge CB2 1RP, United Kingdom

CAMBRIDGE UNIVERSITY PRESS
The Edinburgh Building, Cambridge CB2 2RU, UK   http://www.cup.cam.ac.uk
40 West 20th Street, New York, NY 10011-4211, USA   http://www.cup.org
10 Stamford Road, Oakleigh, Melbourne 3166, Australia

© Cambridge University Press 1999

This book is in copyright. Subject to statutory exception
and to the provisions of relevant collective licensing agreements,
no reproduction of any part may take place without
the written permission of Cambridge University Press.

First published 1999

Printed in the United States of America

Typeset in Melior 10/13.5 pt, in Penta Software Inc. [RF]

A catalog record for this book is available from the British Library

Library of Congress Cataloging-in-Publication Data

Schulkin, Jay.
Neuroendocrine regulation of behavior / Jay Schulkin.
p. cm.
Includes bibliographical references and index.
ISBN 0-521-45385-2 (hbk.). – ISBN 0-521-45985-0 (pbk.)
1. Psychoneuroendocrinology.   2. Steroid hormones – Physiological
effect.   3. Neuropeptides – Physiological effect.   I. Title.
QP356.45.S38   1998
612.8 – dc21   97-41739
                                                              CIP

ISBN 0 521 45385 2 hardback
ISBN 0 521 45985 0 paperback

Accession no.
01111096

Neuroendocrine Regulation of Behavior

LIBRARY

Tel: 01244 375444   Ext: 3301

CHESTER COLLEGE

This book is to be returned on or before the last date stamped below. Overdue charges will be incurred by the late return of books.

CANCELLED
− 3 OCT 2005
− 6 FEB 2006
CANCELLED

In thi
ciple
book i
hormo
then b
animal

The
phic be
tion; th
fourth
fifth is
endogen
emphasi
acting d
variety o

Senior
ogy, psyc
of hormo
medical h

Jay Schul
Biophysic
research a
tional Inst
College of
including

WITHDRAWN

Dedicated to the memory of my maternal grandparents,
Nettie and David Linder,
and my senior colleague, Eliot Stellar

Contents

Acknowledgments

As I was growing up I spent a good deal of time at the Bronx Zoo. Visiting the zoo and the American Museum of Natural History in New York City and playing the clarinet were my favorite preadolescent activities until, as a teenager, my life took a new direction. I well remember those happy hours spent at the zoo, and as an adult I have often visited zoos wherever I have traveled. It is my way of being close to different kinds of animals. What sorts of naturalists would go to zoos? City kids, as I once was.

The origin of this interest perhaps can be traced back to a book my grandparents gave me for Hanukkah when I was seven. It was entitled *All about Animals and Their Young*. My interest was further cultivated by three senior colleagues and mentors, now gone: Alan Epstein, Eliot Stellar, and George Wolf. I feel that I have had some of the best colleagues the world can offer. Bruce McEwen at Rockefeller University has been a great help to me. I thank my colleagues at the National Institute of Mental Health, particularly Phil Gold and Dave Jacobowitz, and at Georgetown University and the American College of Obstetricians and Gynecologists for their instruction and assistance.

My association with Olav Oftedal of the Smithsonian Institution and the Washington Zoo has allowed me to return to my first love. I now go to the zoo here regularly for research, and sometimes I bring my daughter. Our colleagueship is very much appreciated.

As always, my colleagues, friends, and family have been of great help in my scientific pursuits. In particular, I thank Kent Berridge, G. DeVries, Lauren Hill, Micah Leshem, Peter Marler, Adrian Morrison, Ralph Norgren, Dick O'Keeffe, Ellen Parr Oliver, Mike Power, Jeff Rosen, Alan Rosenwasser, Louis Schmidt, and Dana Trevas.

I also wish to thank Sarah Winans Newman. I met her once, talked to her maybe twice. A collaborative visit to Kent Berridge at the University of Michigan more than 10 years ago provided the opportunity for a conversation with her there. She talked about analyzing the hamster's behavioral responses to testosterone and discussed three sexually dimorphic brain nuclei in the forebrain (bed nucleus of the stria terminalis, medial preoptic nucleus, and medial nucleus of the amygdala) that were involved in steroid-induced behavioral effects, thus implanting the thought that these nuclei might also underlie what I was studying: aldosterone-induced sodium appetite.

As I said in my book *Sodium Hunger,* I have been greatly influenced by those with whom I have interacted and in many instances been involved with personally. Science is set in a human context of social relationships. One of the joys of science is participation in the fragile community of inquirers. To those who may feel slighted if I have not properly attended to their work, I apologize. This book, in addition, is not intended to cover all aspects of the neuroendocrine regulation of behavior. It builds and draws on the sodium-hunger perspective for how steroids and peptides influence behavior, and the chapters reflect only a partial purview of the field.

This book is not intended, therefore, as a general textbook in behavioral endocrinology. It does not cover everything nor pretend to. For a comprehensive textbook that recently appeared, I urge readers to look at Randy Nelson's book *An Introduction to Behavioral Endocrinology* or the volume *Behavioral Endocrinology,* edited by J. B. Becker, M. C. Breedlove, and D. Crews.

Many colleagues have been helpful in reviewing this manuscript and making it better. I thank them. Finally, I thank three influential women: my daughter, Danielle, my wife, April, and my mother, Rosalind, for their wonderful love.

Introduction

This is a book about behavioral neuroscience. As a subdiscipline of both biology and psychology, behavioral neuroscience spans the breadth of neuroscience (Gallistel, 1980). Neuroscience itself is a new discipline, a hybrid intellectual entity derived from a number of other disciplines, including physiological psychology (Lashley, 1938; Stellar, 1954), psychobiology (Richter, 1943; Beach, 1948), and ethology (Tinbergen, 1969; Lorenz, 1981), as well as neuroanatomy, neurophysiology, neuroendocrinology, neuropsychology, logic, linguistics, and philosophy.

Specifically, this book focuses on how hormones influence behavior by their actions in the brain. Phylogenetically ancient chemical messengers produced by endocrine glands and the brain, hormones exert their influences throughout the body. Biologists separate them into two major classes: peptide hormones, which typically act on cell membranes to produce their effects, and steroid hormones, which act on the nuclei of cells to promote protein synthesis. The effects of the first class usually are rapid, whereas those of the second class are slow, although these distinctions between peptides and steroids and their rapid and delayed behavioral effects are beginning to become blurred (e.g., Wehling, Eisen, and Christ, 1992). Both peptide and steroid hormones have profound effects on behavior.

Hormones such as insulin, cholecystokinin, and bombesin (essential for feeding behavior and food regulation) are produced in gastrointestinal organs. Aldosterone, angiotensin, and atrial natriuretic factor (hormones involved in mineral and water balance) are produced in the adrenal glands, kidney, and heart. Hormones like vitamin D, parathyroid, thyroid, and melatonin (essential for calcium balance, circadian rhythmicity, and mental health) are produced in the skin and parathyroid, thyroid, and pineal glands. Produced predominantly in the ovaries and testes, testosterone and estrogen are fundamental to gender and gender-specific expressions of sexual, aggressive, and other behaviors. Finally, adrenal corticotropic hormone (ACTH) and corticosterone (needed for glucose transport and adaptation to stress and other challenging situations) are produced in the pituitary and adrenal glands.

Partial lists of hormones produced by endocrine glands and neuropeptides produced in the brain are given in Tables I.1 and I.2. We now know that peptide hormones are produced both peripherally and in the central nervous system. For example, insulin, oxytocin, angiotensin, cholecystokinin, and natriuretic factor are produced both in endocrine cells in the periphery and in the brain. In the periphery, we call them hormones. In the brain, they are known as neuropeptides.

When neuropeptides are activated in the brain, they stimulate neural circuits that influence behavioral responses. Hormones, of course, provide only one part of the story. Experience and genetic disposition also influence the expressions of behavior, hormonal secretion, and brain function.

In all of its chapters, this book makes the point that behavior is a part of biology and as such is often involved in adapting to one's environment. The behavior of sodium ingestion, for example, is the outer end of maintaining the body's fluid balance, whereas kidney functioning is the inner end of conserving or secreting fluid (Denton, 1982; Schulkin, 1991a, b). Both activities are regulated by humoral signals. Behavior serves the same end point as physiology – in this case, homeostatic regulation of fluid and sodium balance.

Hormones influence how an animal responds to its environment, just as the environment influences hormonal responses. Seasons and sensory cues such as temperature, olfaction, and light, for instance, determine hormonal concentrations, which in turn influence behavior. One example is that testosterone levels in a variety of animals are seasonal and will influence sexual responsiveness and often territorial aggression (Wingfield, 1994). But social factors also influence these hormone levels. An animal's place in the social hierarchy, for example, can determine its levels of estrogen, corticosterone, and testosterone. One such example is seen in mice (*Mus musculus*) (e.g., Louch and Higginbotham, 1967) (Figure I.1).

Problem-solving

In order to survive and reproduce, animals spend most of their waking hours trying to solve the problems of constructing a coherent world in which to act (Marler, 1961; Peirce, 1992; Hauser, 1996) and to determine meaning (Smith, 1977; Cheney and Seyfarth, 1990). Key tools in problem-solving, hormones help pass information about the world and its meaning to animals, in addition to passing crucial information among different indi-

Table I.1.

Endocrine Gland	Major Hormones Secreted
Anterior pituitary	Growth hormone
	Prolactin
	Adrenocorticotropin (ACTH)
	Luteinizing hormone (LH)
	Thyroid-stimulating hormone (TSH)
	Follicle-stimulating hormone (FSH)
Neurointermediate lobe/posterior pituitary	Arginine vasopressin
	Oxytocin
	Endorphins, enkephalins
Pineal	Melatonin
Thyroid gland	Thyroxine
	Calcitonin
Parathyroid gland	Parathyroid hormone
Heart	Atrial natriuretic factor
Adrenal cortex	Glucocorticoids
	Mineralocorticoids
	Androgens
Adrenal medulla	Epinephrine
	Norepinephrine
Kidney	Renin
Skin/kidney	Vitamin D
Liver/lung	Preangiotensin/angiotensin
Pancreas	Insulin
	Glucagon
Stomach and intestines	Cholecystokinin
	Vasoactive intestinal peptide
	Bombesin
	Somatostatin
Gonads: ovary	Estrogen
	Progesterone
Gonads: testis	Testosterone
Macrophage, lymphocytes	Cytokines

Table I.2.

Some major neuropeptides and proteins
β-endorphin
Dynorphin
Enkephalin
Somatostatin
Corticotropin-releasing hormone
Urocortin
Atrial natriuretic factor
Bombesin
Glucagon
Vasoactive intestinal polypeptide
Vasotocin
Substance P
Neuropeptide Y
Neurotensin
Galanin
Calcitonin
Cholecystokinin
Oxytocin
Prolactin
Vasopressin
Angiotensin
Interleukin
Thyrotropin-releasing hormone
Gonadotropin-releasing hormone (GRH)
Luteinizing-hormone-releasing hormone (LHRH)
Neurotropin
Calretinin
Leptin

Classic neurotransmitters in the CNS
Catecholamines
   Dopamine
   Norepinephrine
   Epinephrine
Indoles
   Serotonin
   Melatonin
Cholinergic
   Acetylcholine
Amino acids
   γ-aminobutyric acid (GABA)
   Glutamate
   Aspartate

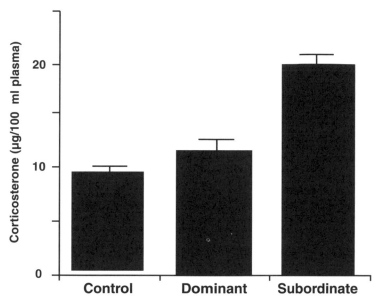

Figure I.1. Mean corticosterone concentrations for control (home alone), dominant, and subordinate mice. (Adapted from Louch and Higginbotham, 1967.)

viduals (Wingfield, 1994). Hormones also induce brain events that prompt animals to behave in ways dictated by environmental clues during problem-solving. When aldosterone is elevated, for example, many animals are able to recall where sources of salt are located and then use that knowledge to satisfy their sodium craving (Krieckhaus and Wolf, 1968), which in turn will restore their extracellular fluid volume. The hormone provides the link between a physiological problem and its behavioral solution.

Maintaining the internal milieu – known as homeostasis – means keeping adequate levels of glucose, sodium, calcium, and other substances within the body using physiological and behavioral strategies. A key to adaptation, behavior often anticipates what the body needs before the item is actually needed. Animals spend some of their cognitive time predicting or anticipating events (e.g., Rescorla and Wagner, 1972; Dickinson, 1980; Gallistel, 1990). Both anticipatory and reactive behaviors are needed to maintain homeostasis (Moore-Ede, 1986; Schulkin, McEwen, and Gold, 1994a).

Central Motive States

An important concept throughout this book is that of central motive states. What is a central motive state? It is a state of the brain (Lashley, 1938) that is knotted to desire and experience and the action that leads to fulfillment of that desire. Two examples will suffice at this outset: The neuropeptide angiotensin is linked to the central state of thirst and sodium appetite, and the neuropeptide corticotropin-releasing hormone is linked to the central state of fear. These neuropeptides, when they are synthesized in appropriate regions of the brain, help orchestrate the behavioral responses. Steroid hormones produced in the periphery influence the expression of neuropeptides (McEwen, 1995).

There are two prominent features of the central motive state of hunger, for example. The first is the appetitive phase – the search for the desired entity – and the second is the consummatory phase – actually ingesting the desired item or otherwise fulfilling a need. This distinction was expressed early on by the American naturalist Wallace Craig (1918) and later by the pragmatist philosopher John Dewey (1989). Within a short period of time it was incorporated within ethology (Tinbergen, 1969) and psychobiology (Beach, 1942) or physiological psychology (Stellar, 1954).

By influencing central states, hormones and their actions in the brain prepare an animal to perceive stimuli and behave in certain characteristic ways; they increase the likelihood of responding to environmental signals (e.g., Gallistel, 1980; Mook, 1987). One example will suffice: In a variety of animals (e.g., rats, *Rattus norvegicus*) estrogen promotes female sexual receptivity (e.g., Pfaff, 1980). Males can try as they may, but if the female is not ready, no sexual contact will occur. Similarly, if testosterone is not elevated in males of a number of species (e.g., Japanese quail, *Coturnix japonica*) (Balthazart, Castagna, and Ball, 1997), sexual behavior will not proceed. Of course, hormone levels in both sexes are linked, and females help induce estrogen production in other females, as do certain environmental stimuli (e.g., Adler, 1981). Behavioral responses do not occur in a vacuum; they are dependent upon environmental events (e.g., Hall, 1990).

When I lived in New York City, for example, for years I watched male pigeons dance – strutting, bulging at the neck, circling, and looking quite elegant. This behavior can be induced with the appropriate stimuli (the female's presence) by elevating testosterone levels. Exposure to estrogen-primed females can promote growth of the testes in males and increase testosterone secretion, as in white-crowned sparrows (*Zonotrichia leucophrys* Gambelii) (Moore, 1983) and common green iguanas (*Iguana iguana*) (Alberts, Pratt, and Phillips, 1992) (Figure I.2).

Both internal and external changes can raise the concentration of a hormone that acts on the brain to induce the readiness for a behavior. Behavior will follow both phases: the appetitive phase (What hurdles will one overcome to gain access to what one wants?) and the consummatory phase (the

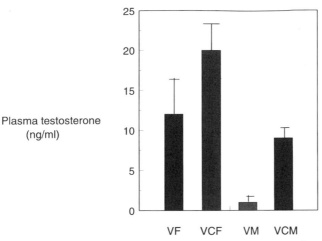

Figure I.2. Plasma testosterone concentrations in groups of juvenile male green iguanas with visual exposure to an adult female (VF), with both visual and chemical exposure to an adult female (VCF), with visual exposure to an adult male (VM), and with visual and chemical exposure to an adult male (VCM). (Adapted from Alberts et al., 1994.)

sexual act) (Goy and McEwen, 1980). By facilitating the expression of neuropeptides (e.g., oxytocin) and/or receptor sites (e.g., progestin), the gonadal steroid hormones lower the threshold at which a sexual response to environmental stimuli will be elicited by inducing central states in the brain (e.g., McEwen, Jones, and Pfaff, 1987).

Similar scenarios describe the courses of most hormonally induced behaviors discussed in this book. Such behaviors include the craving for and ingestion of food, water, and salt, as well as social attachments (including parental behavior) and fear. All these behaviors result from central motive states induced and sustained by humoral mechanisms. They all feature appetitive as well as consummatory phases of behavior.

Regarding the central state of fear, for example, readiness to perceive events as fearful can be induced by a combination of steroid and peptide hormones acting in the brain. The hormones color perception. The adrenal steroid hormones, cortisol or corticosterone, along with the neuropoptide corticotropin-releasing hormone, are linked to the activation and maintenance of fear (e.g., Kalin and Shelton, 1989; Koob et al., 1993). Both kinds of hormones act on limbic sites (e.g., Swanson et al., 1983). When both of these hormones are elevated, fear is one likely outcome.

Among humans, a subset of young children with elevated cortisol levels are more likely to be inhibited and fearful (e.g., Kagan, Reznick, and Snidman, 1988; Gunnar et al., 1989), demonstrating exaggerated startle responses when presented with acoustic stimuli (Schmidt et al., 1997). Perhaps they are prepared to see the world in fearful ways because of cortisol's action in the brain to induce corticotropin-releasing hormone (Swanson and Simmons, 1989; Makino, Gold, and Schulkin, 1994a,b). One neural site underlying fear is the amygdala (e.g., LeDoux, 1987, 1996). Corticosterone's induction of corticotropin-releasing hormone in the amygdala prepares one to perceive danger. The steroid facilitates the expression of the neuropeptide, which increases the likelihood that fear will be experienced.

Table I.3 depicts a number of model systems in which steroids and neuropeptides interact in the regulation of behavior. Some are well established, others much less so.

Functionalism and Evolution

I stress in each chapter that steroids and peptides can influence central states that are typically linked to functional requirements. Consider sodium hunger. The central state of sodium craving is linked to the search for and identification of sources of sodium. The state of sodium craving is initiated by the neuropeptide angiotensin in a number of species (e.g., Buggy and Fisher, 1974). Nerves that recognize sodium (7th nerve) are activated, memories of where sources of sodium may be located (Krieckhaus, 1970) are triggered, and a migration process is expressed for where sodium may be found (Denton, 1982). Forebrain sites orchestrate the behavioral responses whose function is the restoration of the body's sodium balance (Grill, Schulkin, and Flynn, 1986).

But the behavioral state does not exist in a vacuum; it depends upon the environment (e.g., Hinde, 1982). Adaptation reflects an environment in which an animal is trying to cope and provide frameworks of coherence.

"Functionalism" is a characteristic of the problem-solving proclivities of animal life that is reflected by the activation of central states. The evolutionary consideration is apparent (Hinde, 1982). Nature has often linked behavioral responses, in addition to systemic physiological ones, in maintaining bodily health (Richter, 1956). Problem-solving and adaptation are the hallmarks of functionalism amidst evolutionary pressures.

The central state of fear is linked with perceiving and then avoiding dangerous situations (LeDoux, 1995, 1996). The behavioral function of fear is to avoid danger. This response is ancient, and many kinds of land-dwelling mammals, birds, and reptiles demonstrate the phenomenon.

There are two senses of "functionalism" that I want to emphasize. Both senses are rooted in biology. One sense reflects problem-solving internally (e.g., the function of the kidney in maintaining sodium balance). The second is

Table I.3. *Paradigmatic Examples of Steroid Effects on Neuropeptide Expression and Behavior*

Steroid Hormone	Neural Peptides	Behavior
Estrogen Progesterone	Oxytocin LHRH	↑ Female sexual behavior
Testosterone Estrogen	Vasopressin	↑ Flank marking, aggression, territorial defense
Testosterone	Tyrosine hydroxylase (converting enzyme) Substance P	↑ Male sexual behavior
Estrogen Testosterone	Vasotocin	↑ Sex behavior (amphibians), bird song
Estrogen Testosterone	Prolactin Oxytocin Vasopressin	↑ Parental and attachment behaviors
Aldosterone Corticosterone	Angiotensin	↑ Water and sodium appetite
Corticosterone	Corticotropin-releasing hormone	↑ Water and sodium appetite ↑ Fear, anxiety, and depression
Corticosterone	Neuropeptide Y	↑ Food intake

behavioral: the way in which behavior functions in allowing animals to adapt to environmental niches (e.g., Parrott and Schulkin, 1993; cf. Rey, 1997).

Functionalism also recognizes the distinction between (1) what the stuff is made of and where it is located and (2) the fact that, in principle, different kinds of tissue can generate aspects that will initiate the same function (Rey, 1997). That is, one does not have to have a suprachiasmatic nucleus of the hypothalamus to demonstrate circadian rhythmicity (Rosenwasser and Adler, 1986). Single-cell organisms with no hypothalamus can do the same. But the suprachiasmatic region of the hypothalamus is the central circadian pacemaker in mammals and is essential in generating coordinated behavioral and endocrine responses.

Neonatal and Postnatal Developmental Events

A fundamental concept in the neuroscience of hormones and behavior is that of the impact of steroid hormones on the organization of brain morphology during gestation, as well as during the postnatal period, which influences the future expression of a variety of behaviors. In song sparrows, for example, adequate testosterone concentrations during certain stages of development are essential for the full expression of song when the birds become adults; manipulation of the hormone in females can prompt them to render song that otherwise they would not express (e.g., Marler et al., 1988) (Figure I.3). One hypothesis is that song reflects the induction of vasotocin expression in the brain (Voorhuis, De Kloet, and De Wied, 1991). Vasotocin, an analogue of vasopressin in nonmammalian species, when injected into the brain in female white-crowned sparrows, elicits song (Maney, Goode, and Wingfield, 1997). Perhaps the gonadal steroid hormones induce vasotocin gene expression in the brain during early periods of development that are later activated for song production. Steroid hormones, however, can have organizational effects on the brain during gestation, during the postnatal period, and even in adulthood (Nespor et al., 1996; cf. Phoenix et al., 1959; Arnold and Breedlove, 1985; Crews and Moore, 1986).

Neuropeptide Gene Expression in the Brain Is Under the Influence of Steroid Hormones

Negative restraint of hypothalamic hormones is a well-known phenomenon. This is characterized, for example, in the restraint of hypothalamic corticotropin-releasing hormone and its activation of adrenal corticotropic hormone (ACTH) by adrenal steroid hormones (Munck, Guyre, and Holbrook, 1984). But there is another form of regulation for many of these neuropep-

Figure I.3. Drawing of a song sparrow from the work of Margaret Morse Nice (1941), who spent a lifetime in the earlier part of this century studying birds and song.

tides that is not restrained by this mechanism. Steroids can also increase their expression (Harlan, 1988; Kalra, 1993). Moreover, a given neuropeptide can be differentially regulated in different brain regions, as in the case of corticotropin-releasing hormone regulated by glucocorticoids (Swanson and Simmons, 1989; Watts, 1996).

In addition, one role of steroid effects on the brain is to sustain the production of a number of neuropeptides. For example, vasopressin fibers and cell bodies in the medial nucleus of the amygdala are almost fully depleted when testosterone is eliminated and then restored in several species that have been studied (DeVries et al., 1985; DeVries, 1995).

These neuroendocrine events have functional consequences: For example, there is evidence that vasotocin's mediation of courtship behaviors and egg laying in rough-skinned newts (*Taricha granulosa*) is dependent upon the gonadal steroid hormones (Moore, Lowry, and Rose, 1994). In other words, in rough-skinned newts, courtship behaviors are under the control of the gonadal steroid hormones and their influence on central vasotocin expression (Moore et al., 1994). High levels of the gonadal steroid hormones sustain vasotocin expression in the brain that results in male or female sexual behavior in this amphibian (Moore et al., 1994).

Evolution of the Brain

The levels of neural functioning also constitute a fundamental concept in this book. Why? Because something of our evolutionary past is reflected in the organization of the brain. The brain is organized hierarchically for many of its behavioral functions (James, 1952; Jackson, 1958; Rozin, 1976a). Behavior is a piece of biology, and the brain's organization reflects evolution (Porges, 1996). Conversely, inappropriate behaviors reflect the breakdown of neural function (e.g., brain damage, aging). In both cases hormones are critically involved.

Hierarchical organization characterizes the brain in terms of simple central motive states (Von Holst, 1973). Estrogen, for example, acts on forebrain sites that include the ventral medial region of the hypothalamus and the medial region of the amygdala, which then activate brainstem sites that carry out the basic programs for lordosis (Pfaff et al., 1994).

In each chapter, the perspective is that of tracing behavior to hormones and brain circuitry (e.g., Pfaff et al., 1996). Behavior provides one level of explanation, hormones and their actions supply another, and gene transcription furnishes yet another, and all are placed in the context of evolutionary theory. One seeks to understand what the behavior is, what generates it and sustains it, and what makes it possible. These are not abstract questions for philosophers, but concrete questions for experimentalists.

Discussion of the levels of analysis is linked to preserving the behavioral, experiential, and physiological factors as we rush to the molecular. We want to know about second-messenger systems and their regulation, as well as gene manipulation of neuropeptide levels and receptor sites. My focus in this book is to demonstrate common themes in which steroids facilitate behaviors by inducing neuropeptides that help to make the behavioral events possible in suitable environments.

Diversity and Commonality

The great diversity in nature always reminds us that our studies should not be confined to the laboratory rat as the only model. Interestingly, lactation can be found in some male fruit bats, and some fish can change sex.

Diversity, like commonality, is a theme across nature. Differential parental investment reveals various rich reproductive strategies (Brantley, Wingfield, and Bass, 1993), including alternative reproductive strategies that are induced by organizational actions of androgens during critical stages in development (Crews and Moore, 1986).

Commonality is expressed in the utilization of common neural circuits underlying a number of steroids and peptide-facilitated behavioral expressions. Nature uses many of the same neural circuits underlying the behaviors that are described in many of these chapters. For instance, the visceral neural axis includes the solitary nucleus, parabrachial nucleus, and hypothalamic and forebrain limbic regions (Miselis, 1981; Swanson, 1991). These regions are activated during central motive states and participate in regulation of the emotions. Our understanding of this circuit has evolved from the earlier studies of Herrick (1905, 1948), Papez (1937), and MacLean (1949). This circuit is activated when animals are hungry or thirsty, when they crave sodium, when they are afraid, or when they are ready to mate.

A core set of sexually dimorphic regions in the forebrain, moreover, is at the heart of a number of specific steroid- and peptide-induced behaviors. They include the medial amygdala, the medial bed nucleus of the stria terminalis, and the medial preoptic region (Simerly, 1991; Wood and Newman, 1995a,b). These three sites underlie the craving for sodium and other hormone-facilitated motivated behaviors in mammals and birds.

This is not to say that there are not specific anatomic sites subserving specific behavioral functions (Pfaff et al., 1994; LeDoux, 1995), for surely there are. The circuit for angiotensin-induced thirst is not the same as that for angiotensin-induced sodium craving (Fitts and Mason, 1989b). The sensory systems for detection are different, but the motor output, that is, drinking behavior, is the same.

Endocrinology and the Neuroscience of Hormones and Behavior

The study of endocrinology is an enormous field that is evolving rapidly today and cuts across both vertebrate and invertebrate investigations (e.g., Truman, 1992; Masler, Kelly, and Menn, 1993; Gammie and Truman, 1997). Because of the experimental work and basic insights of countless investigators, concepts in the field are changing more rapidly than ever before. For example, endocrinologists once believed that the only way steroids could have an impact was through changes in the genome, a process that takes hours to complete through the mechanism of protein synthesis. Today we know that steroids can also affect cell-surface structure, thus playing roles in membrane-related changes (McEwen, 1991; Joels and De Kloet, 1994) (Figure I.4).

At a behavioral level of analysis, we now know that corticosterone, for example, can reduce sexual behavior in a number of amphibians within minutes; the phenomenon occurs through membrane-related changes (Orchinik, Murray, and Moore, 1994). Another example is that implantation of

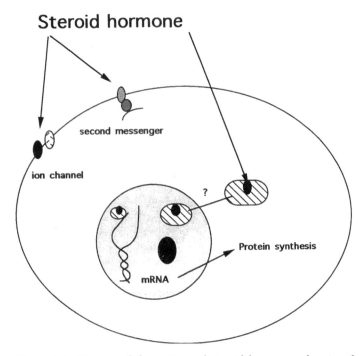

Steroid hormone

second messenger

ion channel

?

Protein synthesis

mRNA

Figure I.4. Diagram of the actions of steroid hormones showing that they exert both genomic and nongenomic effects on a variety of cells in the brain and throughout the body. Cell-surface actions of steroids include effects on receptors with ion channels and on receptors coupled to second messengers via G proteins. A question mark denotes uncertainty as to the extent to which intracellular steroid-hormone receptors reside in the cytoplasm or in the nucleus in the absence of hormone. In the presence of hormone, these intracellular receptors are found in the cell nuclei, where they regulate transcription leading to mRNA production and protein synthesis. (Courtesy of B. S. McEwen.)

aldosterone directly into the brain's medial amygdala can elicit sodium ingestion within minutes. The mechanism seems to reflect membrane-related changes as well as genomic changes in the production of neuropeptide synthesis (Reilly et al., 1993, in press).

Molecular biology has ushered in a new set of techniques that are helping endocrinologists to link hormones to common molecular ancestors. For example, thyroid hormone, corticosteroid hormone, vitamin D (a steroid), and the gonadal steroid hormones share a set of common molecular ancestors (Evans, 1998). Prolactin, growth hormone, and cytokines are members of the same molecular family (Kelly et al., 1991).

A powerful new tool in molecular biology enables researchers to selectively alter one gene that may produce a peptide or a receptor. The technology is allowing researchers to understand how hormones influence behav-

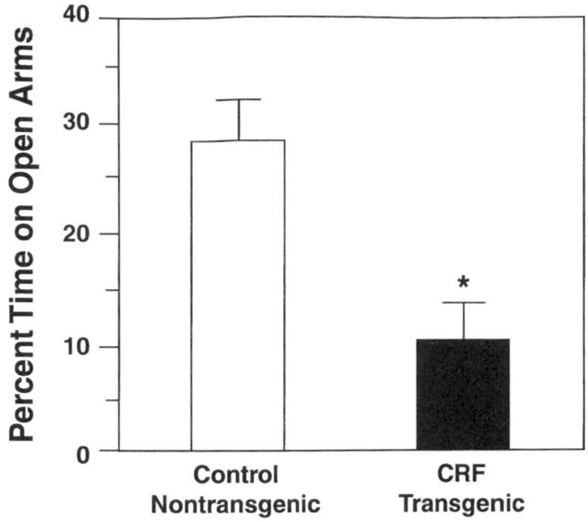

Figure I.5. Means and standard errors for time spent on the open arms of the elevated plus maze by control mice and corticotropin-releasing hormone transgenic mice over a 5-minute test. (Figure 1, Stenzel-Poore et al. "Overproduction of Corticotropin-releasing . . . ," *The Journal of Neuroscience* 14(5):2581 (1994). Reprinted by permission.)

ior, through their actions on the brain, in ways that were once impossible. Antisense oligonucleotides that selectively reduce angiotensin I (AT-I) or oxytocin receptors, for example, have long-term effects on behaviors such as sodium ingestion and female sexual behavior (McCarthy et al., 1993a; Sakai et al., 1994; Phillips and Gyurko, 1997).

Transgenic mice selectively engineered to have impaired glucocorticoid or to be depleted of glucocorticoid receptors (Beaulieu et al., 1994) show increased expressions of fear, and genetic manipulations that increase corticotropin-releasing hormone cell bodies in the brain (Stenzel-Poore et al., 1994) will produce the same reaction. For example, a transgenic mouse that overproduces corticotropin-releasing hormone will exhibit greater fear-like behaviors (e.g., reluctance to go into an open, unfamiliar space).

Model Systems in the Neuroscience of Hormones and Behavior

Guiding much of the research and the insights described throughout this book is the idea of model systems in which to study how hormones induce brain events that result in various behaviors. These model systems, which figure prominently in each chapter (Tables I.3 and I.4, for example), have revealed a number of common themes.

Scope of This Book

The book's six chapters follow a logical order and overlap, hopefully in an integrative fashion. The first chapter focuses on developmental periods and

Table I.4. *A Partial List of Peptide and Neuropeptide Effects on Behavior*

Peptides/Neuropeptides	Behavior/State
Cholecystokinin	Food satiety
Oxytocin	Food satiety, sodium satiety, attachment behavior, sexual behavior
Insulin	Food satiety
Bombesin	Food satiety
Leptin	Food satiety
Corticotropin-releasing hormone	Food satiety, fear
Atrial natriuretic factor	Sodium satiety
Elidoisin	Sodium satiety
Prolactin	Food intake, brooding in birds, behavior quiescence, parental behavior
Vasopressin	Attachment behaviors, paternal behavior, aggression
Endorphins	Pain perception, reinforcement, sexual behavior
Substance P	Pain perception
Melatonin	Sleepiness
Cytokines	Sleepiness

sexual dimorphic behaviors, the second deals with sodium and water appetite and ingestion, the third examines food selection and ingestion, the fourth concerns attachment behaviors (developmental and parental behaviors), the fifth deals with fear and stress, and the last chapter discusses biological clocks.

CHAPTER ONE

Hormones, Development, and Sexual Dimorphic Behaviors

Introduction

For many years, researchers have studied extensively how hormones regulate sexual dimorphic behavior in a range of species (e.g., Beach, 1948; Grunt and Young, 1952). It is well known that estrogen and progesterone stimulate female sexual behavior and that testosterone stimulates male sexual behavior, as reviewed by Goy and McEwen (1980).

Gonadal steroid hormones also regulate a number of other behaviors, such as the singing of birds, territorial behaviors, and spatial abilities in mammals (e.g., Wingfield, Whaing, and Marler, 1994). Although neither testosterone nor estrogen is considered the major hormone of aggression, estrogen can increase the likelihood of aggressiveness related to reproduction in a variety of species. Testosterone has been linked to aggression in males and females given suitable environmental contexts.

Sexual dimorphism in the brain and elsewhere has evolved through both natural selection and sexual selection (Darwin, 1958; Kelley, 1986). Darwin's emphasis on secondary sexual characteristics set the context in which we understand sexual dimorphism in behavioral endocrinology (Darwin, 1958): We look for functional reasons to explain why animals have the characteristics that they have. Although it is possible that not all sexual differences have functional consequences – it is easy to abuse adaptationists' arguments (Gould, 1977) – most no doubt tend to do so. The key to any organism's evolutionary success is sexual reproduction. Perhaps nowhere is this as clearly expressed as in sexually dimorphic characteristics. For example, it is well known that certain regions of the forebrain are sexually dimorphic (e.g., Raisman and Field, 1973; Gorski et al., 1978) and that these same brain regions underlie hormone-influenced behavioral responses. Consider the medial preoptic nucleus. It is larger in males than in females in a variety of species, including humans (e.g., Gorski et al., 1978; Allen et al., 1989) (Figure 1.1).

This chapter introduces many of the themes of the book, principally that steroid and peptide hormones can influence behavior through their actions

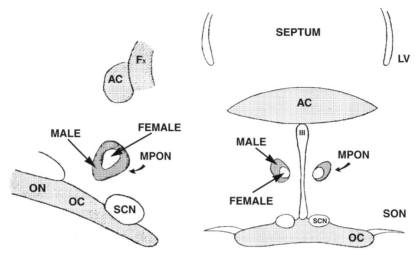

Figure 1.1. Localization of the sexually dimorphic component of the medial preoptic nucleus (MPON) in the sagittal and coronal planes. The nucleus of the female rat is drawn completely within the volume of the nucleus of the male. Abbreviations: AC, anterior commissure; Fx, fornix; LV, lateral ventricle; OC, optic chiasm; ON, optic nerve; SCN, suprachiasmatic nucleus; SON, supraoptic nucleus; III, third ventricle. (Redrawn from *Brain Research*, 148, R. A. Gorski, J. H. Gordon, J. E. Shryne, and A. M. Southam, Evidence for a morphological sex difference within the medial preoptic area of the rat brain, pp. 333–46, copyright 1978, with kind permission of Elsevier Science – NL, Sara Burgerhartstraat 25, 1055 KV Amsterdam, The Netherlands.)

on the brain. We begin with a discussion of the periods during development when hormones influence the structure of the brain and patterns of behavior and proceed to the changing concepts regarding the organizational and activational effects of hormones on brain and behavior. We then discuss a number of sexually dimorphic behaviors that are influenced by steroids and peptides.

Stages in Development and Hormonal Influences

Comparative embryologists have long recognized and studied the stages in development and the impact they have on behavior patterns (Coghill, 1929). In fact, the idea that developmental events are essential in determining behavior has its place in many areas of biological inquiry. The first fact to note is that development and its stages must be placed in the context of the species that is being described.

The importance of the critical developmental stages is reflected, for example, in language acquisition in humans (Pinker, 1994). The brain is more sensitive at some periods than others to learning and speaking languages

(Lennenberg, 1967). In humans, for example, there is a critical period during which one can learn the language of a foreign culture as easily as the language of one's own culture. The result is the ease with which children can achieve syntactic and linguistic competence and express phonetic sounds (Chomsky, 1972). In humans, the critical period for mastering auditory signals is somewhere between 3 and 7 years, and somewhere between 8 and 15 years for syntax (Glietman and Wanner, 1982).

Thus, what can happen in a variety of species is that without appropriate environmental signals – including visual, auditory, and gustatory – neural structure and function may not develop normally during the critical stages. Perhaps the best demonstration of this phenomenon comes from the classic experiments on visual function and structure in cats and monkeys. Depriving these animals of visual stimuli at critical periods during development will impair visual acuity and retard neural development (Hubel and Weisel, 1972). Researchers have reported similar findings from experiments on audition in barn owls (*Tyto alba*) (Knudsen and Brainard, 1991; Knudsen, Knudsen, and Masino, 1993) and gustation in rats (Hill, 1987).

With regard to hormones, during gestation various hormonal events can affect the development of organizational structure (Phoenix et al., 1959). Testosterone, for example, can affect the body size of female gerbils: Female Mongolian gerbils (*Meriones unguiculatus*) born in litters that are predominantly male are likely to be larger than their male littermates and larger than females born in litters that are predominantly female, the reason being the high levels of circulating testosterone. Female gerbils placed between male siblings in the womb can develop brains and behaviors more masculinized than the norm after they are born (Clark, Vom Saal, and Galef, 1992; Clark et al., 1993). As adults, they exhibit more of the typical male scent-marking behavior than is usual for females (Clark and Galef, 1995).

It has been suggested that human females exposed to high levels of testosterone in the womb may develop greater specializations in the cortex than those not so exposed (Diamond, 1991). An important fact that emerges from all of these examples is that physical characteristics can be influenced in utero through the effects of gonadal steroid hormones.

In canaries (*Serinus canarius*), testosterone levels in the egg yolk can determine the extent of both male aggression and female aggression when these birds become mature (Schwabl, 1993). Interestingly, the higher the concentration of testosterone in the yolk, the greater the degree of aggression displayed by the bird as an adult. Dominance among these birds has also been correlated with the testosterone concentrations in the yolks from which they developed (Schwabl, 1993) (Figure 1.2).

In mammals, it is the chromosomes that determine the gender of an ani-

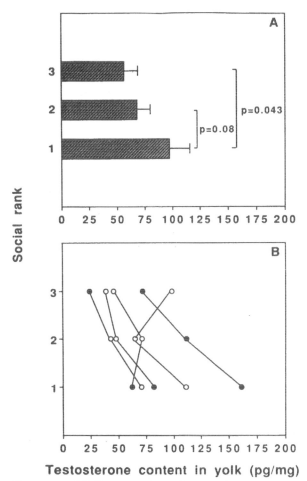

Figure 1.2. (A) Concentrations of maternal testosterone measured in the eggs from which sibling juvenile canaries of different social ranks hatched. High (1), intermediate (2), and low (3) social ranks were assigned from observations of access to food. Levels of significance between ranks are indicated next to brackets. (B) Testosterone concentrations in the eggs and the social ranks of the individual birds in each of these cohorts. (From H. Schwabl, 1993, Yolk is a source of maternal testosterone for developing birds, *Proceedings of the National Academy of Sciences U.S.A.* 90:11446–50. Copyright 1993 National Academy of Sciences, U.S.A. Reprinted by permission.)

mal. In several reptile species that have been studied in some detail, an environmental factor – namely, temperature – determines sex (Crews, 1993; Crews et al., 1996). In the garter snake (*Thamnophis sirtalis parietalis*), for example, each animal is genetically prepared to develop as either sex; it is the temperature at which the eggs are incubated that determines the outcome. Sex changes linked to temperature can also occur in adult snakes.

Investigations have shown that both humoral signals and the physical environment contribute to gender determination.

Consider another example: Temperature, in addition to estradiol concentration, can determine the sex of the red-eared slider turtle (*Trachemys scripta*) during critical periods of development (Wibbels, Bull, and Crews, 1991). Estrogen treatment during embryonic development leads to the development of female hatchlings, whereas testosterone leads to males (Wibbels et al., 1991). Thus in several reptile species, the temperature at which eggs develop – in conjunction with estrogen and testosterone concentrations – determines the gender of offspring (Crews, Bulls and Wibbels, 1991; Wibbels et al., 1991, 1992).

The diversity in nature can be as interesting as are the commonalities. I have sought here to emphasize, where possible, some commonalities regarding hormonal effects on the brain and behavior; but having said that, we must keep in mind the diversity of expression and how with evolution, specifically, it is clearly the case that sexual behavior is in humans less closely linked to the concentrations of gonadal steroid hormones than in rabbits or turtles.

Altering the concentrations of testosterone and other gonadal steroid hormones postnatally can dramatically change behavior. Since the nineteenth century it has been well known that removal of the testes will alter a rooster's crow as well as its physical prowess. Ranchers have long practiced castration of male animals to better channel the development of their herds. During the Middle Ages, young choirboys were castrated so as to maintain their high-pitched voices. In China and the Middle East, eunuchs once were created to provide a class of docile males bound to the service of the ruling class.

Important insights led to the discovery (e.g., Phoenix et al., 1959) that manipulation of gonadal steroids during certain stages of development would alter the sexual behavior of both male and female guinea pigs (*Cavia porcellanus*). For example, Phoenix et al. (1959) demonstrated that testosterone was essential for the organization and later activation of male sexual behavior in guinea pigs. Similarly, Barraclough (1961, 1962) observed that a single injection of testosterone to female rats 5 days postnatally would disrupt normal female cyclicity when they became adults.

Manipulation of gonadal steroid hormones during certain stages in postnatal development can also affect body weight, pituitary function (Pfeiffer, 1936), and hormone secretion, such as secretion of luteinizing hormone in ground squirrels (*Spermophilus lateralis*) (Smale et al., 1986). Whereas male rats, for example, typically are larger and heavier than females, gonadal manipulations during critical stages of development can alter that situation (e.g., Reilly et al., 1993). Conversely, female rats treated with testosterone and deprived of estrogen during the first 14 days of postnatal development are heavier than normal females, as reviewed by Goy and McEwen (1980). In the behavioral area, estrogen and testosterone deprivations during certain periods of postnatal development can impair the sexually dimorphic behaviors of rats as adults (e.g., Rosenblatt, 1970, 1994a,b).

Some species exhibit greater variability than others in hormone-influenced characteristics during critical periods of development, characteristics that will later influence their reproductive patterns, such as body size, territorial behavior, and color changes in tree lizards (*Urosaurus ornatus*) (Crews and Moore, 1986; Moore, 1991).

The distinction between organizational and activational effects is not as clear-cut in regard to their timing as was once believed (Arnold and Breedlove, 1985). When the ruling concepts were introduced, they were initially confined to critical stages of neonatal development during which behavior is determined by the gonadal steroids. We now know that a wide variety of organizational effects can operate during gestation, as cited earlier. Moreover, organizational and activational effects can also occur later in life (Arnold and Breedlove, 1985). For example, Breedlove has shown that regions of the spinal cord (bulbocavernosus) can be altered by androgens at times that lie outside the classic critical window of development. This is expressed in the spinal neurons that control erection in male rats (Breedlove, 1992).

Sexually Dimorphic Behaviors: Water, Sodium, and Calcium Ingestion

Gonadal steroid hormones control much more than sexual behavior. For example, they affect the propensities for rough play (higher in males) and for saccharin ingestion (higher in females) (e.g., Wade and Zucker, 1969). Manipulations of estrogen and testosterone concentrations during critical stages of development can alter the patterns of these behaviors later in life (Wade and Zucker, 1969).

A very clear example of a sexually dimorphic behavior is hormonally induced water and sodium ingestion. As detailed in Chapter 2, the mechanisms that regulate water and sodium ingestion are closely linked to those regulating extracellular fluids and maintaining the body's internal milieu. In general, females ingest greater amounts of sodium and water than do males (Krecek, 1972, 1974; Kaufman, 1980; Wolf, 1982). They also have greater concentrations of aldosterone circulating systemically, and they may perhaps experience greater induction of angiotensin sites centrally (L.Y. Ma, unpublished observations).

Virgin female rats and rhesus monkeys (*Macaca mulatta*), for example, ingest more water than males when they are depleted of extracellular fluids (Wolf, 1982; Schulkin et al., 1984). The same effect is seen when they are injected with angiotensin (Kaufman, 1980). This does not occur in response to other dipsogenic signals (e.g., intracellular induced thirst or carbachol-induced thirst) (Kaufman, 1980).

Combinations of the hormones of female sexual reproduction will also induce sodium ingestion. Administration of a combination of progesterone and estrogen to rabbits, for example, will induce sodium ingestion (Shulkes et al., 1972). The effects are even greater when the hormones are combined with oxytocin and prolactin (Shulkes et al., 1972). Administration of either prolactin or oxytocin can increase sodium intake in rabbits (Shulkes et al., 1972). In rats, prolactin can induce water intake (Kaufman, MacKay, and Scott, 1981), although its effect decreases during lactation (Kaufman et al., 1981).

It is well known that females have higher requirements for both sodium and calcium during pregnancy and lactation (Richter, 1956; Denton, 1982). The former is tied to extracellular fluid balance, whereas the latter is fundamental for intracellular events. In the laboratory, Richter demonstrated that female rats ingested more of both minerals during pregnancy and lactation. Interestingly, extracellular depletion in female rats during pregnancy also results in elevated sodium intake by their offspring (Nicolaidis, Galaverna, and Metzler, 1990). Perhaps the elevated hormonal sequelae that result from depletion alter the response to salt in the offspring.

Since Richter's work, others have reported the phenomenon of increased sodium intake during pregnancy for several other species (Denton, 1982). Females of many species migrate to salt licks during reproduction. Moreover, the greater the size of a rabbit litter, the more minerals ingested (Denton, 1982).

Several classic manipulations that are known to affect sexuality and sexually dimorphic behaviors in lower animals also affect sodium ingestion. Gonadectomy, for example, can either decrease sodium ingestion (removal of the ovaries in females) or increase it (removal of the testes in males) if the procedure is done within a critical window during development (before 20 days of age) (Krecek, 1972, 1973a,b, 1974; Chow et al., 1992). Testosterone given to females during this period changes their long-term pattern of ingestion of sodium (Krecek, 1973a,b). These same neonatal manipulations determine aldosterone concentrations (L.Y. Ma, unpublished observations).

Thus manipulation of steroid hormones can determine a sexually dimorphic behavior as well as its underlying hormonal state (e.g., fluid intake during the estrous cycle) (Kucharczyk, 1984; see Chapter 6, this volume).

In rats, calcium ingestion (like sodium ingestion and other sexually dimorphic behaviors) is affected by the gonadal steroid hormones. Manipulating these hormones during critical stages of development can alter calcium ingestion by both male and female rats when they become adults. In one experiment, rats were either gonadectomized or not during development. The females were also treated with testosterone at that time. When they became adults, the animals were given access to a calcium solution, and their ingestive behavior was monitored. The results were clear: Females ingested more calcium than males. Moreover, males could be made to act like females, and vice versa, by manipulation of the gonadal steroid hormones during development (Reilly and Schulkin, 1993) (Figure 1.3).

Like sodium ingestion, calcium ingestion increases during pregnancy and lactation (Richter, 1956; Denton, 1982). Whereas female rats always ingest more calcium than males, multiparous females ingest more calcium than do

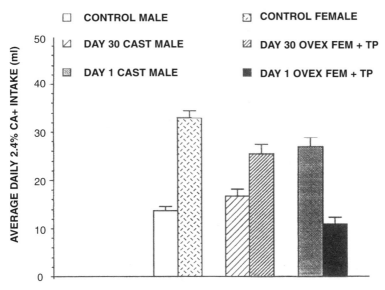

Figure 1.3. Means and standard deviations for 24-hour intake of 2.4% calcium by control male and female rats, gonadectomized rats treated with testosterone at 1 day of age or 30 days of age (CAST, castrated; OVEX, ovariectomized; +TP, plus testosterone pellet), and gonadectomized male rats treated at 1 day or 30 days of age. (From J. J. Reilly and J. Schulkin, 1993, Hormonal control of calcium ingestion: the effects of neonatal manipulations of the gonadal steroids both during and after critical stages in development on the calcium ingestion of male and female rats, *Psychobiology* 21:50–4. Reprinted by permission.)

virgins. The hormones of reproduction increase calcium ingestion as well as sodium ingestion (Shulkes et al., 1972). In an experiment with common marmosets (*Callithrix jacchus*) we found that calcium ingestion was greatest during pregnancy and lactation in females (Power et al., in press).

Like the hormones of sodium homeostasis, the hormones of calcium homeostasis may play a role in behavior. The steroid hormone vitamin D, for example, may be involved in calcium hunger. Circulating concentrations of 1,25-dihydroxyvitamin D_3 are greater in females during reproduction, because of the greater strain on calcium reserves (DeLuca, 1987). Vitamin D acts to conserve and redistribute sources of calcium, which is analogous to aldosterone's role in sodium balance. Along with parathyroid hormone, vitamin D is the principal hormone involved in calcium retention and involved with calcitonin in promoting calcium excretion. Vitamin D is derived from a common molecular ancestor that also gave rise to aldosterone, corticosterone, the gonadal hormones, and thyroid hormones (Evans, 1988). Vitamin D acts as a classic steroid on receptor sites and on ligand transcription factors (DeLuca, 1987; Hollick, 1994). Vitamin D receptors are localized in the same regions of the brain that underlie sodium appetite (see Chapter 2) and that also may underlie calcium appetite (medial amygdala).

Vitamin D is known to activate calcium-binding proteins (Sonnenberg, Pansini, and Christakos, 1984), as does calcium deficiency (Strauss, Jacobowitz, and Schulkin, 1994). Perhaps the same circuit that underlies a number of steroid- and peptide-induced behaviors also underlies vitamin D's role in calcium ingestion. One intriguing hypothesis suggests that vitamin D induces parathyroid hormone in the brain, which results in the central state of calcium hunger. Raising the concentrations of vitamin D in the common iguana augments calcium ingestion (Oftedal, Chen, and Schulkin, 1997). Ultraviolet light, which increases dermal production of vitamin D, also seems to be correlated with calcium ingestion: The greater the intensity of the light, the greater the calcium ingestion, as well as the circulating concentrations of vitamin D metabolites (Oftedal et al., 1997). In addition, systemic injections of vitamin D in calcium-replete rats increases their calcium ingestion (Leshem, DelCanho, and Schulkin, 1996).

Sexual Dimorphism in the Brain

In most mammals studied to date, the brain in either sex is feminine until it is converted to a masculine form in males through the action of testosterone (Goy and McEwen, 1980). This masculinization, or defeminization, of the brain affects not only sexual behavior but also a wide range of behaviors relating to aggression, play, and ingestion, as well as a number of key physiological processes. Masculinization in rats, for example, can be reversed by removing testosterone from the male during the first 15 days after birth. Conversely, the female brain can be made to look more like the male's if testosterone is administered during this period. One result is that when a variety of species of either sex are given testosterone during adulthood, male behavior is elicited. If they are not exposed to testosterone during early development, the likelihood of such behavior is reduced.

Specifically, it is the conversion of testosterone to estradiol in the brain – by an enzymatic process called aromatization – that changes the female brain to the male form (McEwen et al., 1977). Though it may be counterintuitive that a female hormone is ultimately responsible for development of the male brain, the data reported thus far demonstrate that such is the case.

The metabolic pathway by which aromatization occurs is depicted in Figure 1.4. For sexual differentiation, the process typically takes place during certain periods of development. If the hormonal events occur outside this window, the feminine brain does not convert to the male form. Similarly, a male rat that is deprived of testosterone during this critical window will turn out to have a brain that remains much like a female's, including the types of synapses and the neurochemicals or neuropeptides that the brain produces (e.g., cholecystokinin) (Simerly, 1991).

Three critical sexually dimorphic forebrain nuclei are induced during early periods in development. The preoptic nucleus is sexually dimorphic, as is the medial bed nucleus of the stria terminalis and the medial amygdala (e.g., Lehman, Winans, and Powers, 1980; Lehman, Powers, and Winans, 1983). When the medial amygdala, bed nucleus, or preoptic region is deprived of circulating testosterone, sexual behaviors decline, as do neurotransmitters and neuropeptides (Simerly, 1995).

These same three sexually dimorphic forebrain sites are also affected by the removal of testosterone during adulthood (Simerly, 1995; DeVries, 1995), typically leading to reductions in neuropeptide concentrations (e.g., Simerly, 1995). Innervated by olfactory signals, these dimorphic forebrain regions constitute a circuit in the regulation of both male and female behaviors in response to gonadal steroid hormones (Simerly, 1995). The three sites constitute a common circuit underlying a number of hormone-induced behavioral responses (Wood and Newman, 1995a, b).

Within the medial preoptic nucleus, the numbers of cholecystokinin-producing cells differ in males and females; males have more of these cells than do females (Simerly, 1991). There are also divisions within nuclei that differ between the sexes. In many cases, the sexually dimorphic region is defined by the number of cells of a neuropeptide that is synthesized there (e.g., vasopressin in the bed nucleus of the stria terminalis) (DeVries, 1995).

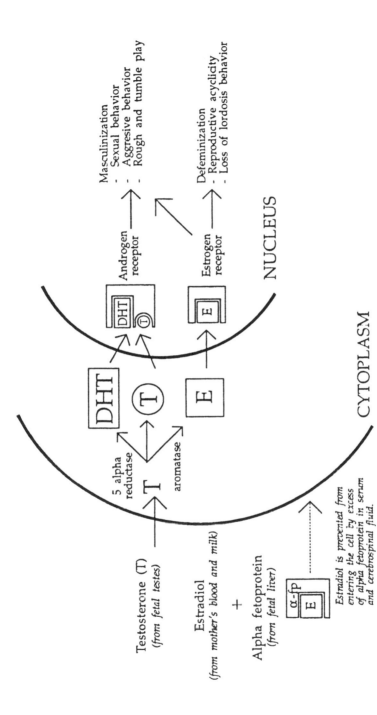

Figure 1.4. A depiction of the metabolic process for the gonadal steroid hormones. DHT, dihydroxytestosterone. (Courtesy of B. S. McEwen.)

These phenomena hold true not only for mammals but also for reptiles and a variety of birds.

Hormonal and Neural Control of Sexual Behaviors

Frank Beach (1942) devoted his career to uncovering the neural and hormonal mechanisms of sexual behavior. In his experiments with mammals treated with gonadal steroid hormones it was observed that testosterone influenced male sexual behavior and that estrogen and progesterone influenced female sexual behavior. He also demonstrated convincingly in a variety of animals that the brain is the critical end point in influencing such behavior. Beach (1942) concluded that the "central excitatory state" is induced by both hormonal signals and external events. Gonadal hormones acting on the brain reduce the threshold to respond sexually. After having determined some of the behavioral effects of gonadal hormones, Beach went on to study which portions of the brain are involved in orchestrating sexual behavior. Using a classic Jacksonian framework, he disconnected or aspirated various parts of the brain in rats. He found that decerebration (surgically disconnecting the forebrain from the brainstem) (e.g., Bard, 1940) reduced the male's behavioral responses to gonadal steroid hormones. Decortication (removal of the neocortex) eliminated one set of responses, and isolation of the hypothalamus eliminated another set (Beach, 1948; Meisel and Sachs, 1994).

Female Hormones and the Induction of a Sexual Response

Sexual behaviors that are under the control of steroid hormones are not the same for all species of animals. There is evidence, for example, that in primates sexual behavior has become somewhat independent of hormone levels (Wallen, 1990), whereas rodent behavior is closely linked to hormonal levels (Beach, 1948; Wallen, 1990).

Still, gonadal-steroid-treated animals demonstrate both appetitive and consummatory sexual behaviors. The gonadal steroid hormone estrogen activates the dopamine reward system located in the nucleus accumbens (e.g., Mermelstein and Becker, 1995). Of course, being deprived of estrogen during critical stages of development may change sexual behavior later in life (e.g., Young et al., 1939; Feder and Whalen, 1965).

One of the best-worked-out systems in the analysis of steroid hormones and peptides is that which facilitates lordosis, or female receptivity to sex. Lordosis is triggered peripherally by systemic administration of estrogen and progesterone, and then by centrally delivered oxytocin (McEwen et al., 1987). The combination of these hormones triggers the potential for lordosis (e.g., Boling and Blandau, 1939; Feder and Whalen, 1965; Feder, Blaustein, and Nock, 1979; Pfaff, 1980).

The anatomic sites for estrogen are depicted in Figure 1.5 by autoradiography (Pfaff, 1980). These sites have also been substantiated by in situ hybridization studies (Simerly et al., 1990). Holding true among fish, amphibians, reptiles, and mammals (Pfaff, 1980; Young et al., 1994), this distribution of sites has been preserved throughout evolution (Figure 1.5).

Treatment with estrogen and progesterone induces motivated behavior. Estrogen-treated female rats, for instance, will bar-press to gain access to males and will even cross an electric grid to gain access to males. They will choose locations in which sexually active males are to be found. These sorts of appetitive states are dose-dependent. The consummatory phase is the time of actual sexual receptivity when the female comes into contact with the male. Both reflect the central state of motivation (Beach, 1948; Pfaff, 1980).

How does this system work? Estrogen, in conjunction with progesterone, acts on the brain to induce sexual receptivity (e.g., Mani, Blaustein, and O'Malley, 1997). This includes the stimulation of sensory systems (Adler, Tesko, and Goy, 1970; Blaustein et al., 1994) associated with lordosis, such as receptivity to relevant male behavior. Such behavior can also trigger this state. The sensory fields for sexual receptivity are enhanced in estrogen- and progesterone-treated rats (Pfaff, 1980). Estrogen, for example, enhances the acuity of olfactory cues for males. This information is transduced to the medial amygdala via the vomeronasal organ by increasing neuropeptide expression within this region and other sexually dimorphic sites (Simerly and Swanson, 1988).

Whereas progesterone often inhibits male sexuality, in several species of reptiles the hormone can actually potentiate androgen stimulation of sexual behavior (Young, Greenberg, and Crews, 1991; Lindzey and Crews, 1992, 1993). The reason may be that the male pattern of sexual behavior is derived from that of the female. This could mean that female hormones are also capable of promoting the expression of male-like behavior (Crews, 1993).

Brain Mechanisms for Sexual Responsiveness: Estrogen, Progesterone, and Oxytocin Actions in the Brain

Decerebrated rats do not demonstrate lordosis behavior (e.g., Bard, 1940; Pfaff, 1980). Within the brain, estrogen and progesterone receptors are lo-

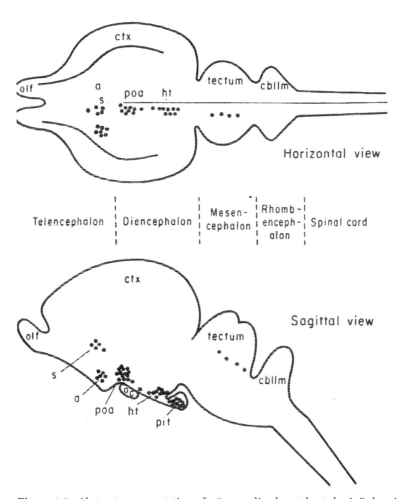

Figure 1.5. Abstract representation of a "generalized vertebrate brain" showing locations of estrogen- and androgen-concentrating cells common to all vertebrates studied. The schematic drawings show a horizontal view and a sagittal view. Black dots represent groups of steroid-concentrating cells. Features of the distribution of estrogen- and androgen-addressed neurons that are common across vertebrates include labeled cells in the limbic forebrain (e.g., septum, amygdala), preoptic area, tuberal hypothalamic nuclei, and specific subtectal loci in the mesencephalon. Abbreviations: a, amygdala or archistriatum; cbllm, cerebellum; ctx, cortex; ht, nuclei in the tuberal region of the hypothalamus; oc, optic chiasm; olf, olfactory bulb; pit, pituitary; poa, preoptic area; s, septum. (From D. W. Pfaff, S. Schwartz-Giblin, M. M. McCarthy, and L.-M. Kow, 1994, Cellular and molecular mechanisms of female reproductive behaviors, in *The Physiology of Reproduction,* 2nd ed., ed. E. Knobil and J. D. Neill, New York: Raven Press, fig. 3. Reprinted by permission.)

calized in sites that include a number of sexually dimorphic regions, as well as in the ventral medial hypothalamus (Parsons et al., 1982). The ultrastructural characteristics of these receptors have been determined (e.g., Silverman, Don Carlos, and Morrell, 1991) and are differentially regulated in different brain regions for different animal species (Young, Nag, and Crews, 1995).

Although the basic reflexes for sexual responses are mediated by the brainstem (e.g., Pfaff, 1980), forebrain sites initiate the expression of motivated behaviors for sexual contact. Classic studies have indicated the role of the hypothalamic-pituitary-gonadal axis in the regulation of sexual behavior.

When estrogen is implanted in the ventral medial hypothalamus, the threshold for lordosis is reduced in ovariectomized rats. This effect is dose-dependent. Although implantation of estrogen into other regions of the brain can also result in lordosis, the dose required is larger. Estrogen implants in the ventral medial hypothalamus generate greater responsiveness to progesterone in eliciting lordosis. This occurs in part through the induction of progestin receptors that project to the medial amygdala and medial preoptic nucleus, as reviewed by Pfaff et al. (1994). The movement of an estrogen implant can be traced by radiolabeling the hormone. Such experiments have further confirmed that the site of action and the locus of the spread are in and around the lateral region of the ventral medial hypothalamus (McEwen et al., 1987).

Estrogen implants can also elicit sexual receptivity in other mammalian species such as the guinea pig (Feder et al., 1979; Blaustein et al., 1994), as well as in some nonmammals. Local implants of estradiol into the ventral medial hypothalamus in whiptail lizards (Cnemidophorus inornatus), for instance, can elicit sexual receptivity (Wade and Crews, 1991) (Figure 1.6).

Induction of c-fos immunoreactivity in oxytocin cells in estrogen- and progesterone-treated rats has shown activation in the paraventricular nucleus and preoptic nucleus and within the ventral medial region of the hypothalamus (Flanagan et al., 1992a). These findings have strengthened earlier anatomic and functional observations that these regions underlie female sexual behavior.

Estrogen and progesterone induce protein synthesis (as well as structural changes related to protein synthesis) in the lateral ventral medial hypothalamus (McEwen et al., 1987). Local implants of protein-synthesis inhibitors into this part of the brain inhibit the lordosis response to systemic estrogen and progesterone. Because it is well known that steroids promote protein synthesis, these findings suggest that local synthesis of proteins that synthesize neurotransmitters underlies the behavioral responses to the estrogen

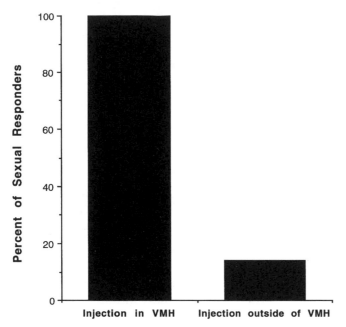

Figure 1.6. Sexual receptivity for two species of lizards implanted with estrogen directly into the ventral medial hypothalamus or outside this region. (Adapted from Wade and Crews, 1991.)

and progesterone stimulation of lordosis (e.g., McEwen et al., 1987).

Estrogen also induces oxytocin receptors in the ventral medial hypothalamus, and progesterone induces oxytocin receptors in estrogen-primed rats (Caldwell et al., 1988; Schumacher et al., 1989). The two hormones appear to act synergistically, with estrogen stimulating oxytocin in the hypothalamus and progesterone inducing it in the surrounding area containing oxytocin receptors (Figure 1.7). The major result is that oxytocin doses that do not by themselves elicit lordosis behavior do so when given in conjunction with estrogen and progesterone in rats.

Interestingly, progesterone treatment of estrogen-primed rats can produce rapid effects on oxytocin receptors that facilitate lordosis behavior (Schumacher et al., 1990), and those receptors are located in the same regions of the brain that induce maternal behavior (Caldwell et al., 1988). These effects were reported to have occurred 30 minutes after treatment and appeared to be mediated by the caudal end of the lateral ventral medial hypothalamus (Figure 1.8).

Estrogen regulates a number of genes for brain peptides, including prolactin and tachykinins. But a number of neuropeptides or neurotransmitters

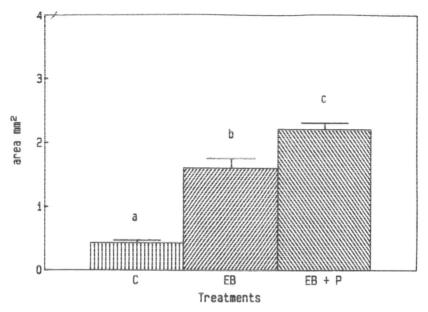

Figure 1.7. Areas of specific oxytocin-receptor binding in the ventral medial nucleus of the hypothalamus demonstrated by computerized quantitative neurotransmitter-receptor autoradiography in control (C), estrogen-treated (EB), and estrogen-and-progesterone-treated (EB + P) rats. Bars show the extent of spread of oxytocin receptor. (Reprinted with permission from M. Schumacher, H. Coirini, D. W. Pfaff, and B. S. McEwen, 1990, Behavioral effects of progesterone associated with rapid modulation of oxytocin receptors, *Science* 250:691–4. Copyright 1990 American Association for the Advancement of Science.)

also influence sexual behavior. They include endorphins, gonadotropin-releasing hormone (GRH), luteinizing-hormone-releasing hormone (LHRH), substance P, cholecystokinin (CCK), and dopamine (e.g., Micevych and Hammer, 1995).

Specific proteins are induced by estrogen in the brain (e.g., HIP-70). These proteins, which include both inhibitory and excitatory proteins [γ-aminobutyric acid (GABA), N-methyl-D-aspartate (NMDA) receptors] (McCarthy et al., 1993a,b), are functionally related to sexual behavior, as is LHRH release from the pituitary (Pfaff et al., 1994). Structural changes in these proteins may not necessarily depend on genomic changes (see Chapter 2).

Using antisense oligodeoxynucleotides, researchers have found, for example, that infusion of antisense for oxytocin receptors or progesterone receptors (Mani et al., 1994) into the ventral medial hypothalamus reduces lordosis behavior in estrogen-primed rats (McCarthy et al., 1993a, b); the rats were treated for several days with doses of estrogen that ordinarily

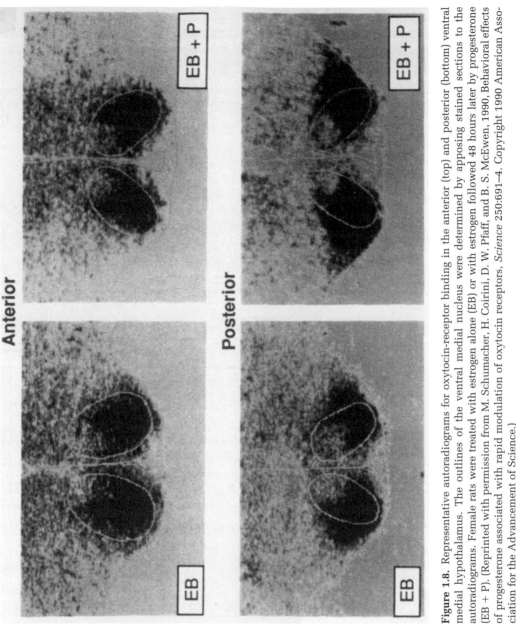

Figure 1.8. Representative autoradiograms for oxytocin-receptor binding in the anterior (top) and posterior (bottom) ventral medial hypothalamus. The outlines of the ventral medial nucleus were determined by apposing stained sections to the autoradiograms. Female rats were treated with estrogen alone (EB) or with estrogen followed 48 hours later by progesterone (EB + P). (Reprinted with permission from M. Schumacher, H. Coirini, D. W. Pfaff, and B. S. McEwen, 1990, Behavioral effects of progesterone associated with rapid modulation of oxytocin receptors, *Science* 250:691–4. Copyright 1990 American Association for the Advancement of Science.)

would have induced sexual receptivity. Interestingly, lordosis induced by progesterone was not affected, suggesting a mechanism different from that of estrogen.

Nongenomic effects have been demonstrated in facilitating sexual behavior in rats that were primed with estrogen and progesterone at the level of the ventral tegmental region, but not the ventral medial hypothalamus (Frye, Mermelstein, and DeBold, 1992). That is, whereas estrogen-facilitated lordosis at the level of the ventral medial hypothalamus depends upon genomic actions, it may not at the level of the ventral tegmental region. Prolonged sexual receptivity may require nongenomic activity at the level of the brainstem, whereas initial receptivity may be triggered at the level of the forebrain or the hypothalamus.

Figure 1.9 depicts an anatomic circuit that underlies sexual receptivity in female rats (Pfaff et al., 1994). One of this book's central themes – that of steroids increasing the production of neuropeptides to generate central motive states and influence behavior – is quite striking in this example.

Testosterone and Male Sexual Responses

It has long been known that testosterone influences the central motive state in males to search for females and then to try to mount them under suitable conditions (e.g., Beach, 1948), as reviewed by Meisel and Sachs (1994). Both appetitive and consummatory behaviors are affected by testosterone in a wide variety of species (e.g., Everitt and Stacey, 1987; Balthazart et al., 1995). Not only does testosterone generate motivated behaviors, but also the hormone itself is rewarding to male rats. Males injected with testosterone in one location returned to that place more frequently than to others (Alexander, Packard, and Hines, 1994).

The consummatory phase of male sexual behavior in rats includes pelvic thrusting, mounting, neck gripping, and ejaculation (Baum, 1992). Male rats treated with testosterone can tell the difference between females in estrus and those that are not (Baum, 1992). The vomeronasal organ (relaying olfactory information) activates several sexually dimorphic brain sites that respond to females in estrus (Wood and Newman, 1995a, b).

Development of the peripheral system (penile erection) is of course required for success at reproduction. Testosterone has organizational effects on brainstem and spinal-cord sites essential for penile erection (Breedlove, 1992). For example, testosterone alters dendrites in motor neurons that act directly on target muscles for penile erection (Rand and Breedlove, 1992). These regions are larger in males than in females, and testosterone enhances neuronal survival by activation of neurotrophic factors [e.g., spinal nucleus

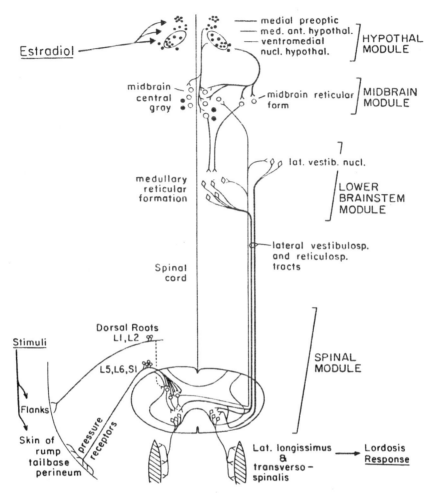

Figure 1.9. Summary diagram of neural circuit for activating lordosis behavior. The entire circuit is shown, from the somatosensory stimuli required to activate the behavior, through the sensory and ascending pathways, to the ventromedial hypothalamus, where estradiol and progesterone act to facilitate the behavior. Descending pathways from hypothalamus to midbrain, from midbrain to medullary reticular formation, and from reticular formation to lumbar spinal cord are also shown. All stimuli, neural pathways, hormone actions, and responses are bilateral, but are shown here on one side for convenience of illustration. The neural modules proven to participate in the neural and endocrine control of this behavior are outlined on the right, and they also serve to organize the text in this part of the chapter. (From D. W. Pfaff, S. Schwartz-Giblin, M. M. McCarthy, and L.-M. Kow, 1994, Cellular and molecular mechanisms of female reproductive behaviors, in *The Physiology of Reproduction,* 2nd ed., ed. E. Knobil and J. D. Neill, New York: Raven Press, fig. 16. Reprinted by permission.)

of the bulbocavernosus (SBN)] (Forger et al., 1993). Lesions of the SBN reduce penile display (Breedlove, 1992).

But it is the forebrain that orchestrates the underlying male sexual behaviors. Androgens bind to sexually dimorphic brain regions such as the medial preoptic area. This knowledge was first based on autoradiographic studies of androgen and estrogen binding properties, as later verified by in situ hybridization studies (e.g., Simerly et al., 1990; Wood and Newman, 1995a,b). Androgen receptors have been constant across evolution (Young et al., 1994). The medial preoptic area is particularly important for male sexual behavior (Simerly and Swanson, 1988; Meisel and Sachs, 1994; Liu, Salamone, and Sachs, 1997b).

Lesions of the medial preoptic area can reduce or abolish testosterone-induced sexual behaviors in rats (e.g., Everitt and Stacey, 1987). Similarly, implants of testosterone within this region can induce male sexual behavior (Davidson, 1966). This pattern holds true for a variety of species (Kingston and Crews, 1994). Testosterone is known, for example, to induce sexual behavior in the Japanese quail (*Coturnix japonica*) (e.g., Adkins and Pniewski, 1978; Pannzica, Pannzica, and Balthazart, 1996). The medial preoptic region is a critical site for male sexual behavior related to elevated concentrations of testosterone (Meddle et al., 1997), with testosterone implants facilitating the behavior, and lesions abolishing it. Moreover, aromatase inhibition aborts the activation of male sexual motivation (Balthazart et al., 1997). The relationship between testosterone and the metabolic transformation of testosterone is nicely depicted for the quail in Figure 1.10.

Pathways from the medial preoptic area, including projections to the retrorubal field or ventral lateral periaqueductal field, are also involved in male sexual behavior in rats (Finn and Yahr, 1994), as are other sexually dimorphic regions such as the medial nucleus of the amygdala (Yahr and Jacobsen, 1994). Amygdala lesions reduce the motivation to gain access to females, and preoptic-area lesions reduce both the motivation to gain access to females and the male's sexual performance in their presence (Everitt and Stacey, 1987).

All three sexually dimorphic nuclei (medial preoptic, bed nucleus, and medial amygdala) are involved in this behavior and are induced by testosterone. Interference with these nuclei disrupts male sexual behavior (Lehman et al., 1980, 1983). Anatomic markers such as c-fos for early gene expression have confirmed that this circuit is activated when androgens are elevated (Wood and Newman, 1995a, b). Specifically, testosterone facilitates substance-P expression in the brain, as well as other neuropeptides or neurotransmitters. For example, tyrosine hydroxylase (TH) neurons in the medial amygdala are elevated during mating when testosterone is elevated

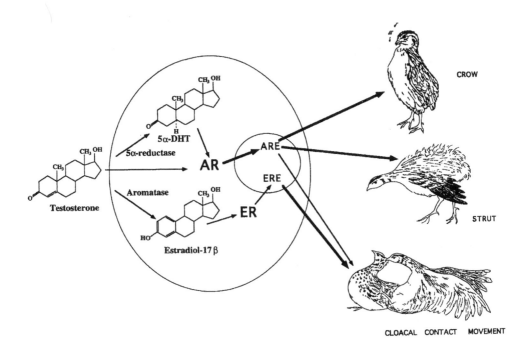

Figure 1.10. Enzymatic transformations of testosterone (T) in the quail brain and their relevance for the activation of male reproductive behavior. In its target cells, T can be reduced into 5α-dihydrotestosterone (5α-DHT) or aromatized into estradiol (E); T and 5α-DHT bind to androgen receptors (AR) and activate androgen-responsive elements (ARE) in the nucleus; E binds to estrogen receptors (ER) and activates estrogen-responsive elements (ERE). Androgenic action is necessary for the activation of crowing and strutting in quail. The activation of copulatory behavior is mostly under the control of estrogen, but there is a synergism with androgens. (From G. C. Pannzica, C. V. Pannzica, and J. Balthazart, 1996, The sexually dimorphic medial preoptic nucleus of quail: a key brain area mediating steroid action on male sexual behavior, *Frontiers in Neuroendocrinology* 17:51–125. Reprinted by permission.)

(Asmus and Newman, 1993). Both substance P and TH neurons are linked to male sexual behavior (Wood and Newman, 1995a,b).

Local implants of testosterone in the medial preoptic area or the medial bed nucleus of the stria terminalis or the medial nucleus of the amygdala will elicit male sexual behavior in Syrian hamsters (*Mesocricetus auratus*) (Wood and Newman, 1995a,b).

Olfaction is a primary sense that is essential for successful mating behaviors in a number of mammals that depend upon this sense for primary sensory input. Interestingly, many of the androgen- and estrogen-containing sites are linked to olfactory sites essential for mating (Wood et al., 1992; Wood and Newman, 1993a,b). The medial nucleus of the amygdala receives

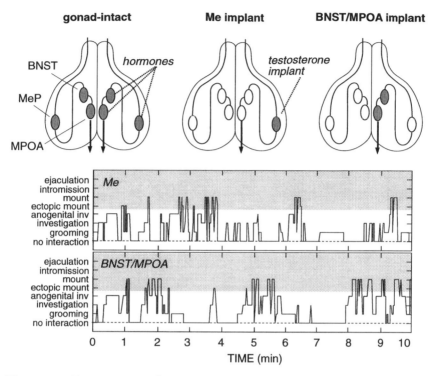

Figure 1.11. Testosterone implants in regions of the forebrain [medial amygdala (Me), bed nucleus of the stria terminalis/medial preoptic region (BNST/MPOA)] and male hamster copulatory behavior. (From R. I. Wood and S. W. Newman, 1995, The medial amygdaloid nucleus and medial preoptic area mediate steroidal control of sexual behavior in the male Syrian hamster, *Hormones and Behavior* 29:338–53. Reprinted by permission.)

olfactory input from two sources: the vomeronasal organ and the olfactory tract and its interactions with androgen sites. When testosterone is elevated, the sensory threshold at which environmental events will trigger approach behavior is reduced. Pathways from the amygdala to other sexually dimorphic sites (amygdalafugal pathways) interact with olfactory cues in driving the central state (Meisel and Sachs, 1994).

Gonadal Steroid Regulation of Amphibian Sexual Behavior

In the rough-skinned newt, testosterone and vasotocin regulate male sexual behavior. Injections of vasotocin or gonadotropin-releasing hormone will stimulate sexual behavior in both males and females (Moore, Wood, and

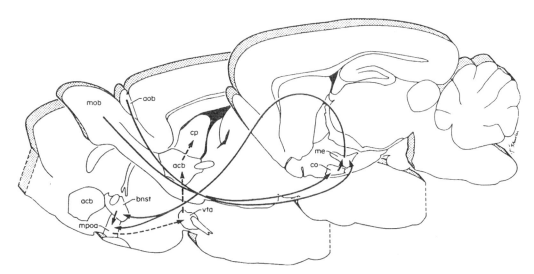

Figure 1.12. Some of the neural pathways regulating male copulatory behavior. Only a few of the relevant inputs to the preoptic area are indicated, primarily those arising from the medial amygdala. Abbreviations: acb, nucleus accumbens; aob, accessory olfactory bulb; bnst, bed nucleus of the stria terminalis; co, cortical amygdala; cp, caudate-putamen; me, medial amygdala; mob, main olfactory bulb; mpoa, medial preoptic area; vta, ventral tegmental area. (From R. L. Meisel and B. D. Sachs, 1994, The physiology of male sexual behavior, in *The Physiology of Reproduction,* 2nd ed., ed. E. Knobil and J. D. Neill, New York: Raven Press, p. 19. Reprinted by permission.)

Boyd, 1992). This behavior is dependent upon the gonadal steroid hormones (Moore et al., 1994). It can be inhibited by corticosterone or by activation of the stress axis (see Chapter 4 and 5).

Testosterone influences vasotocin expression in the brain and by doing so increases the likelihood of male sexual behavior (Moore et al., 1992). Gonadectomy eliminates male sexual behavior by reducing central vasotocin expression in regions of the brain, such as the amygdala, that underlie behavioral responses (Boyd and Moore, 1991).

In the rough-skinned newt, two distinct behaviors with common motor outputs are regulated by gonadal steroid hormones (Moore et al., 1992): female egg-laying behavior and male-like courtship. Clasping behavior is involved in both.

Female Hyenas, Androgens, and Dominance

Aggressive display is a common characteristic across evolution; it has been linked to the concentrations of androgens that circulate during critical per-

40

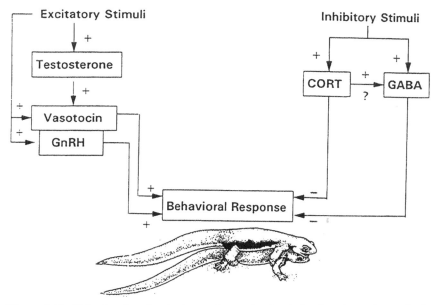

Figure 1.13. Behavioral endocrinology of reproductive behaviors in male rough-skinned newt. (From F. L. Moore and M. Orchinik, 1991 Multiple molecular actions for steroids in the regulation of reproductive behaviors, *Seminars in the Neurosciences* 3:489–96. Reprinted by permission.) Both excitatory and inhibitory stimuli are indicated in the figure. Note that activation of the stress hormone corticosterone (CORT) reduces the response (see Chapter 5).

iods of development. That is, it is well known that during critical stages in neonatal development, testosterone induces organizational effects in the brain that later are expressed in terms of aggressive behavior. The idea that adult aggression is facilitated by the actions of androgens during critical developmental stages has been supported by a wealth of experimental evidence in a number of species. For example, females injected with testosterone during critical stages develop male-like aggressive behavior when they become adults. They also express male-like characteristics during play-fighting and other sexually dimorphic behaviors.

In one unusual case, that of the spotted hyena (*Crocuta crocuta*), female aggression is particularly pronounced. Both male aggression and female aggression in this species are determined by high testosterone concentrations in utero. Aggression among spotted hyenas is fierce: Pups are born with sharp teeth, and infanticide can reach up to 25% (Glickman et al., 1992a,b). Females are both larger and more aggressive than males as a result of being exposed to higher concentrations of testosterone and its metabolites in the womb. Those females exposed to the greatest amounts of testosterone in

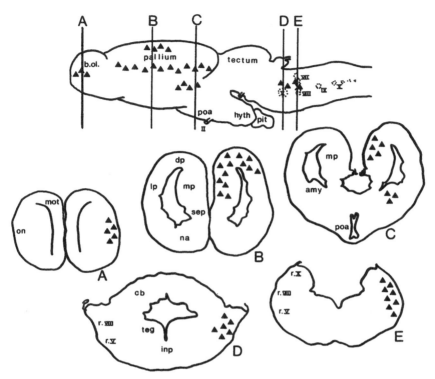

Figure 1.14. Schematic diagram of the newt brain with regions containing AVT receptors indicated by closed triangles. The top figure is the exterior side view. Parts A – E are typical frontal sections at the levels indicated on the side view. Abbreviations: amy, amygdala; b.ol., olfactory bulb; cb, cerebellum; dp, dorsal pallium; hyth, hypothalamus; inp, interpeduncular nucleus; lp, lateral pallium; mot, medial olfactory tract; mp, medial pallium; na, nucleus accumbens; on, olfactory nerve terminal field; pit, anterior pituitary; poa, preoptic area; sep, septal nucleus; teg, tegmentum V and VII; II–X, cranial nerves. (Reprinted from *Brain Research*, 541, S. K. Boyd and F. L. Moore, Gonadectomy reduces the concentrations of putative receptors for arginine vasotocin in the brain of an amphibian, pp. 193–7, copyright 1991, with kind permission of Elsevier Science – NL, Sara Burgerhartstraat 25, 1055 KV Amsterdam, The Netherlands.)

utero are likely to be the most aggressive. Although testosterone declines in females more than in males upon reaching maturity, the hormone's early impact means a lifelong tendency toward aggressive behavior (Glickman et al., 1992 a,b).

In addition to being larger than the male, the female spotted hyena has external male-like genitalia. Androgens operative during critical stages of development (in this case, fetal) also determine these physiological characteristics. Females also exhibit rougher play behaviors than do males (Glickman et al., 1992 a,b). The process through which this masculinization occurs in female hyenas is thought to be the following: Androstenedione secreted by the ovary is changed to testosterone within the placenta, directly affecting the fetus and the later expression of its genitalia. High concentrations of androstenedione also play a key role in the development of aggressive behavior. Whereas testosterone levels are about the same in males and females, androstenedione levels remain high in females. Aromatization processes that convert androstenedione to estrogen are weakly expressed in this species (Glickman et al., 1992a,b).

We do not know the hyena's underlying neural circuit for these behaviors, but in rodents there is a neural circuit linked to aggression that includes the medial preoptic and anterior hypothalamic areas, the medial bed nucleus of the stria terminalis, and the medial nucleus of the amygdala (Albert, Jonik, and Walsh, 1992; Ferris et al., 1997). This forebrain steroid circuit seems to underlie testosterone-dependent aggression. In particular, the bed nucleus seems to be the vital link in organizing aggressive behavior (Albert et al., 1992). All three are important, however. Lesions in any of these areas will reduce aggression, and all three are affected by developmental events (determined by the gonadal steroid hormones) that organize the structure of the circuit underlying aggression (Albert et al., 1992). Perhaps the more dominant hyena has greater concentrations of vasopressin in these three structures, as will be discussed later. There is evidence to suggest that testosterone, by modulating vasopressin receptors in the brain, facilitates aggression (Delville, Mansour, and Ferris, 1996).

Gonadal Steroids, Vasopressin, and Flank-marking Behaviors

A nicely described system in which steroid hormones and peptides interact to affect behavior is that of scent marking by hamsters (Wood and Newman, 1995a,b). This behavior is linked to dominance (Yahr and Jacobsen, 1994), determining territorial boundaries and attachments to conspecifics (Ferris et al., 1990; Wingfield et al., 1994). Scent marking is facilitated by testosterone or estrogen (e.g., Ulibarri and Yahr, 1996).

Vasopressin is both a posterior peptide hormone and a neuropeptide in the brain. Central infusions of vasopressin facilitate scent markings by golden hamsters (*Mesocricetus auratus*) (Ferris, 1992). For example, infusions of vasopressin in brain regions such as the bed nucleus of the stria terminalis facilitate scent marking (see Chapter 4). A background of testosterone along with the central vasopressin will increase the behavioral expression of scent marking (Albers and Cooper, 1995). This suggests that the normal secretion of testosterone linked to reproductive events facilitates

vasopressin synthesis, which in turn triggers the behavior of marking territorial claims.

Interestingly, in female Syrian hamsters (*Mesocricetus auratus*), in contrast to most species, there is greater flank marking than by males, and
females tend to be more dominant; this behavior is not due to differences in
vasopressin-facilitated flank marking (Hennessey, Huhman, and Albers,
1994). In this species, estrogen, by increasing vasopressin gene expression,
facilitates the likelihood of flank marking (Huhman and Albers, 1993)

The medial bed nucleus and the medial nucleus of the amygdala are
essential for flank-marking behavior. Vasopressin concentrations in the medial amygdala, bed nucleus, and lateral septum are higher in males than in
females for a number of mammal species (DeVries et al., 1985; Wang et al.,
1995). Vasopressin concentrations are reduced by castration, particularly
during development (Wang et al., 1994, 1995; Johnson, Barberis, and Albers,
1995). Testosterone, as indicated earlier, is necessary to maintain vasopressin in these sites (DeVries, 1995). Estrogen injections also stimulate vasopressin gene expression in males (DeVries, 1995).

These three brain sites have a number of neuropeptides whose expressions are differentiated as functions of steroid activation during critical
stages of development. Vasopressin in the bed nucleus, for example, is determined by androgens during this critical period (DeVries, 1995). Testos-

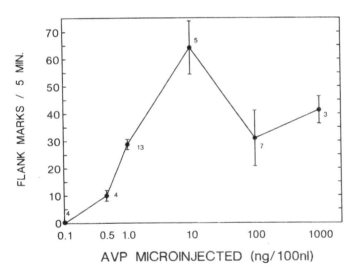

Figure 1.15. Dose-dependent induction of hamster flank-marking behavior by injection of
arginine vasopressin (AVP) into the medial preoptic region. The numbers of flank markings
observed during the 5-minute period are indicated on the vertical axis. (Adapted from Albers et al., 1981.)

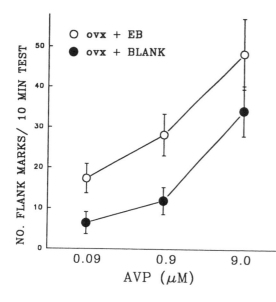

Figure 1.16. Flank-marking reaction to central administration of AVP in estrogen-treated (EB) and nontreated (BLANK) ovariectomized (ovx) female hamsters. (Reprinted with permission from *Peptides* 14, K. L. Huhman and H. E. Albers, Estradiol increases the behavioral response to arginine vasopressin (AVP) in the medial preoptic-anterior hypothalamus, pp. 1049–54, 1993, Elsevier Science Inc.)

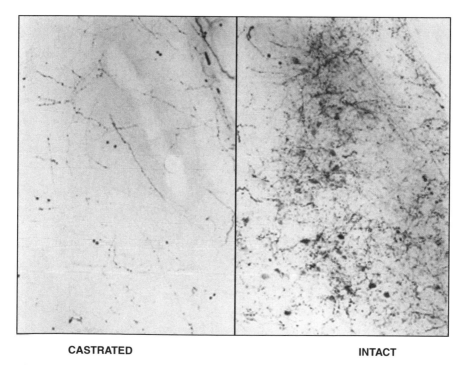

CASTRATED   INTACT

Figure 1.17. Vasopressin-immunoreactive cells and fibers in the medial nucleus of the amygdala in castrated and intact rats. Notice on the left side the dramatic effects on vasopressin expression in castrated rats. (Courtesy of G. J. DeVries, 1995, with permission.)

CHESTER COLLEGE LIBRARY

terone (and, to some extent, estrogen) will both sustain and increase vaso-pressin mRNA in the lateral septum, medial amygdala, and bed nucleus in several species (DeVries, 1995). The circuit that underlies scent marking includes the medial preoptic/anterior hypothalamic region, in addition to the bed nucleus of the stria terminalis and the periaqueductal gray (Bam-shad and Albers, 1996).

Frogs, Birds, Gonadal Steroids, and Song

The male South African clawed frog (*Xenopus laevis*) produces an elaborate song that is tied to reproduction (Kelley, 1992). The song facilitates com-municative signaling in problem-solving. Though the female frog also emits sound, it is in the form of what are called "ticks," as opposed to full song (Kelley, 1992). Like bird song (Marler et al., 1988), the frog song shows a syntax, or set of rules. A hormonally controlled behavior, the singing of clawed frogs is orchestrated by androgen sites that researchers have now mapped.

The juvenile muscles that ultimately will produce frog song in adults are particularly sensitive to testosterone during development (Kelly, 1986; Kel-ley et al., 1988). Specifically, testosterone induces cell proliferation in the larynx, an event that later plays a role in the expression of acoustic signals. The result of testosterone elevation is that the male larynx is three times heavier and has eight times more muscle fibers than the female larynx. Although males and females start out with similarly sized larynges, testos-terone during early development also acts to prevent the fiber loss that naturally occurs in females (Sassoon, Segil, and Kelley, 1986; Sassoon, Gray, and Kelley, 1987).

If male frogs are castrated early during development, they will grow up to resemble females, both anatomically and behaviorally. Conversely, if the ovaries are removed from a female early during development and she is treated with testosterone, she will have large larynx muscles and will sing like a male in adulthood (Kelley, 1992). A similar phenomenon holds true for songbirds (see Chapter 6).

Researchers have observed changes in the neural circuits that control the frog's song – notably that the cranial motor nerves experience the same hormonal regulation as the larynx during development (Kelley, 1992). Using a combination of methods, investigators have uncovered the sites in the brain that underlie the singing behavior of the frog. These regions include a vocal-motor-pattern generator in the tegmentum, the sensory region of the thalamus, the preoptic region, and the striatum (Kelley, 1992). More re-cently, researchers have studied the development of the androgen receptors

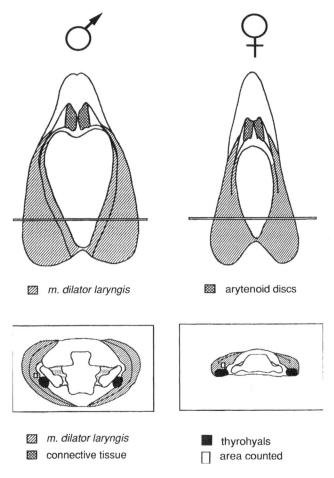

Figure 1.18. Dorsal view (top) and cross section (bottom) of adult male and female larynges of frogs. For the dorsal view, anterior is up. For the cross sections, dorsal is up. Cross sections were taken from the anterior–posterior level indicated in the laryngeal dorsal views. Fibers were counted from an area (small rectangle in cross sections) within the inner bipennate muscle just dorsal and lateral to the thyrohyal cartilage. The characteristic shape of the lumen of the larynx at this anterior–posterior level is illustrated in the cross sections. (From M. L. Marin, M. L. Tobias, and D. B. Kelley, 1990, Hormone-sensitive stages in the sexual differentiation of laryngeal muscle fiber number in *Xenopus laevis, Development* 110: 703–12. Reprinted by permission of Company of Biologists, Ltd.)

that are essential for frog song, as well as the gene expression that makes this possible.

We come now to one of my favorite model systems for illustrating hormonal effects on brain and behavior – singing in some species of birds. Not all birds sing, but the zebra finch (*Poephila guttata*) is one that does (Gurney and Konishi, 1980). It is quite a beautiful species, and the syntax and semantics of its song compose a well-studied piece of ethology (e.g., Marler et al., 1988). Typically, the male birds of this and a variety of other species sing melodic songs in order to attract females (though there are species variations in the scenario that follows).

Bird song depends on two events: the activation of testosterone and the perception of song during critical stages of development. Testosterone also potentiates the recognition of a conspecific's song (Cynx and Nottebohm, 1992a,b). Although birds deprived of testosterone during critical stages of development may be able to sing, their song will be muted. Similarly, birds deprived of the ability to sing during this time by occlusion of the auditory canal will also sing a muted song when they become adults.

Song perception is a highly specialized ability tied to communicative and territorial competence. Like human language, song production is lateralized in the brain; the left side is dominant in these behaviors (Nottebohm, 1993). Song production, like language production, appears to follow syntactic rules (Marler et al., 1988; Chomsky, 1972). Song resembles speech coupled to syntax and local dialectics.

The songbird (e.g., zebra finch) sings during spring, when the high-vocal-center (HVC) nucleus enlarges as testosterone concentrations become elevated (e.g., Smith, Brenowitz, and Wingfield, 1997). Testosterone implanted in this region facilitates song expression (Nottebohm, 1993), whereas lesions of the HVC nucleus abolish or impair song production in males, as well as song reception in females (Brenowitz, 1991).

Song production induced by testosterone depends on protein synthesis. Immunohistochemical studies have indicated that a number of neuropeptides are synthesized in the circuitry underlying song production (Ball et al., 1988). Testosterone and the metabolic conversion process that transforms it to estradiol induce synaptogenesis for song production (Nottebohm, 1993).

Many examples of sexual differentiation (including singing) are initiated by estrogen (Gurney, 1982; Konishi and Gurney, 1982; Arnold and Schlinger, 1993). Importantly, the brain itself is a source of the estrogen synthesis and aromatization activity that regulates song (Schlinger and Arnold, 1992). Aromatization activity is particularly extensive in the caudal neostriatum (Shen et al., 1995).

Estrogen implants into the high vocal center will masculinize part of the zebra finch's song repertoire (Grisham, Mathews, and Arnold, 1994). In fact, estrogen, testosterone, or converting enzyme will induce masculinization at this nucleus (Grisham et al., 1994).

It is important to remember that neither estrogen nor testosterone alone

Figure 1.19. (a) Testosterone activates song and courtship in adult estrogen-treated females (E-females). This female zebra finch (right) was treated with estrogen as a chick, and then as an adult received a Silastic pellet containing 100 μg of testosterone. When courting a Bengalese finch (left), she approached it with pivoting movements, then straightened to an erect posture, fluffed her throat feathers, and rapidly repeated her short song phrase in a behavioral sequence that closely resembled the courtship behavior of a normal male. This female was raised by Bengalese finches and became sexually imprinted upon that species. (b) An example of the stereotyped song developed by this E-female after 28 days of continued exposure to testosterone. (Reprinted with permission from M. E. Gurney and M. Konishi, 1980, Hormone-induced sexual differentiation of brain and behavior in zebra finches, *Science* 208:1380–3. Copyright 1980 American Association for the Advancement of Science.)

controls bird song; more important is how the two hormones are linked (Gurney and Konishi, 1980; Gurney, 1982). For example, Gurney and Konishi showed that a female zebra finch treated with estrogen during critical periods in development also needed treatment with testosterone for song expression when she was mature (Gurney and Konishi, 1980; Gurney, 1982).

There is now evidence to suggest that changes in vasotocin expression facilitated by testosterone in the frog or songbird may underlie song (Voorhuis et al., 1991; Marler, Chu, and Wilczynski, 1995). For example, vasotocin infusions into the third ventricle in white-crowned sparrows elicit vocalization; none of the animals tested sang without this treatment. The effect was particularly strong in estrogen-primed females (Maney et al., 1997).

Sexual Dimorphism in the Human Brain and Behavior

In humans, the evidence suggests that adrenal hyperplasia in neonates results in girls preferring boys' toys more than do control girls (Berenbaum and Snyder, 1995). This finding and others make it more than plausible that high estrogen or testosterone concentrations help determine sexually dimorphic behavior in humans, just as they do in other animals. The open question is to what extent.

Throughout her long career, Margaret Mead, the great anthropologist, went to great lengths to describe the diversity of sexual behaviors and gender roles that human beings exhibit (Mead, 1974). Simultaneously, she was quite responsive to new developments within psychobiology. Nonetheless, perhaps nowhere in the fields of biology are issues as controversial as those surrounding sexual dimorphism in humans. Whereas there is no dispute about the findings of differences between men and women in regard to mineral ingestion and nutrient preferences, when it comes to differences in sexuality and other behaviors, discussion of the issues can become quite heated. Because politics inevitably becomes involved, scientists must be aware of the implications of their findings. Whereas the biological differences between men and women are real, how such differences are interpreted must be handled with care.

That the brains of human males and females are different is now an uncontested fact (e.g., Gorski et al., 1978; Hines, 1982; De Lacost-Utamsing and Holloway, 1982). We still must work out what that means, however. For example, parts of the corpus callosum (a bundle of nerve fibers that connects the two cerebral hemispheres) are larger in women than in men.

This may suggest that women are better able to integrate information across the cortex. In fact, common sense and some science suggest that women are better at dealing with diverse information and appear to be more flexible (Kimura, 1992). This behavior may be an adaptation for dealing with the host of diverse information encountered during child-rearing.

Perhaps most troubling for some people is the fact that regions of the hypothalamus not only differ between men and women but also may differ between homosexual men and heterosexual men. In one interesting study, LeVay (1991a,b) reported that males who died from AIDS (acquired immune deficiency syndrome) incurred through sexual behavior had different brain structures than those who acquired AIDS through the use of intravenous drugs (i.e., heterosexual men), including several hypothalamic structures such as the anterior hypothalamus (Hines, Allen, and Gorski, 1992). The anterior commissure is also larger in women than in men and is larger in homosexual men than in heterosexual men. These regions of the brain are sexually differentiated during critical early stages in development. These findings suggest that sexual orientations may reflect the concentrations of testosterone that are circulating during critical stages in development and their subsequent effects on neuropeptide expression in the brain (e.g., Swaab and Hofman, 1995).

It is generally recognized that females, on average, are better at verbal tasks, and men, on average, are better at spatial-mathematical reasoning. Testosterone concentrations are correlated with performance: The higher the testosterone concentration, the poorer the verbal performance by men (Kimura, 1992). Interestingly, high testosterone concentrations in women are correlated with greater ability in spatial tasks. High estrogen concentrations tend to decrease performance at tasks that usually favor men (deductive reasoning, spatial ability) and increase performance at tasks that favor women (e.g., articulation, manual speed and dexterity, and perceptual speed) (Hampson, 1990; see Chapter 6, this volume). Men with low testosterone concentrations still do better at spatial reasoning than do women, and this holds true for young children (Gladuc and Beatty, 1990).

But high levels of estrogen (e.g., during estrogen therapy) are now known to preserve memory functions (Sherwin, 1996). One of the roles of estrogen in the brain is to facilitate some functions (memory, well-being) that are not linked to reproduction. Women, in general, have greater abilities in verbal-recall tasks (Hampson, 1990). Estrogen seems to preserve these functions (Kimura, 1995). Estrogen also seems to enhance the state of well-being in older women (Sherwin, 1996). An interesting finding is estrogen's induction of the genes that encode brain neurotrophin factor (Swaab et al., 1994).

Conclusion

This chapter has described several systems in which steroid hormones have profound effects on sexual dimorphism in the brain and on behavior. The fact that critical periods of brain development are affected by the actions of hormones is of paramount importance to our biology. Whereas hormones act primarily during critical periods of development to determine the brain structure and the neural circuitry that will last for a lifetime, they can also have an impact on behavior later in life (Arnold and Breedlove, 1985).

Steroid hormones induce central motive states by activating neuropeptides within neural circuits. We have considered several instances in which steroids facilitate neuropeptides in the brain. This results in behaviors that are of functional significance. Because nature is economical, the same neural circuits control a number of other steroid-induced behaviors that are described throughout this book. In all cases, the end result is activation of a central state leading to behavior that achieves an important biological end point.

CHAPTER TWO

Hormonal Regulation of Sodium and Water Ingestion

Introduction

In what follows we shall review the hormonal control of water and sodium appetite. The greatest emphasis will be on sodium ingestion, for it reflects my own research interests. The first major section is on angiotensin-induced water and sodium appetite, with discussion of the sites of action. The second major section discusses corticosteroid-induced sodium appetite and the sites of action. The third major section deals with stress-induced (or corticotropin-releasing-hormone- and corticosterone-induced) sodium intake and the sites of action. The fourth major section concerns angiotensin- and corticosteroid-induced water and sodium appetite and the sites of action. The fifth section deals with atrial natriuretic peptide and inhibition of water and sodium ingestion. The final section depicts an anatomic circuit that may underlie the cravings for water and sodium.

Sodium Hunger

Sodium appetite, in addition to thirst, provides the backbone of extracellular fluid regulation at a behavioral level of analysis. It has been known since the turn of the century that loss of extracellular fluid can generate thirst; it occurs when one bleeds or sweats under natural circumstances (Fitzsimons, 1979; Denton, McKinley, and Weisinger, 1996). Later it was discovered that sodium hunger is also expressed under these conditions (Wolf, 1969b; Denton, 1982). The hormones that regulate extracellular fluid balance have been preserved across evolution in terrestrial vertebrates (Denton, 1965). Sodium ingestion is linked to extracellular fluid balance. Thus the hunger for sodium provides a window into extracellular fluid regulation (Denton, 1982).

Sodium hunger, or sodium craving, is a phenomenon of nature. Several species are known to travel great distances to ingest salt at mineral licks and other sources of salt when the sodium content of their diet is reduced (Denton, 1982). The range of this phenomenon is expressed in a variety of species of omnivorous and herbivorous birds and mammals. Sodium appetite is a phenomenon found in humans, typically following body-fluid depletion and elevation of natriorexegenic hormones (Beauchamp et al., 1990). It is less clear whether or not carnivores experience sodium hunger, for sodium is ingested during consumption of their prey.

Salt licks contain a number of minerals, and ingestion of salt-lick deposits can satisfy a number of mineral requirements concurrently (Schulkin, 1982, 1991a). Elegant studies during the 1960s by Derek Denton's group demonstrated relationships involving the activities of the renin-angiotensin-aldosterone and corticosterone systems (Blair-West et al., 1963), the diet being consumed and its reduced sodium content, with the ingestion of sodium in several herbivorous mammals (e.g., *Oryctolagaus cuniculus*).

Numerous other studies have documented enlargement of the adrenal gland, aldosterone production, and ingestion of salt during "naturally occurring sodium deficiencies" (e.g., Hoffman and Robinson, 1966; Weeks and Kirkpatrick, 1976). Typically these are seen when the food sources on

Figure 2.1. Photograph of a kangaroo at a salt peg. (Courtesy of D. Denton.)

which animals are grazing are depleted. The behavioral activity serves the same end as the physiological end, namely, to maintain adequate sodium in the body. Nature uses both behavior and physiology to do so.

Under natural conditions, mineralocorticoid hormones in conjunction with angiotensin arouse the appetite for sodium in the brain of a sodium-depleted animal (Epstein, 1982b; Fluharty and Epstein, 1983; Sakai, 1986). Either hormone alone can arouse the appetite, but both are elevated during sodium depletion. The combination of the actions of the two hormones when they are elevated concurrently is larger than the sum of their individual actions. The adrenalectomized rat's avidity for sodium may be an example of the singular action of cerebral angiotensin in the brain. Without the presence of mineralocorticoid hormone, as in the case of the adrenalectomized rat, the appetite depends entirely upon angiotensin. In support of this, Epstein and his colleagues suggested that central (not peripheral) angiotensin-receptor blockade abolished salt appetite in the adrenalectomized rat, suggesting that the appetite was dependent upon cerebral angiotensin (Sakai and Epstein, 1990). However, the story is more complicated than that, because recent evidence in the rat implicates systemic angiotensin in the generation of the behavior (Thunhorst, 1996), and sodium hunger in sheep has been found to be related to angiotensin of peripheral origin (Weisinger et al., 1987).

Richter believed that the hunger for sodium resulted in changes in the brain and that the brain orchestrated this behavior in response to the hormones of sodium homeostasis. His primary focus was on putative changes within the gustatory neural axis (Richter, 1956). Indeed, salty tastes are perceived differently in sodium-hungry rats (cf. Contreras, 1977; Berridge et al., 1984; Jacobs, Mark, and Scott, 1988; Nakamura and Norgren, 1995; McCaughey and Scott, in press) and in humans placed on sodium-deficient diets or treated with diuretics (Beauchamp et al., 1990).

Behaviorally, the hormones that generate an appetite for sodium do so by eliciting a craving for a salty substance (Schulkin, 1982). A sodium-hungry animal searches for salty substances, recalls where they were encountered when it did not need them (Krieckhaus and Wolf, 1968), or what they were associated with (Berridge and Schulkin, 1989), or at what time of day they would appear (Rosenwasser, Schulkin, and Adler, 1988), and utilizes that knowledge to satisfy its hunger for sodium, as reviewed by Wolf (1969a,b) and Schulkin (1991a).

Sodium appetite is a paradigmatic example of a hormonally induced motivated behavior. Sodium-hungry animals express both the appetitive and consummatory phases of motivated behavior. For example, mineralocorticoid-treated rats will run down a runway for sodium-taste rewards (Schulkin, Arnell, and Stellar, 1985). This is the appetitive phase of motivated behavior.

The consummatory phase consists in ingestion of the desired substance – in this case, salty commodities. For example, when hypertonic NaCl is infused into the oral cavity, sodium-hungry rats emit species-specific stereotyped oral-facial responses to the salt (3% NaCl) as if they were ingesting something like sucrose; when not sodium-hungry, they behave as if it had an aversive taste, like quinine water (Berridge et al., 1984; Berridge and Schulkin, 1989). Thus both the appetitive and consummatory phases of motivated behavior are altered when the hormones of sodium homeostasis are elevated and act on the brain.

Recall that the conditions for sodium hunger generally emerge from perturbations of extracellular fluid homeostasis, including hypovolemia (loss of extracellular fluid and extreme sodium loss). In all these situations the concentrations of plasma renin, aldosterone, and corticosterone are elevated (Stricker et al., 1979; Denton, 1982). In each case a hunger for sodium is aroused. This is expressed in a number of species (e.g., hamster) (Fitts et al., 1983).

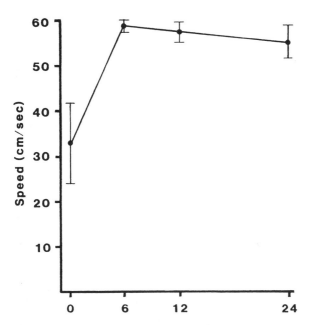

Figure 2.2. Running speed to reach sodium in mineralocorticoid-treated rats. Note that the highest concentration is roughly eight times the concentration in seawater, which the rats will not ingest, though they still express appetitive behavior toward it. (Adapted from Schulkin et al., 1985.)

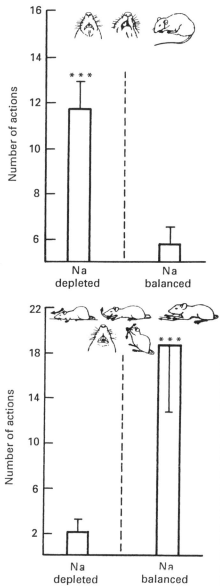

Figure 2.3. Taste-reactivity profiles of rats to intraoral infusions of hypertonic NaCl when the hormones of sodium homeostasis were elevated (Na depleted) and when they were not (Na balanced). Top: Combined mean (±SEM) number of ingestive actions (rhythmic tongue protrusions, nonrhythmic lateral tongue protrusions, and paw licks). Bottom: Combined mean (±SEM) number of aversive actions (chin rubs, head shakes, paw treads, gapes, face washes, and forelimb flails). (From K. C. Berridge and J. Schulkin, 1989, Palatability shift of a salt-associated incentive drive during sodium depletion, *Quarterly Journal of Experimental Psychology* 41B: Fig.2. Copyright © 1989. Reprinted by permission of The Experimental Psychology Society.)

It is difficult, but not impossible, to generate sodium hunger in humans. McCance (1936, 1938), in the late 1930s, attempted to deplete himself of sodium by combining a sodium-deficient diet with excessive sweating, and thereby reduce his extracellular volume. What he found was a blunting of sodium taste responses. This effect has been seen by others (Beauchamp et al., 1990). A sodium appetite was expressed in a child who had an adrenal

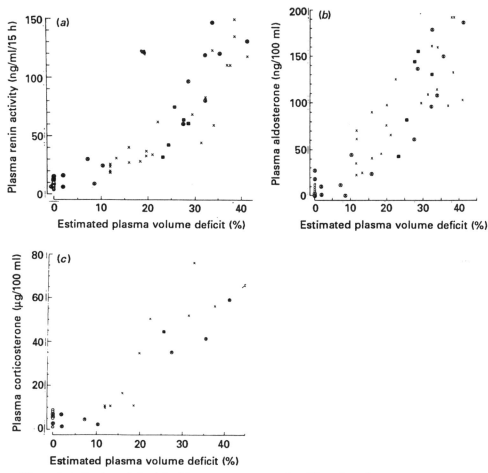

Figure 2.4. Renin activity (*a*), aldosterone concentration (*b*), and corticosterone concentration (*c*) as functions of estimated plasma volume deficit. (From Stricker et al., 1979, with permission.)

malfunction and chronic sodium loss (Wilkins and Richter, 1940), and experimental depletion of sodium with furosemide can induce a modest sodium appetite in adults (Beauchamp et al., 1990), though the results in humans are not that strong. Because humans are salt gluttons, it is difficult for a sodium appetite to be elicited.

In the laboratory, hunger for sodium is expressed in a variety of species. They range from pigeons (Massi and Epstein, 1990) to sheep, rabbits, mice (Denton, 1982), hamsters (Fitts et al., 1983), rhesus monkeys (Schulkin et al., 1984), and baboons (Denton et al., 1993).

Angiotensin-induced Water and Sodium Appetite

Angiotensin is one of the primary hormones involved in regulation of extracellular fluid balance (Fitzsimons, 1979). Behaviorally, angiotensin is known to promote ingestion of water and sodium (Buggy and Fisher, 1974; Chiaraviglio, 1976; Fluharty and Manaker, 1983). Although not all drinking responses are mediated by renin-angiotensin, the water and sodium ingestions that are generated by angiotensin are essential to maintain extracellular fluid volume (Stricker and Wolf, 1969; Fitzsimons, 1979).

It is a remarkable phenomenon to witness animals abruptly drink when angiotensin is injected in the brain. They drink water, and then exhibit a delayed motivation to ingest sodium. The phenomena are clear, though natriuresis does occur over time (e.g., Buggy and Fisher, 1974; Fluharty and Manaker, 1983).

Both ingestive responses are expressed in neonates well before they are able to independently ingest water or sodium (e.g., Leshem and Epstein, 1989), suggesting that the appetite is innate and that the hormone activates a neural circuit designed to orchestrate these behaviors.

Moreover, the appetite that is expressed following central administration of angiotensin results in persistent increases in sodium consumption even after the infusion is ended (Byrant et al., 1980). Angiotensin-induced increases in water intake have been reported in the Japanese eel (*Anguilla japonica*) (Takei, Hirano, and Kobayashi, 1979), the spadefoot toad (*Scophiopus couchii*) (Propper, Hillyard, and Johnson, 1995), and a number of reptiles, birds, and mammals (Fitzsimons, 1979), and sodium appetite has been expressed in different species (e.g., pigeons, mice, sheep, monkeys) (Rowland and Fregly, 1988; Denton et al., 1990; Weisinger et al., 1990a,b), with the ingestion of sodium being further augmented when angiotensin is combined with sodium depletion in sheep (Weisinger et al., 1987, 1997). Finally, central injections of angiotensin increase the appetitive phase of this motivated behavior (e.g., running down a runway for water or sodium) (Zhang, Stellar, and Epstein, 1984).

Both inhibitory and excitatory signals interact in regulating behavioral events. It may be the case that oxytocin and angiotensin reflect such inhibitory and excitatory signals in the regulation of body fluids. Each behavior is activated by common and separate mechanisms (Stricker and Verbalis, 1987). Hormonal mechanisms appear to complement and perhaps stimulate the behavioral expression. Oxytocin is an inhibitory signal on ingestive behavior (see Chapter 3). Central infusions of oxytocin inhibit hypovolemic and angiotensin-induced sodium appetite (Blackburn et al., 1992), while also promoting sodium excretion. Both behavioral expressions are laboring for the same end: maintaining extracellular fluid balance.

In addition to the behavioral actions of angiotensin, this hormone also participates in the physiological regulation of body-fluid homeostasis. Hypovolemia is one of several stimuli that can activate the renin-angiotensin system (Peach and Chiu, 1974; Denton, 1982). This can occur naturally during blood loss or from water and sodium deprivation. The rate-limiting step in the synthesis of angiotensin is renin release from the juxtaglomerular cells of the kidney. Renin converts the plasma globulin protein angiotensinogen to angiotensin I, which is subsequently converted to angiotensin II by a carboxyl dipeptidase known as angiotensin-converting enzyme. Circulatory angiotensin has numerous peripheral actions, including vasoconstriction, aldosterone release, augmentation of sympathetic-nervous-system function, and renal conservation of sodium and water (Philips, 1978). Circulating angiotensin also has important central-nervous-system effects, even though, like most other peptide hormones, it has restricted access to cerebral structures because of the blood–brain barrier (Simpson and Routtenberg, 1973; Philips, 1978). By acting on forebrain circumventricular organs (e.g., subfornical organ), blood-borne angiotensin can regulate pituitary function, stimulate central pressor responses, and elicit thirst.

In the brain there are two major types of angiotensin receptors, referred to as AT1 and AT2 (Millan et al., 1991; Rowe, Saylor, and Speth, 1992). Whereas the dipsogenic response to angiotensin may be mediated solely by AT1 receptors (Rowland et al., 1992), expression of the appetite for sodium appears to depend upon both types of angiotensin receptors (AT1, and perhaps AT2) (Galaverna et al., 1992; Rowland et al., 1992).

Many of angiotensin's cellular actions in its peripheral target organs, as well as the brain, involve intracellular mobilization of calcium. This response is mediated by AT1 receptors and involves membrane-associated phosphoinositide hydrolysis and subsequent production of IP3. In neuron-like cultures, evidence suggests that a phosphoinositide-specific phospholipase C is coupled to angiotensin receptors (Mah et al., 1992) via a G protein (Siemens et al., 1991). A second cellular action of angiotensin in neurons is modulation of cGMP content (Sumners et al., 1990; Sumners and Myers, 1991), which may involve both increases and decreases in the concentrations of this cyclic nucleotide. When angiotensin stimulates the production of cGMP, it appears to do so through an intermediate formation of nitric oxide, a response that involves both AT1 and AT2 receptors. Moreover, the interactions among the two receptor types, nitric oxide formation, and second-messenger production may require interactions between angiotensin receptors expressed on both neurons and glial cells. Decreases in cGMP concentrations appear to involve exclusively AT2 receptors and activation of a cGMP-specific phosphodiesterase; this arrangement may exist only in neurons (Sumners and Myers, 1991). In both cases of cGMP regulation, the end result of the second-messenger action appears to involve modulation of ionic conductances that determine membrane excitability (Sumners et al., 1991b; Sumners and Myers, 1991). These cellular changes ultimately underlie the behavioral and physiological changes elicited by the peptide's action in the brain (Fluharty and Sakai; 1995).

Transgenic rats with increased concentrations of central vasopressin exhibit increased water ingestion when angiotensin is centrally injected (Moriguchi et al., 1994). Moreover, antisense oligonucletides for angiotensin-I receptors infused into the third ventricle were reported to reduce angiotensin-II-induced water or sodium drinking when angiotensin II was also infused into the third ventricle (Sakai et al., 1994). The rats were infused for several days prior to the angiotensin treatment. Angiotensin-II-induced water intake was reduced, but carbachol-induced water drinking was not affected. Moreover, only angiotensin-I receptor sites were reduced (Sakai et al., 1994). Other findings have demonstrated that the antisense inhibition of AT1 receptors can reduce the hypertension that is genetically and neurally determined (e.g., Gyurko, Wielbo, and Phillips, 1993).

Sites of Action for Angiotensin-induced Water and Sodium Appetite

Angiotensin receptors and cell bodies are localized in circumventricular organs [e.g., subfornical organ, organum vasculosum of the lamina terminalis (OVLT)], in the median preoptic nucleus, and in magnocellular cells of the paraventricular hypothalamus, supraoptic nucleus, zona incerta, lateral hypothalamus, parabrachial nucleus, and solitary nucleus (Miselis, 1981; Lind, 1988). Angiotensin-containing neurons and terminal fields have also been found in many of these same regions, including the amygdala and the bed nucleus of the stria terminalis (Lind, Swanson, and Ganten, 1985; Lind, 1988; Reagan et al., 1994). These same regions are activated by both systemic and central angiotensin infusions using c-fos to determine early gene expression (Herbert, 1993; Rowland et al., 1994).

Subfornical neurons are responsive to changes in blood pressure (Nicolaidis et al., 1983), and systemic angiotensin acts on the subfornical organ (a circumventricular region of the brain) to elicit the ingestion of water (Simpson and Routtenberg, 1973). Very small doses of angiotensin, when injected into the subfornical organ, will elicit a water appetite. It is the most sensitive site in the brain for the dipsogenic response. But angiotensin-induced

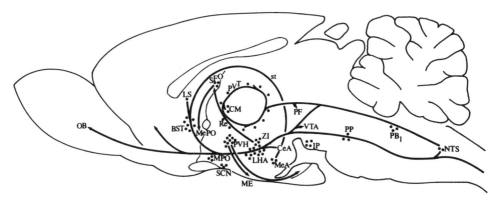

Figure 2.5. Schematic midsagittal view of rat brain illustrating the major angiotensin-II-immunoreactive cell groups and fiber pathways studied. Arrows are drawn to indicate the orientation of projection in cases where this is obvious. Note the NTS, nucleus of the solitary tract; PB, parabrachial region; LHA, lateral hypothalamic region; SFO, subfornical area; BST, bed nucleus of the stria terminalis; CeA, MeA, central and medial amygdala. (From Lind, Swanson, and Ganten, 1985. Reproduced with permission of S. Karger AG, Basel.)

thirst is also elicited when it is injected elsewhere in the brain, including the bed nucleus of the stria terminals and the preoptic region (Swanson and Sharpe, 1973; Swanson, Kucharczyk, and Mogenson, 1978).

Through the subfornical-organ connectivity to the anterior third ventricular region (AV3V, including the median preoptic nucleus and the OVLT), the two sites regulate the ingestion of water to angiotensinergic signals (e.g., Buggy and Johnson, 1977; Philips, 1978; Camacho and Philips, 1981; Miselis, 1981; Lind, 1988). However, neither the subfornical organ nor its efferent projections (Schulkin, Eng, and Miselis, 1983; Thunhorst et al., 1987; Fitts and Mason, 1989a) are essential for a number of forms of hormonally induced sodium appetite (cf. Weisinger et al., 1990b; Thunhorst, Ehrlich, and Simpson, 1990), although lesions can reduce the expression of this appetite (Weisinger et al., 1990b). Therefore, the anatomic sites that regulate water ingestion and sodium ingestion to an angiotensinergic signal appear to be different.

One site potentially important for the angiotensinergic arousal of a sodium appetite is the AV3V region, because of its involvement in body-fluid and cardiovascular regulation (Buggy and Johnson, 1977; Johnson, 1985). Osmotic and sodium sensors monitor intracellular changes and facilitate the release of vasopressin, in addition to eliciting thirst. Lesions of this region impair both intracellular- and extracellular-induced thirst (Buggy and Johnson, 1977). There is evidence that lesions of this region reduce the salt appetite in rats on a sodium-deficient diet (Johnson, 1985) or following sodium depletion (Chiaraviglio and Perez Guaita, 1984).

Water ingestion ceases following lesions of the AV3V region (Buggy and Johnson, 1977). This region integrates angiotensinergic signals from peripheral angiotensin with centrally delivered angiotensin in eliciting thirst or water ingestion (Johnson, 1985). Lesion studies of the AV3V region of the brain resulted in the abolition of renin-angiotensin-induced sodium appetite, whereas sodium appetite induced by deoxycorticosterone acetate (DOCA) remained intact (Fitts, Tjepkes, and Bright, 1990; Fitts and Mason, 1990; DeLuca et al., 1992). Importantly, infusions of angiotensin within this region increased sodium ingestion, whereas the same doses near the subfornical organ produced no such effect (Fitts and Mason, 1990). It is still unclear what roles peripheral angiotensin and centrally derived angiotensin play in this behavior (e.g., Sakai and Epstein, 1990; Thunhorst, 1996).

Corticosteroid-induced Sodium Appetite

The distinction between the roles of corticosterone and aldosterone is not as clear in fish (Vinson et al., 1979; Denton, 1965) as it is in land-dwelling mammals and perhaps reptiles (Denton, 1965). Both can increase sodium-retaining capacities, but aldosterone is 500 times more powerful than corticosterone in reducing the sodium appetite that occurs following adrenalectomy (McEwen et al., 1986). In fact, the ability to retain sodium may have been an important physiological adaptation to land. In the sea the issue is to get rid of sodium, but on the land the issue is to retain it. Nature selected both physiological and behavioral means to ensure this end.

Richter, and later others (e.g., Braun-Menendenz and Brandt, 1952; Braun-Menendez, 1953), noted that deoxycorticosterone, a mineralocorticoid hormone, would increase the ingestion of sodium by adrenally intact rats (Richter, 1941; Rice and Richter, 1943); in adrenalectomized rats, it first reduced and then increased the appetite, depending upon the dosage of mineralocorticoid that was administered. Years later it was discovered that aldosterone, the naturally occurring mineralocorticoid hormone, could either increase or decrease the appetite for sodium in adrenalectomized rats (Fregly and Waters, 1966; Wolf, 1964) and could elicit an appetite for sodium in adrenally intact rats (Wolf, 1964; Wolf and Handel, 1966).

When a rat's aldosterone becomes elevated for the first time, the result is sodium ingestion. In a group of rats raised on a sodium-rich diet since birth (therefore never having been in a sodium-deficient state and without elevated concentrations of aldosterone), the first time they were given aldoster-

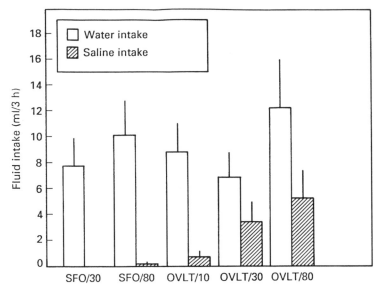

Figure 2.6. Water and salt intake of rats injected with angiotensin either within the subfornical organ (SFO) or within the OVLT. It is clear from the figure that only water intake increased following angiotensin injected in the SFO, but both water and sodium ingestions increased following OVLT injections. (From D. A. Fitts and D. B. Mason, 1990, Preoptic angiotensin and salt appetite, *Behavioral Neuroscience* 104:643–50. Copyright 1990 by the American Psychological Association. Reprinted with permission.)

one they ingested sodium. On the basis of this and other evidence, we have argued that the appetite aroused by this hormone is innate (Schulkin, 1978).

Sodium ingestion consequent to mineralocorticoid hormone administration can be experimentally demonstrated in neonates, before adult ingestive behavior is actually expressed (Thompson and Epstein, 1991). Mineralocorticoid-induced sodium appetite is expressed in several mammalian species: mice (Denton et al., 1990), hamsters (Fitts et al., 1983), gerbils (Wong and Jones, 1978), sheep (Denton, 1982; Hamlin et al., 1988), pigeons (Epstein and Massi, 1987). Mineralocorticoids also promote motivated behavior; rats will hurry down a runway for small rewards of sodium (Schulkin et al., 1985) or will bar-press in an operant chamber to obtain sodium (Quartermain and Wolf, 1967).

But mineralocorticoid activation obviously is not necessary for expression of the appetite, because adrenalectomized rats express a sodium appetite (Richter, 1936). Moreover, hypovolemia can induce a sodium appetite in adrenalectomized rats treated therapeutically with mineralocorticoid (given enough mineralocorticoid hormone to hold sodium, but unable to increase their mineralocorticoid output in response to extracellular fluid loss and sodium loss) (Stricker and Wolf, 1969). Perhaps, in this case, central angiotensin is activating the behavioral function.

Consideration of the question of how a given hormone can both decrease and increase the appetite for sodium in adrenalectomized rats may provide valuable insights into the hormone's role. Aldosterone has several important physiological effects. At the level of the kidney, it reduces sodium excretion; it mobilizes sodium transport out of sodium reserves such as bone and redistributes sodium from salivary glands and the alimentary tract, all for the physiological end of maintaining extracellular fluid volume (Blair-West et al., 1963; Denton, 1965, 1982). In the case of adrenalectomized rats, restoring the original mineralocorticoid hormone concentration reverses the chronic sodium loss (Denton, 1965). The behavioral adaptation of ingesting sodium to compensate for the loss of sodium can then be terminated. But with higher concentrations of circulating mineralocorticoid, the rat's brain is activated to promote the search for and ingestion of sodium as if it needed sodium. In other words, regions of the brain that regulate the behavioral end of sodium homeostasis in the body are activated, and the animal acts as if it were in need of sodium. The hormone therefore maintains sodium homeostasis under normal circumstances by promoting physiological responses to conserve and redistribute sodium, and then acts on the brain to elicit the behavior of sodium ingestion (Wolf, 1965; Schulkin, 1991a,b). Both the physiological and behavioral responses have the same end point: to maintain extracellular sodium balance.

As already indicated, both mineralocorticoids and glucocorticoids, in addition to angiotensin, are playing roles in the regulation of the body's sodium homeostasis, and during sodium depletion, or extracellular fluid loss, both mineralocorticoid and glucocorticoid hormones are elevated (Stricker et al., 1979; Denton, 1982). In rats, glucocorticoid activation does not typically result in increased sodium ingestion by itself (e.g., Braun-Menendenz, 1953; Wolf, 1965). But when glucocorticoids are combined with mineralocorticoids, the ingestion of sodium is greater than when mineralocorticoids are given alone (Wolf, 1965; Coirini, Schulkin, and McEwen, 1988; Zhang, Epstein, and Schulkin, 1993; Ma et al., 1993). This phenomenon was noted when corticosterone was combined with aldosterone (Coirini et al., 1988; Zhang et al., 1993; Ma et al., 1993), when aldosterone was combined with dexamethasone (Ma et al., 1993), when deoxycorticosterone was combined with dexamethasone or with corticosterone (Wolf, 1965; Coirini et al., 1988; Rodd et al., in press), and when aldosterone was combined with RU 28362, which is a glucocorticoid agonist (Devenport and Stith, 1992; Ma et al.,

1993). The enhanced appetite seems to be independent of sodium loss and is therefore primary to the corticosteroid activation in the brain (Ma et al., 1993).

Why should glucocorticoids potentiate mineralocorticoid-induced sodium ingestion? First, the two hormones evolved from a common molecular ancestor, and the distinction between the two hormones is less clear in lower organisms (Arriaza et al., 1988). Second, and importantly, glucocorticoids can increase the number of aldosterone-preferring type-1 corticosteroid sites in brain (Coirini et al., 1988). This increase in binding sites could then increase the natriorexegenic efficacy of the mineralocorticoid hormones. Indeed, because there are many fewer aldosterone-preferring type-1 corticosteroid sites than glucocorticoid-preferring type-2 corticosteroid sites in brain, this mechanism would ensure the uptake of the mineralocorticoid, an event that occurs through an enzyme (11β-hydroxy-steroid dehydrogenase) that degrades and thereby reduces glucocorticoid binding to aldosterone-preferring type-1 corticosteroid sites (Funder et al., 1988; Funder, 1993). Therefore, glucocorticoid hormones may have an important role in the expression of sodium appetite.

Perhaps we can now understand one of the puzzling facts about mineralocorticoid-induced sodium appetite, namely, that deoxycorticosterone produces greater ingestion of sodium in rats and mice (Ma et al., 1993; Blair-West et al., 1995) than does aldosterone, the naturally occurring mineralocorticoid hormone. Why? Aldosterone is a pure mineralocorticoid (whereas deoxycorticosterone is not), and it has glucocorticoid properties in addition to mineralocorticoid properties. Hence the reason that deoxycorticosterone may be more natriorexegenic is because it is like giving the animal glucocorticoid hormone in addition to mineralocorticoid hormone.

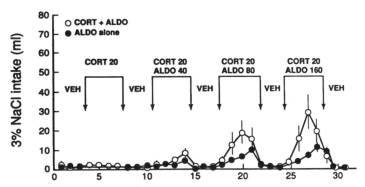

Figure 2.7. Intake of NaCl over 24 hours by rats treated with vehicle (VEH), corticosterone (CORT), and various doses of aldosterone (ALDO). (From Ma et al., 1993.)

Sites of Action for Corticosteroid-induced Sodium Appetite

There is evidence that the phenomenon of adrenal-steroid-induced sodium appetite is centrally induced, for the following reasons: Mineralocorticoid antagonists are known to block mineralocorticoid-induced sodium ingestion (Wolf, 1969a; Ma et al., 1992), while leaving intact sodium ingestion due to other causes (pregnancy-induced sodium appetite). Peripheral blockade of mineralocorticoid receptors does not affect mineralocorticoid-induced sodium appetite, but central blockade does (Ma et al., 1992). What would be the likely sites for such effects in the brain?

Several studies of receptor binding in the brain have indicated that the amygdala is a site of action for adrenal steroids (e.g., Coirini et al., 1988). Although the hippocampus contains aldosterone-preferring sites (e.g., Ermisch and Ruhle, 1978; Birmingham, Sar, and Stumpf, 1984), it is not importantly involved in regulating homeostatic behaviors like sodium appetite (e.g., Magarinos et al., 1986). The medial amygdala is involved in other steroid-mediated behaviors (e.g., Lehman and Winans, 1980; see Chapters 3 and 4, this volume). Importantly, the amygdala is known to be involved in sodium appetite (e.g., Nachman and Ashe, 1974). In addition, one site in the brain that contains both type-1 and type-2 corticosteroid sites for these receptors (Coirini et al., 1988; Arriaza et al., 1988; S. Makino, P. W. Gold, and J. Schulkin, unpublished observations) is the amygdala, specifically the medial nucleus, which caudally becomes the amygdala–hippocampal transition zone (Canteras, Simerly, and Swanson, 1992). We therefore looked at the role of this nucleus in corticosteroid-induced behavior.

Electrolytic lesions of the medial nucleus of the amygdala reduced or abolished aldosterone- or deoxycorticosterone-induced sodium appetite, and the effect was specific for adrenal-steroid-induced appetite (Schulkin, Marini, and Epstein, 1989; Nitabach, Schulkin, and Epstein, 1989). Ibotenic acid, which destroys cell bodies, leaving fibers intact within the medial nucleus of the amygdala, abolished corticosteroid-induced sodium appetite, whereas angiotensin-induced sodium appetite was unaffected (Zhang et al., 1993).

In addition to lesion studies, there is evidence that central administration of mineralocorticoid hormone to the medial amygdala elicits sodium ingestion, whereas angiotensin infusions do not elicit this response, nor do implants of aldosterone placed elsewhere in the brain (M. Nitabach and J. Schulkin, unpublished observations; Reilly et al., 1993). Moreover, mineralocorticoid antagonists (RU 28318) or mineralocorticoid-receptor antisense oligonucleotides applied to this region abolish the sodium appetite induced by intravenous aldosterone (Ma et al., 1992, 1997). Intracerebral injections of mineralocorticoid receptor antisense into the medial nucleus of the amygdala abolishes mineralocorticoid-induced sodium appetite (L. Y. Ma, unpublished observations). This evidence, coupled with the receptor localization studies and lesion studies, argues that the medial nucleus of the amygdala is one site importantly involved in the arousal of sodium appetite by corticosteroid hormones.

Evidence reported over the past several years clearly indicates that aldosterone, like other steroid hormones, not only has genomic effects but also has membrane effects (Wehling et al., 1992; Joels and De Kloet, 1994). Our recent evidence suggests that under some conditions one can see relatively rapid natriorexegenic effects (i.e., within minutes) when the hormone is applied to the medial nucleus of the amygdala (Reilly et al., 1993, in press), suggesting that to some extent these effects may not reflect genomic actions via classic intracellular changes, but rather may be mediated by the actions of neuroactive steroids, generated by metabolism of deoxycorticosterone in brain (Paul and Purdy, 1992). In fact, there is evidence that sodium appetite is elicited within 15 minutes when aldosterone or TH-aldosterone is applied to the medial amygdala. The effect is specific for sodium, because neither sucrose ingestion nor water ingestion is increased. These rapid effects suggest that these hormones may be acting via GABA-A or glutamate receptor systems (Reilly et al., 1993, in press).

But both corticosteroid hormones also bind to receptors and promote protein synthesis by DNA-related RNA synthesis (Edelman, 1978). The mineralocorticoid activates the sodium cellular electrogenic pump via protein-dependent changes (Edelman, 1978). Interestingly, there are sodium-transport changes in the amygdala following mineralocorticoid treatment (Grillo et al., 1989), and sodium transport is known to influence sodium ingestion (Michell, 1976; Denton, 1982). Both events depend upon cellular mechanisms for protein synthesis (Edelman, 1978). This suggests that one mechanism by which mineralocorticoid-induced sodium ingestion may be elicited is via changes in sodium-transport-dependent protein synthesis in the medial amygdala.

Stress-induced (Corticosterone- and Corticotropin-releasing Hormone-induced) Sodium Intake

There are three distinct instances in which sodium hunger is generated in response to humoral signals. The first and perhaps most important is during pregnancy and lactation (see Chapter 1). The second is what I have de-

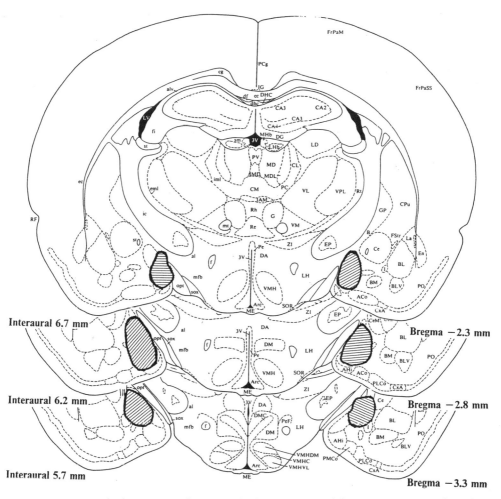

Figure 2.8 Hatched areas are those in which corticosteroid hormones were implanted within the medial nucleus of the amygdala.

scribed in this chapter, namely, when the hormones that are essential to maintain sodium balance and extracellular fluid volume are elevated. The last instance is when the hormones of stress are elevated (corticotropin-releasing hormone, ACTH, corticosterone).

Stressful conditions can lead to increased sodium intake. For example, when rabbits are constrained by jackets that limit their movements, one result is greater NaCl ingestion (Denton, 1982). This treatment also results in activation of the hypothalamic-pituitary-adrenal axis.

In Chapters 4 and 5 we shall discuss in detail the relationships among corticosterone, corticotropin, and corticotropin-releasing hormone. Adren-

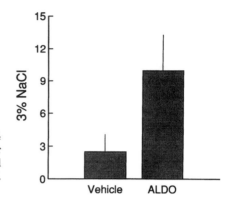

Figure 2.9. Ingestion of sodium over the 24 hours following administration of vehicle or aldosterone directly delivered to the medial region of the amygdala (R. Goldman and J. Schulkin, unpublished observations).

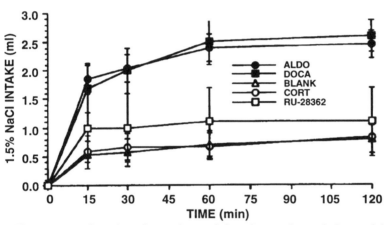

Figure 2.10. Bilateral implants of steroid directly into the medial amygdala produce rapid increases in the intake of 0.25-M NaCl. Intake of saline solution was measured immediately after steroid implantation. The steroids of greatest efficacy were DOCA and ALDO. Implants of CORT and RU 28362 were ineffective, showing no differences from blank implants. (Adapted from Reilly et al., 1993.)

alectomized or hypophysectomized rats cannot tolerate stress; in fact, they often die. Hypophysectomized rats have compromised natriorexegenic responses to depletion of body fluid (Jalowiec, Stricker, and Wolf, 1970; Schulkin et al., 1989), as do sheep (Denton, 1982). But when ACTH is given to rats or sheep, water intake and sodium intake are increased (Weisinger et al., 1978; Denton, 1982; Blair-West et al., 1996). In sheep, the increased ingestion of sodium is independent of adrenal output, because adrenalectomized sheep maintained with sufficient amounts of aldosterone, so as not to lose sodium, still generate a sodium appetite when treated with corticotropin. The same holds for rabbits (Blaine et al., 1975), but not for rats (Weisinger et al., 1978).

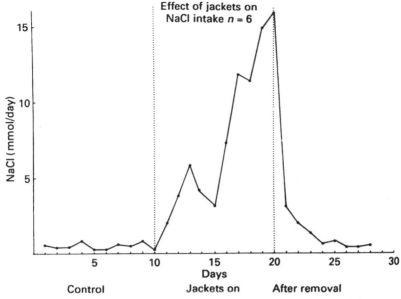

Figure 2.11 Sodium ingestion by rabbits following stress (by restraint). (From D. A. Denton, 1982, *The Hunger for Salt,* Berlin: Springer-Verlag. Reprinted by permission.)

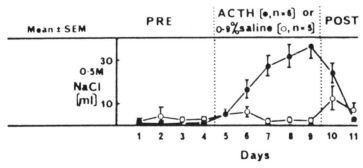

Figure 2.12. Mean intake of salt following ACTH injection in rats (90 IU/day). (Reprinted with permission from *Pharmacology Biochemistry & Behavior* 8, R. S. Weisinger, D. A. Denton, M. J. McKinley, and J. F. Nelson, ACTH induced sodium appetite in the rat, pp. 339–43, 1978, Elsevier Science Inc.)

Whereas glucocorticoids given alone to rats are not natriorexegenic, when given to rabbits or sheep they are (Blaine et al., 1975). In fact, in rabbits, when both hormones of stress were given together systemically, the appetite was greater than when either hormone was given alone (Blaine et al., 1975).

Given the foregoing, it may not be surprising that conditions that mimic the release of ACTH and glucocorticoid hormones can also result in

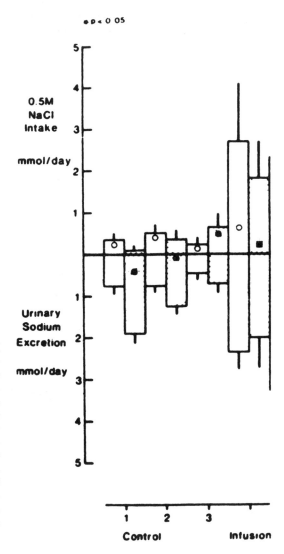

Figure 2.13. Daily intake of 0.5-M NaCl solution, urinary sodium excretion, and daily sodium balance (sodium ingested with 0.5-M NaCl and food, minus sodium excreted in the urine) of wild rabbits in rabbits infused with corticotropin-releasing hormone. Values are means (± SEM). (Reprinted from *Brain Research Bulletin*, 26, E. Tarjan and D. A. Denton, Sodium/water intake of rabbits following administration of hormones of stress, pp. 133–6, copyright 1991, with permission from Elsevier Science.)

increased sodium ingestion. There are two instances that come to mind. The first is immobilization-induced stress. When rats are released from this condition, their intake of sodium is increased. Of course, they also will have lost sodium during the immobilization. The second concerns conditions of aggression and crowding among male mice.

Centrally infused corticotropin-releasing hormone typically decreases food appetite (Glowa et al., 1992; Spina et al., 1996; see Chapter 3, this volume). But this does not hold for sodium ingestion, for infusions of corticotropin-releasing hormone elicit sodium ingestion (Tarjan and Denton, 1991). The ingestion is independent of sodium loss.

Interestingly, corticotropin-releasing-hormone mRNA is decreased in the central nucleus of the amygdala following hypertonic-saline infusions, but it is also decreased in the paraventricular nucleus (PVN), which is linked to corticosterone-regulated negative feedback (Watts, 1996). The central nucleus of the amygdala is at least one site that integrates signals from both angiotensin and corticotropin-releasing hormone in generating a sodium appetite.

Anatomic Circuit

An anatomic circuit by which corticotropin-releasing hormone ACTH and corticosterone would participate in stress-induced sodium intake should have the central nucleus of the amygdala playing a role in this response. This is because of its role in the regulation of stress and other corticotropin-releasing-hormone-induced behaviors tied to cardiovascular regulation and stress.

Angiotensin- and Corticosteroid-induced Water and Sodium Appetite

There have been indications that the hormones of sodium homeostasis may be working together to arouse a sodium appetite, as reviewed by Fregly and Rowland (1985). The "synergy hypothesis" was first tested using deoxycorticosterone in combination with central angiotensin (Fluharty and Epstein, 1983; Zhang et al., 1984; Fitzsimons and Fuller, 1985). At doses of angiotensin and mineralocorticoid hormone that when given individually would not raise the appetite for sodium, the two given in combination did increase appetite. Moreover, at higher doses, when sodium appetite would be increased by the dose of angiotensin, the effect was even greater when it was combined with mineralocorticoid hormone.

The same phenomenon was later demonstrated using aldosterone in concert with central angiotensin (Sakai, 1986). Stricker (1983) examined sodium ingestion among rats on a sodium-deficient diet and treated with an agent that depleted extracellular fluid and concluded that the exaggerated appetite that developed resulted from a synergistic interaction of aldosterone and angiotensin. Moreover, the two hormones when combined produced motivated behavior (scurrying down a runway) (Zhang et al., 1984) that reflected the concentrations of hormones; that is, the greater the amount of hormone administered, the faster the running speed and the higher the sodium ingestion. Finally, expression of synergistically increased sodium appetite appears early in the life of neonates, before independent ingestion of sodium (Thompson and Epstein, 1991), as also demonstrated in other species (e.g., pigeons) (Massi and Epstein, 1990).

The hormonal treatment effects are central, although there is a peripheral contribution to the behavioral expression of sodium ingestion (Thunhorst, 1996). But the combination of peripheral angiotensin and aldosterone does not produce such effects on sodium appetite (Sakai and Epstein, 1990). Moreover, when both hormones are blocked in the brain, the arousal of sodium appetite that results from sodium depletion is abolished (Sakai, Nicolaidis, and Epstein, 1989); when the hormones are blocked peripherally, that does not occur (Sakai and Epstein, 1990); when either hormone is blocked alone, the appetite is expressed, but in muted form (Weiss, Moe, and Epstein, 1986; Sakai et al., 1989). Thus the infusion studies of the hormones, taken in conjunction with the blocker studies, provide a case that these hormones of sodium homeostasis raise the appetite by their actions in the brain, and when they are blocked there is no sodium appetite expressed to sodium deprivation (however, see Thunhorst, 1966).

Other studies have demonstrated functional relationships between corticosteroids and the central regulation of angiotensin. For instance, mineralocorticoids are known to increase the number of angiotensin receptors in the brain (Wilson et al., 1986; King, Harding, and Moe, 1988). The same treatment results in upregulation of angiotensin receptors in primary cultures of neonatal rat brain and is mediated by aldosterone-preferring type-1 corticosteroid receptors (Sumners et al., 1991a,b). The same doses of mineralocorticoids combined with central angiotensin resulted in greater sodium ingestion than when angiotensin was given alone (Fluharty and Epstein, 1983; King et al., 1988). In those studies, the mineralocorticoid hormone that was used was DOCA, which contains both adrenal steroids (i.e., aldosterone and corticosterone). Interestingly, if one compares sodium ingestion among rats treated with either aldosterone or deoxycorticosterone in combination with intraventricular infusions of angiotensin, the intake by the deoxycorticosterone group is greater (cf. Fluharty and Epstein, 1983; Sakai, 1986). In addition, we know that subsequent studies looking at mRNA for angiotensinogen have shown that corticosterone, not aldosterone, resulted in increased activation of angiotensinogen mRNA in the brain (Angulo, Schulkin, and McEwen, 1988).

In addition, corticosterone increases angiotensin-induced water drinking as well as mineralocorticoid-induced sodium appetite. With regard to angiotensin potentiation by corticosterone, it occurs via AT2 angiotensin receptors (Sumners et al., 1991a,b).

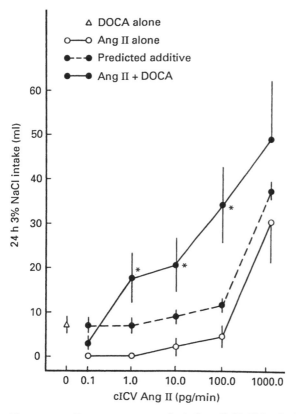

Figure 2.14. Dose–response analysis for 3% NaCl intake (means ± SE) following systemic DOCA at 2 mg/day or continuous intracerebroventricular infusion of angiotensin II (Ang II), or both, in rats treated with sesame oil. The intakes were recorded every 24 hours, and the duration of the infusion was 3 days. Also plotted are hypothetical curves in which 3% NaCl intakes were predicted by adding the separate effects of Ang II and DOCA when each was administered alone. (From S. J. Fluharty and A. N. Epstein, 1983, Sodium appetite elicited by intracerebroventricular infusion of angiotensin II in the rat: II. Synergistic interaction with systemic mineralocorticoids, *Behavioral Neuroscience* 97:746–58. Copyright 1983 by the American Psychological Association. Reprinted with permission.)

Additional insights into the interactions of both adrenal steroid hormones on central angiotensin have emerged from work on cultured neuronal cell lines. Specifically, in several neuroblastoma cell lines, corticosterone and the glucocorticoid agonist RU 28362 upregulate both angiotensin AT1 and AT2 receptors; the same effect has been demonstrated in brain tissue (Maki et al., 1992). Aldosterone, by contrast, only seemed to upregulate the AT2 angiotensin sites, to amplify stimulatory cGMP responses, and to regulate vasopressin-stimulated pressor responses. Type-2 corticosteroid agonists, such as RU 23986, increase levels of both receptor subtypes and enhance angiotensin-stimulated calcium mobilization and cGMP production. The endogenous steroid corticosterone produces the same pattern of receptor–effector changes. In this regard, phospholipase C (PLC-α), which appears coupled to angiotensin receptors in neuronal-like tissue, is also upregulated by steroid action.

This upregulation of angiotensin appears to depend upon genomic effects, because it can be blocked with actinomycin, a protein-synthesis blocker, and because injections of RNA isolated from steroid-treated cells into oocytes result in twofold to threefold increases in angiotensin-receptor expression (S. J. Fluharty, unpublished observations). When taken together, these results suggest that the gene(s) encoding for the multiple types of angiotensin receptors possess a steroid-responsive element that may underlie adrenal steroid facilitation of angiotensin in brain and its effects on sodium and water ingestion.

The increase in corticosteroid activation, along with angiotensin, mimics the endocrine profile of the body during sodium and extracellular fluid depletions (Stricker et al., 1979; Denton, 1982). The result is ingestion of sodium. Adrenal steroid hormones have genomic actions that prime the brain for angiotensin by upregulating angiotensin receptors and increasing gene transcription. Thus both adrenal steroid hormones, in concert with angiotensin, arouse the hunger for sodium.

Corticosterone appears to be doing two important things in the arousal of a sodium appetite and thirst: The first is to induce greater binding of aldosterone-preferring type-1 corticosteroid receptors in the brain (Coirini et al.,

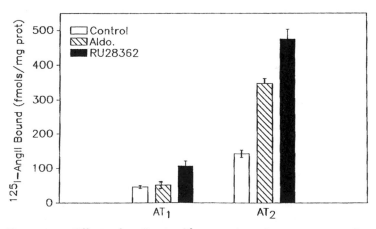

Figure 2.15. Effects of corticosteroids on angiotensin receptor types in cultured neuronal cell lines. Cells were treated with steroids for 20 hours, and angiotensin receptors were measured in membrane preparations (S. J. Fluharty, unpublished observations; Schulkin and Fluharty, 1993).

1988). Second, corticosterone increases the concentration of the angiotensin precursor angiotensinogen (Angulo et al., 1988). This precursor is found only in glial cells, suggesting that this interaction ultimately should involve neuron/glial interactions, as is true for the second-messenger system briefly described earlier. In other words, it appears that there are multiple cellular mechanisms that underlie the amplification of the action of angiotensin induced by adrenal steroids in the brain. On the one hand, type-II occupancy increases angiotensin production, thus providing more ligand to the receptors. In addition, both type-I and type-II genomic actions result in an increased density of receptors to bind the ligand and amplification of the cellular signaling mechanisms that transduce the binding of angiotensin to cellular actions.

Sites of Action of Angiotensin- and Corticosteroid-induced Water and Sodium Appetite

The previously described functional studies have implicated the AV3V region as the site of action for angiotensin-induced sodium appetite, and the medial amygdala for adrenal-steroid-induced sodium appetite, but which sites are responsive to both adrenal steroid and angiotensin hormonal signals? The central nucleus of the amygdala appears a plausible candidate.

The central nucleus of the amygdala receives the densest projections from brainstem gustatory sites and more general visceral information (Herrick, 1905; Norgren, 1976, 1984). It concentrates both hormones and is rich in angiotensin nerve terminals (e.g., Lind, 1988). The central nucleus of the amygdala is involved in cardiovascular regulation (Galeno et al., 1982). Moreover, as indicated earlier, recent investigations using c-fos as an anatomic marker have found that angiotensin infusions activate early gene expression in the central nucleus of the amygdala; that also occurs during sodium depletion (Herbert, 1993; Rowland et al., 1994). These areas include, in addition to the central nucleus of the amygdala, the AV3V region, the subfornical organ, the bed nucleus of the stria terminalis, the PVN, and the supraoptic nuclei.

The body's sodium homeostasis is intimately tied to cardiovascular regulation, and gustation is the specific sensory channel for identifying sodium in the environment (Schulkin, 1982; Norgren, 1984). Lesions of the central nucleus of the amygdala disrupt sodium appetite. We have found that both thirst and sodium appetite induced by central administration of renin-angiotensin and mineralocorticoid-induced sodium appetite are abolished by central-nucleus lesions, though angiotensin-induced thirst remains relatively intact (Galaverna et al., 1992).

The central nucleus of the amygdala may also be importantly involved in yohimbine-induced sodium appetite (featuring rapid changes in blood pressure that are linked to angiotensin) (Thunhorst and Johnson, 1994). After all, it is known that norepinephrine can potentiate angiotensin-induced sodium appetite (Chiaraviglio, 1976).

As indicated earlier, it has been known for some time that injections of angiotensin within the medial bed nucleus of the stria terminalis can elicit thirst (Swanson et al., 1978). The medial and central nuclei of the amygdala are anatomically tied to the bed nucleus of the stria terminalis (in fact, are continuous with it) (Johnston, 1923; Alheid and Heimer, 1988), and the bed nucleus has been called the "extended amygdala" (see Chapter 5). Lesions of either the medial or central nucleus of the amygdala affect sodium ingestion, and perhaps sites within the bed nucleus of the stria terminalis may also be important. Lesions of the bed nucleus are known to reduce sodium-depletion-induced sodium appetite (Zardetto-Smith, Beltz, and Johnson, 1994) and mineralocorticoid-induced sodium appetite (Reilly et al., 1994). It is possible that oxytocin inhibition may be via the bed nucleus of the stria terminalis, where it might interact with angiotensin or corticosteroids. Moreover, tachykinins, which inhibit both thirst and sodium appetite, exert their actions in both the medial amygdala and the medial bed nucleus of the stria terminalis (Pompei et al., 1992). Studies are needed featuring application of these hormones directly to this region to determine whether or not the appetite for sodium is aroused.

Atrial Natriuretic Peptide and the Inhibition of Water and Sodium Ingestion

Atrial natriuretic peptide (ANP) hormone has behavioral effects on water and sodium ingestion that follow logically from its known involvement in the body's sodium homeostasis and its regulation of the renin-angiotensin-aldosterone system (McCann and Antunes-Rodrigues, 1996). We know that centrally administered ANP decreases the sodium appetite in sodium-depleted rats (Fitts, Thunhorst, and Simpson, 1985) and rabbits (Tarjan, Denton, and Weisinger, 1988). Peripherally administered ANP shows no such effects (Fitts et al., 1985; Tarjan et al., 1988).

ANP also reduces the motivated behavior of the sodium-hungry rat. For both sodium-depleted rats and adrenal-steroid-treated rats, running speed for sodium salts was reduced following centrally administered ANP (P. Arnell and E. Stellar, unpublished observations). Moreover, thirst and sodium appetite induced by centrally administered angiotensin are reduced by central administration of ANP (Masotto and Negro-Vilar, 1985; McCann, Francis and Antunes-Rodrigues, 1989). Other dipsogenic responses are not reduced by ANP treatment (e.g., carbachol-induced or intracellularly induced thirst) (Fitts et al., 1985). Moreover, ANP does not seem to reduce other forms of ingestive behavior (e.g., sucrose consumption) (P. Arnell and E. Stellar, unpublished observations). Taken together, these findings suggest that ANP selectively acts on the hormones of sodium-hunger elicitation in reducing sodium ingestion.

The physiological background for the role of ANP in behavior is briefly the following: For years there was thought to be a natriuretic hormone, based on studies of sodium excretion (McCann and Antunes-Rodrigues, 1996). Moreover, we now know that there are multiple forms of atrial natriuretic factor (ANF) (Saper, 1995). Morphological studies have demonstrated that cells in the atria resemble endocrine cells; the numbers of these cells can be increased by sodium loading and decreased by sodium restriction. It has been demonstrated that crude atrial extract exhibits some of the most potent natriuretic-diuretic properties known (DeBold et al., 1981). Cardiac-volume receptors are coupled to the endocrine function of the atria, and thus an increased plasma volume stimulates atrial receptors by distension, in addition to the increased sodium concentration releasing ANP (Cantin and Genest, 1985). Once in the circulation, ANP activates a variety of target organs, including kidney, adrenal cortex, vasculature, and brain, in such a way as to reduce plasma volume. The phenomenon is general, as it has been expressed in a wide variety of species (e.g., fowl).

One of the ways in which ANP participates in the body's fluid homeostasis is by interacting with other hormonal regulators such as the renin-angiotensin-aldosterone system. This hormone is the antagonist of the renin-angiotensin-aldosterone system, For instance, synthetic ANP has been

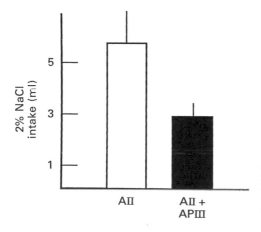

Figure 2.16. Effects of atropeptin III (AP III) on angiotensin-II-induced sodium intake. (Adapted from Masotto and Negro-Vilar, 1985.)

shown to inhibit in vitro and in vivo renin release (Brands and Freeman, 1988). The effect appears to reflect a direct action of ANP on the juxtaglomerular cells that may be mediated in part by changes in cyclic-nucleotide production, particularly a decrease in cAMP and an increase in cGMP. It appears to cause a tonic inhibition of renin release. In addition to limiting the production of angiotensin, ANP opposes many of the physiological actions of angiotensin and the release of aldosterone (Brands and Freeman, 1988). The observation that ANP inhibits aldosterone production elicited by dibutyryl-cAMP suggests that the site of action for ANP is distal to the hormone receptors and may involve inhibition of adenylate cyclase and/or stimulation of guanylate cyclase activity in adrenal cortical membranes and in brain (Israel et al., 1988).

Sites of Action for ANP Inhibition of Water and Sodium Appetite

Although the name implies that ANP is confined to the heart, it is expressed in the brain (e.g., Skofitsch and Jacobowitz, 1985a,b). Manipulations of the blood volume perfusing the atrium or impairments of atrial function will interfere with water ingestion in response to volume depletion (Fitzsimons, 1979). It is now clear that one of the most interesting extracardiac sites is in the brain. In general, it has been shown that immunoreactive cell bodies and fibers are widely distributed throughout the brain (Skofitsch and Jacobowitz, 1985a, b). In the telencephalon, immunoreactivity for ANP has been localized in the bed nucleus of the stria terminalis and both the central and medial amygdala, and the diencephalon sites show very strong immunoreactivity in the AV3V region, in addition to hypothalamic nuclei. Immunoreactivity has been observed in brainstem sites, including the parabrachial and solitary nuclei (Gibson et al., 1986). In addition to the apparent presence of ANP-containing neurons, autoradiographic techniques have demonstrated the existence of receptors for this peptide(s) in a variety of brain regions, including circumventricular organs and brainstem sites such as the solitary nucleus (Zamir et al., 1986). These receptors are specific for ANP, are saturable, and exhibit high affinity.

The presence of ANP-binding sites in circumventricular organs provides a route through which circulating ANP might regulate some aspects of the central-nervous-system control of the body's sodium homeostasis. ANP infused directly into the subfornical organ reduces angiotensin-induced water drinking (Ehrlich and Fitts, 1990). Interestingly, ANP injected into the AV3V region (OVLT) does not reduce the sodium ingestion consequent to angiotensin-related captopril treatment (Fitts, 1993). These data suggest that

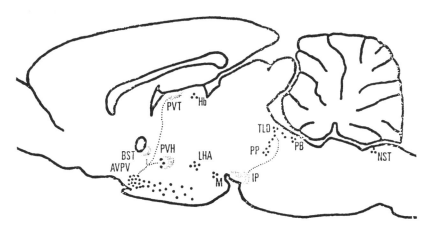

Figure 2.17. ANP sites in the brain of the rat, including the solitary nucleus (NST), parabrachial region (PB), lateral hypothalamic region (LHA), bed nucleus of the stria terminalis (BST), paraventricular nucleus of the hypothalamus (PVH), and anterior ventricular posterior region (AVPV). (Reprinted from *Trends in Neuroscience* 8, D. G. Standaert, C. B. Saper, and P. Needleman, Atriopeptin: potent hormone and potential neuromediator, pp. 509–11. Copyright 1985, with permission from Elsevier Science.)

ANP may be acting on one site in the brain to decrease the angiotensin signal for thirst, with another site for sodium, though ANP's site of action for sodium intake remains unclear. The possible functions of ANP in other brain sites are not known, but like other peptides, it may be acting as a neurotransmitter or neuromodulator within these circuits. Interestingly, central oxytocin inhibition of angiotensin-induced sodium ingestion may also be linked to central ANP (Blackburn et al., 1992).

Anatomic Circuit for Sodium Appetite in Response to the Hormones That Regulate Extracellular Fluid and Sodium Balance

Initiation of the appetite for sodium depends upon forebrain structures for its expression. When the hormones of sodium homeostasis are elevated, decerebrated rats will not increase their ingestion of sodium when sodium is infused directly into the oral cavity (Grill et al., 1986), whereas they will increase their food intake when metabolically deprived (Miller and Sherrington, 1915; Grill, 1980; see Chapter 3, this volume). Moreover, infusions of angiotensin into the fourth ventricle will not elicit sodium ingestion (Fitts and Mason, 1989b). Perhaps this difference in neural levels of control (Jackson, 1958) reflects the fact that the behavioral regulation of the body's mineral and fluid homeostasis was selected for when animals like ourselves emerged from the sea, which is why the appetites for both water and sodium are more encephalized within the nervous system, as compared with hunger.

Sites in the forebrain responsive to the hormones of sodium homeostasis include the medial and central nuclei of the amygdala and regions of the bed nucleus or extended amygdala, in addition to the AV3V region. Also depicted in Figure 2.19 are other sites known to be involved in the body's fluid and sodium homeostasis (e.g., lateral hypothalamus, zona incerta) (Wolf and Schulkin, 1980; Grossman, 1990). They include gustatory sites in the solitary and parabrachial nuclei in the brainstem (Flynn et al., 1992; Scalera, Spector, and Norgren, 1995). These brainstem sites send visceral afferents and receive efferents from the amygdala and bed nucleus and the AV3V region and are known to be involved in sodium appetite and body fluid balance. This ventral pathway from the parabrachial nucleus into the forebrain may be the afferent limb of viscerally generated motivated behavior (Pfaffmann, Norgren, and Grill, 1977; Ohman and Johnson, 1986; Spector, 1995). Moreover, though the stria terminalis pathway is not essential for such hormonally induced behavior (Black et al., 1992), the amygdalafugal pathway is; transection of this pathway reduces both adrenal-steroid-

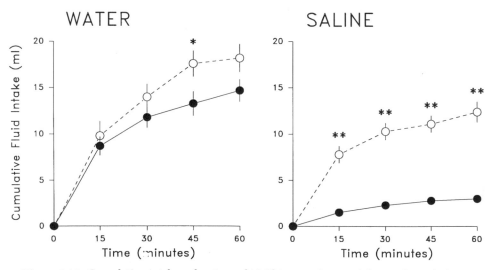

Figure 2.18. Cumulative intakes of water and NaCl in rats after receiving angiotensin intraventricularly during a 1-hour drinking test. Rats were treated 30 minutes earlier with ventricular injections of either an oxytocin-receptor antagonist (open circles) or vehicle (filled circles). (From Blackburn et al., 1992, with permission.)

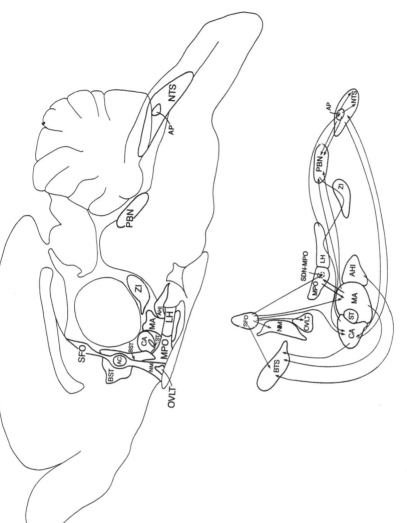

Figure 2.19. Characterization of rat brain showing areas that underlie body sodium homeostasis and the major connections. Abbreviations: AC, anterior commissure; AHI, amygdala-hippocampal regions; AP, area postrema; BST, bed nucleus of the stria terminalis; CA, central nucleus of the amygdala; LH, lateral hypothalamus; MA, medial nucleus of the amygdala; MPO, medial preoptic nucleus; NM, nucleus medianus; NTS, nucleus of the solitary tract; OVLT, organum vasculosum of the lamina terminalis; PBN, parabrachial nucleus; SDN-MPO, sexually dimorphic nucleus of the medial preoptic area; SFO, subfornical organ; ST, stria terminalis; ZI, zona incerta. (From Schulkin, 1991a, with permission.)

induced sodium ingestion and angiotensin-induced sodium ingestion, and it is known to contain angiotensin fibers (Chiaraviglio, 1971; Alheid, Schulkin, and Epstein, 1993). This pathway may be an efferent limb from the amygdala to midline sites in generating the appetite for sodium in response to the hormones of sodium homeostasis.

Thus, during sodium depletion and extracellular fluid depletion, renin-angiotensin-aldosterone and corticosterone are activated. Peripherally they act to conserve and redistribute sodium to maintain sodium balance. Centrally they act to generate a hunger for sodium. Both behavior and physiology have the same end point: the maintenance of sodium and extracellular fluid balances. Anatomically, consider the following working hypothesis for where the hormones of sodium homeostasis are acting in the brain: Angiotensin acts within midline structures surrounding the third ventricle (AV3V), where ANP exerts its inhibition on the appetite and perhaps sends inhibitory signals from the bed nucleus of the stria terminalis (oxytocin, tachykinins), in addition to elevated concentrations of sodium interacting with these signals to contribute to the inhibition of appetite. The medial amygdala mediates the corticosteroid activation of the appetite, and the central nucleus of the amygdala and perhaps the extended amygdala (bed nucleus) are responsive to both natriorexegenic signals.

Conclusion

Studies of sodium appetite and thirst are fundamental for understanding the regulation of the internal milieu (Bernard, 1957) and homeostatic regulation (Cannon, 1932). Curt Richter built on the insights of Bernard and Cannon and encouraged attention to the behavioral end point; both behavior and physiology are serving the same end: maintenance of the internal milieu. Such research is also of medical import, for the hormones of sodium and water homeostasis are tied to both extracellular fluid balance and cardiovascular regulation. The hormones of sodium and water homeostasis induce a central state, the state of sodium or water craving. The central state involves both appetitive and consummatory behaviors in the search for salt and its consumption. The anatomic circuits depicted are involved in a number of steroid- and peptide-regulated behaviors. In the case of sodium hunger or thirst, they are like other central states, which include sexual desires, food desires, the state of fear, and the desire to be attached to others.

CHAPTER THREE

Hormonal Regulation of Food Selection

Introduction

Wild animals spend much of their time foraging for food. Underlying these behavioral events are hormonal changes. Together, behavioral and endocrine mechanisms serve the end point of nutrient homeostasis. What is homeostasis? Broadly conceived, it is defined as maintenance of the internal milieu – keeping adequate amounts of glucose, sodium, calcium, and other substances within the body through physiological as well as behavioral means. Homeostasis is key to maintaining the well-being of all living organisms (Oftedal, 1991).

The concentrations of a number of hormones are elevated or lowered when animals are hungry and are searching for food. For instance, corticosterone (secreted by the adrenal gland) and neuropeptide Y (synthesized by neurons in the brain) together generate the central motive state of searching for and ingesting food (e.g., Dallman et al., 1993; Leibowitz, 1995). Other hormones are increased when an ingested meal is utilized. For example, increases in insulin (fundamental in glucose utilization and energy and fat metabolism) (Woods et al., 1979) and two other satiating hormones (cholecystokinin and oxytocin) (Gibbs, Young, and Smith, 1973; Verbalis et al., 1993; Smith, 1997) and a protein (leptin) (Seeley et al., 1996) signal the termination of a meal. These hormones also influence food selection and food avoidance.

Food selection involves appetitive (search) and consummatory (ingestion) behaviors that reflect the central state of hunger (Stellar and Stellar, 1985). Consider, for example, the appetitive and consummatory phases of motivation in hungry rats. Rats will run toward food when hungry. This phenomenon has long been a convenient tool with which to study motivated behavior in the laboratory. The rat's degree of hunger is reflected in the speed of its running (Hull, 1943; Miller, 1959–71), which depends upon the incentive of the reward (Stellar and Stellar, 1985). Running to a food source is an example of the appetitive phase of motivated behavior. The consummatory phase is the ingestion of the food. Measuring the relationship between these two events is one way researchers study motivated behaviors (Stellar and Stellar, 1985).

Hormonal changes underlie both appetitive and consummatory behaviors (Craig, 1918). Humoral signals that initiate appetitive ingestive behaviors (e.g., corticosterone, neuropeptide Y) and those that initiate consummatory ingestive behaviors (e.g., cholecystokinin, insulin, oxytocin) reflect the two phases of motivated behaviors. But these hormonal signals also underlie the distinction between short-term regulation and long-term regulation of energy balance and homeostasis (e.g., Strack et al., 1995).

This chapter describes the role that behavior plays in food selection (and therefore in maintaining homeostasis), as well as the steroid and peptide hormones associated with various behaviors. Many of the peptides are synthesized within the gustatory neural axis, which is fundamental in food selection and ingestion. We shall focus on several model systems in which we understand how hormones influence nutrient intake, beginning with the behavior of food selection.

Behavioral Strategies: Novel-Food Cautiousness

Curt Richter (1943, 1956) championed the idea that animals optimize their behavioral responses to ensure adequate diet. Except for occasional work with wild rats in the field, his work was mostly in the laboratory.* Whereas Richter emphasized the role of behavior – in addition to physiological and endocrine processes – in maintaining homeostasis, he emphasized reactive behavior (e.g., eating when glucose levels drop) as opposed to anticipatory behavior (e.g., eating in preparation for hibernation) (Moore-Ede, 1986). Yet anticipation governs many behaviors as well as the hormonal signals underlying them.

Bait shyness, or novel-food cautiousness, is an important behavioral adaptation among a wide variety of animals. Richter, among others (Barnet, 1963), demonstrated the phenomenon: the reluctance to sample new and unfamiliar food sources, sometimes accompanied by visceral illness (Richter, 1956; Rozin, 1967a,b). Expressed in nearly all omnivorous animals forced to sample new foods (Rozin, 1967a), it is one of the most pronounced behavioral adaptations used in foraging.

In the laboratory, rapid and long-lasting learning takes place when animals ingest something that is noxious (e.g., Garcia, Hankins, and Rusiniak,

---

*We shall refer to Richter's work in nearly every chapter. I believe that when Paul Rozin (1976b) described Richter as the "compleat psychobiologist" he was right (Schulkin, Rozin, and Stellar, 1994b).

1974), a phenomenon that has been called "taste-aversion learning" (Rozin, 1976a). Rats, for example, that ingest a food and then become sick are reluctant to eat that food again (e.g., Rodgers, 1967; Richter, 1976). They learn quickly, and the memory is long-lasting and profoundly imprinted upon the brain. The selection pressure for such behavior is clear: A food that has poisoned an animal must be avoided in the future if the animal is to survive and reproduce. A similar mechanism operates in dietary selection. Diets that make animals sick, when they are recognized as such, are avoided by animals (Rozin, 1976a). The behavior – which is specialized to gastrointestinal-related malaise – is expressed in a number of herbivorous and omnivorous vertebrates.

Another example of food-avoidance behavior comes from laboratory studies of induced thiamine deficiency in rats (Rozin, 1967b): After being deprived of thiamine during development over a 3-week period, rats are offered the choice to continue the deficient diet or begin a different diet. They almost always choose the novel diet. Because there is no specific taste for thiamine, the animal must learn to associate its ingestion with the restoration of homeostatic balance (Rozin, 1967b). The rats eat the new diet because they have learned that the old diet made them ill. As part of the behavioral adaptation of taste-aversion learning, this type of avoidance mechanism is general and operates in a number of contexts (Rozin, 1976a).

Behavioral Strategies: Self-selection of Diet

As far back as 1915, Evaard demonstrated that pigs (*Potamochoerus S. scrofa*), when offered nine choices of foods, could select those that would constitute a diet adequate for growth. By 1918 it had been shown that rats offered choices of protein or nonprotein foods preferred the protein (Osborne and Mendel, 1918). By 1925 it was commonly known that cattle would choose to consume phosphorus when they needed it. Years later, Denton (1982) and his group in Australia reported a striking example of this phenomenon: Cattle that were rendered phosphorus-deficient ingested the mineral immediately when they were exposed to it. The appetite for bone may be innate in phosphorus-deficient cattle. All of these early observations contributed to the concept of bodily wisdom – that animals "knew" what it was they needed (Cannon, 1932).

In a series of experiments that could never be repeated today, researchers demonstrated a similar ability among human infants (Davis, 1939). In a cafeteria setting, the experimenter offered neonates a choice of different purified foods, which were their only possible sources of nutrition. The infants chose a balanced diet. Since then, debate has raged as to the extent

Figure 3.1 Phosphate-deficient cows ingesting bone. (Courtesy of D. A Denton.)

to which those results were innate or learned (e.g, Rozin and Schulkin, 1990) versus artificial (Galef, 1991). Despite this criticism, humans do seem to have some innate knowledge of what their bodies need (e.g., sodium, calcium, carbohydrates, water). In all animals, what is needed for food recognition and selection is an internal detector to monitor changes in requirements, as well as an external detector to determine what in the environment can meet these metabolic needs (Rozin and Schulkin, 1990).

Two ways of looking at food choice have been described. One emphasizes that behavior reflects innate preferences: The mechanisms of choice are wired into the animal. Sodium hunger in the rat is a paradigmatic example of a hormone-dependent behavior that is innately organized (see Chapter 2). The other view is that preferences are learned. Food-selection research has often pitted the two views against one another (Rozin and Schulkin, 1990), but in reality both are operative in food choice.

Feeding and Feeling

In animals, including humans, whether to approach or avoid a food source is often mediated by hedonic (palatability) judgments. Preference, however, is not the same thing as hunger. One can prefer something but not need it

(Young, 1949); or one can need something but not prefer it, or even like it (e.g., Berridge, 1996). Hedonic mechanisms may in turn reflect changes in humoral signals (e.g., Berridge, Grill, and Norgren, 1981). Two kinds of hedonic judgments about food seem to underlie ingestive responses: one linked to acceptance and approach toward food sources, and one linked to avoidance (Berridge and Grill, 1984; Stellar and Stellar, 1985).

One emotion linked to food avoidance – which is essential in food selection (Rozin and Fallon, 1987), as noted by Darwin (1965) – is that of disgust. The disgust response orients an animal away from an object; the hedonic sense is negative. The human emotion of disgust is closely linked to ingestive behavior (Rozin and Fallon, 1987). We tend to associate inedible objects with disgust; contaminants like rotten food are paradigmatic examples of disgusting objects, as are feces (Rozin, 1996).

The facial responses to foods that are aversive differ from the responses to foods that are attractive in a variety of mammalian species (e.g., Steiner, 1977; Grill and Norgren, 1978a,b). Many species are known to avoid noxious stimuli and approach attractive ones. Sometimes the intensity of the stimulus determines the behavioral response – whether to approach or avoid the object (Schnierla, 1959).

The functional significance of palatability judgments ties hedonic assessment to the regulation of the internal milieu (Cabanac, 1971). Homeostatic needs can change the palatability of a taste (e.g., seawater when hungry for sodium) (see Chapter 2) (Berridge et al., 1984). A clear example of a change in hedonic assessment occurs following taste-aversion learning. When lithium chloride, which can induce taste aversion (Nachman, 1963), is combined with sucrose, that will change the favorable facial response to sucrose

DR. ELIOT STELLAR

"Brain Mechanisms in Hunger and Other Hedonic Experiences"

Figure 3.2. Caricature of a sated rat. (Courtesy of Eliot Stellar.)

to a response resembling the aversive response to quinine. The sucrose with lithium chloride, which now tastes bad, also affects hormones. For example, insulin secretion – normally stimulated by oral infusions of glucose – decreases following taste aversion (Berridge et al., 1981).

Sensory Influences on Food Selection

In birds, species in which the primary sensory system is vision, visual stimuli are easily linked to the animals' responses to ingesta (Hernstein, Loveland, and Cable, 1976). Chickens, for example, learn to associate gastrointestinal relief with visual cues linked to the ingesta. Young chicks also innately associate the visual properties of water with the body's fluid needs (Stricker and Sterritt, 1966).

Taste plays a major role in the food preferences of many species. In herbivores, omnivores, and carnivores alike, determining the gustatory characteristics of a potential food – whether it tastes salty, sweet, or bitter – is an important part of foraging patterns. Sweet tastes usually indicate that a food is nutritious, and bitter that it may be dangerous (Janzen, 1977).

Gustatory sensibility is phylogenetically ancient and has been preserved throughout evolution (Halpern, 1983). The anatomy of the gustatory system in vertebrates shows that the anterior tongue is maximally responsive to sweet and salty tastes, whereas the back of the tongue is responsive to bitter tastes (Nowlis, 1977). The front of the tongue initiates acceptance of a gustatory source, and the back initiates rejection. The front is innervated by the seventh nerve, and the back by the ninth cranial nerve in addition to the tenth nerve, which send gustatory afferents to the brain (Norgren, 1984). The seventh nerve is also linked to the perception of sweet tastes, and the ninth is linked to bitter tastes (Nowlis, 1977; Bartoshuk, 1988).

These three nerves project to the anterior portion of the solitary nucleus (Pfaffmann et al., 1977). Dietary manipulations (e.g., taste aversion, sodium hunger) affect the gustatory region of the solitary nucleus, as do hormones that initiate and stop food ingestion (e.g., glucagon) (Scott and Mark, 1986; Giza and Scott, 1987). The gustatory region integrates internal signals (sickness) with external events (ingesta). The nucleus of the solitary tract projects to the medial parabrachial region (Norgren, 1984), which is essential for taste-aversion learning (Flynn et al., 1992; Spector, 1995). This anatomic pattern holds true in organisms from fish to mammals (Herrick, 1905; Finger and Kanwal, 1992); and this neural circuit subserves both specific gustatory information and more general visceral information.

Gustatory afferents bifurcate as they ascend from the parabrachial nucleus through the ventral posterior region of the thalamus to the insular cortex

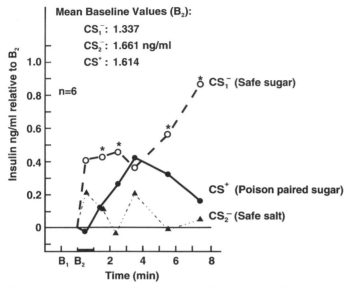

Figure 3.3. Relative preabsorptive insulin release following taste-aversion conditioning: (mean insulin responses (equated for baseline) to oral infusions of CS^+ (15% maltose or fructose), CS_1^- (15% maltose or fructose), and CS_2^- (0.15% NaCl). (From K. C. Berridge, H. J. Grill, and R. Norgren, 1981, Relation of consummatory responses and preabsorptive insulin release to palatability and taste aversions, *Journal of Comparative and Physiological Psychology* 95:363–82. Copyright 1981 by the American Psychological Association. Reprinted with permission.)

(Norgren and Wolf, 1975). The second route follows a ventral pathway that courses through the lateral hypothalamus and reaches the central nucleus of the amygdala and the bed nucleus of the stria terminalis (Norgren, 1976).

Gustatory neuroanatomy is particularly interesting because, first, it extends into the forebrain and, second, it is involved in a number of motivated behaviors beyond foraging and feeding (Herrick, 1905; Pfaffmann et al., 1977). Underlying several hormone-induced behaviors, this basic circuit is part of the visceral neural axis in the central nervous system. The circuit is replete with steroid-regulated neuropeptides that project from the parabrachial nucleus to ventral forebrain sites, namely the amygdala, bed nucleus, and paraventricular nucleus. These peptides include neuropeptide Y, cholecystokinin, corticotropin-receptor hormone, enkephalin, neurotensin, and galanin (e.g., Moga, Saper, and Gray, 1989). Corticosterone has receptor sites in many of these same regions (McEwen, 1992), indicating that the taste–visceral neural axis contains neuropeptides linked to the regulation of many motivated behaviors.

Figure 3.4 shows that the gustatory system, as part of the limbic circuit

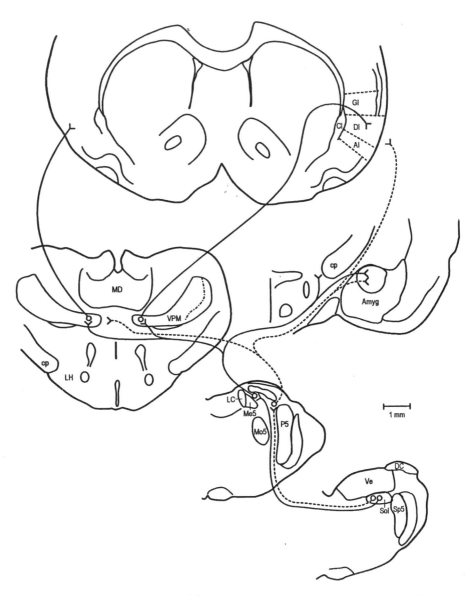

Figure 3.4. Schematic summary of the gustatory system in rat brain. Outlines of coronal sections through the rostral medulla (lower right), pons, diencephalon, hypothalamus and amygdala, and cerebral cortex, covering a rostrocaudal distance of about 12 mm. The solid lines connecting the panels represent axons known to convey gustatory information; dashed lines are axons associated with the taste system, but without documented sensory function. None of the lines follow actual pathways, nor do the bifurcations necessarily imply collateral projections. Abbreviations: AI, agranular insular cortex; Amyg, amygdala; Cl, claustrum; cp, cerebral peduncle; DC, dorsal cochlear nucleus; DI, dysgranular insular cortex; GI, granular insular cortex; LC, locus ceruleus; LH, lateral hypothalamus; MD, mediodorsal nucleus; Me5, mesencephalic trigeminal nucleus; Mo5, motor trigeminal nucleus; P5, principal sensory trigeminal nucleus; Sol, nucleus of the solitary tract; Sp5, spinal trigeminal nucleus; Ve, vestibular nuclei; VPM, ventral posteromedial nucleus. (From R. Norgren, 1995, Gustatory system, in *The Rat Nervous System,* Orlando: Academic Press, pp. 751–71. Reprinted by permission.)

underlying several hormone-induced behaviors, has centrifugal control over behavioral and physiological events (Powley, 1977; Norgren, 1984; Altschuler, Rinaman, and Miselis, 1992). Forebrain regions project down to hindbrain sites in regulating basic (taste reactivity) and physiological responses (gastric motility).

Anticipatory Secretion of Insulin: Gustatory and Conditioning Factors

Anticipatory responses predominate in the behavioral and physiological regulation of the internal milieu. Pavlov (1927) introduced the concept of cephalic phase, or cephalic influence, to explain these anticipatory responses (Powley, 1977). As mentioned previously, sweet tastes elicit preabsorptive insulin in laboratory rats. Whereas glucose is the best insulin stimulant, other sweet tastes will also cause an animal to secrete insulin, which in turn prepares the body to absorb nutrients (Grill, Berridge, and Ganster, 1984; LeMagnen, 1985). Oral factors that regulate insulin secretion also help regulate ingestive behavior by signaling the absorption of glucose and its subsequent utilization and the regulation of adiposity; these are fundamental roles for peripheral insulin (e.g., Friedman and Ramirez, 1987).

Elevated insulin concentrations facilitate the uptake, storage, and use of nutrients. Linked to the cephalic phase, this hormone figures prominently in foraging and feeding behavior. Powley reintroduced the concept of the cephalic phase in understanding the role of the ventral medial hypothalamus in elevated insulin concentrations and enhanced feeding (King et al., 1983; Powley and Berthoud, 1985). Damage to the ventral medial hypothalamus increases insulin secretion, whereas vagal damage reduces it (Powley, 1977). An elevated insulin concentration is secondary to the change in central state, namely, the increased hunger for food. The elevated concentration of insulin may reflect the greater anticipatory utilization of glucose (Berthoud and Powley, 1990).

This anticipatory secretion of insulin is linked to learning. Tying together insulin, learning, and homeostasis is the interesting observation that if rats are fed at the same time every day, they will begin to secrete insulin in anticipation of feeding (Woods, Hutton, and Makous, 1970). Rats that are fed only once each day (i.e., deprived of food for 23 hours) will begin to secrete insulin right before the anticipated availability of food at 24 hours, perhaps reflecting the activities of a circadian clock (see Chapter 6) (Dallman et al., 1993). But they can also learn to associate odors and other sensory cues not linked to time of day, but linked to a food source, that prompt

93

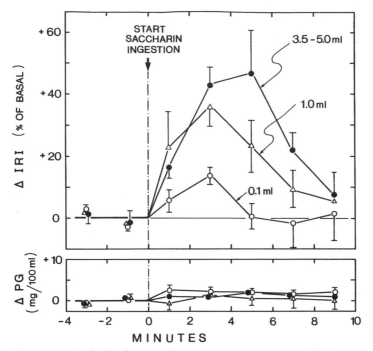

Figure 3.5. Cephalic-phase insulin responses to ingestion of different volumes of a solution of 0.15% saccharin in four freely moving male rats bearing chronic transjugular vena cava catheters and deprived of food for 16 hours. The animals had been trained to drink the saccharin solution readily. Blood was continuously sampled (100 μl/min) by means of a peristaltic pump. (From Powley and Berthoud, 1985, with permission; © Am. J. Clin. Nutr. American Society for Clinical Nutrition.)

insulin secretion (Woods et al., 1977). Presumably, insulin is secreted to facilitate digestion and metabolic use of the anticipated food.

The foregoing examples show how cephalic and gustatory influences facilitate insulin secretion, either by activation of the oral cavity or by learning an association between feeding and time of day. In both cases, anticipation triggers the secretion of insulin, which will facilitate the metabolic utilization of fuels.

Glucocorticoids and Food Ingestion

Glucocorticoid hormones secreted by the adrenal gland are fundamental in glucose metabolism. Food deprivation is known to activate the hypothalamic-pituitary-adrenal axis (Dallman et al., 1993). Released by the adrenal glands, glucocorticoids are involved in energy homeostasis and food intake

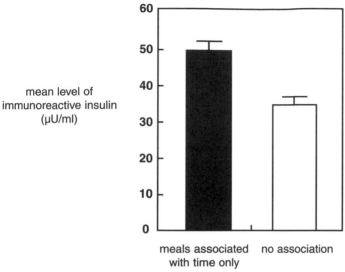

Figure 3.6. Mean levels of immunoreactive insulin for control and experimental rats. (Adapted from Woods et al., 1977.)

and thus are fundamental to the regulation of systemic physiology (Dallman et al., 1993, 1995; Borer et al., 1992). Low concentrations of corticosterone, for example, will stimulate food intake in rats, and high concentrations will reduce food intake (Dallman et al., 1993). Whereas adrenalectomy will reduce food consumption and increase fat utilization (e.g., Hamelink et al., 1994), moderate amounts of corticosterone replacement will restore normal patterns (e.g., Tempel and Leibowitz, 1994).

It is well known that lesions of the ventral medial hypothalamus result in hyperphagic, overweight animals (e.g., Powley, 1977). Corticosterone is essential for ventral-medial-hypothalamic-related hyperphagia (King et al., 1988; Dallman et al., 1993); adrenalectomy abolishes the hyperphagia, and corticosterone replacement reinstates it. This occurs through the activation of type-2 corticosteroid receptor sites (see Chapter 2 for a discussion of type-1 and type-2 corticosteroid receptor sites) (Thomas, Devenport, and Stith, 1994).

Though it may be taking advantage of laboratory artifacts to attempt to separate various metabolic requirements, researchers have found that systemic corticosterone injections affect carbohydrate intake, as judged by the choices rats make under cafeteria-type selections (Tempel and Leibowitz, 1994; cf. Kamara, Kamara, and Castonguay, 1992). Protein intake is unaffected by corticosterone treatment, though fat intake is altered by such treatment (Castonguay, Dallman, and Stern, 1986; Bligh, Douglass, and Castonguay, 1993). To the extent that glucocorticoids are fundamental in energy homeostasis and glucose metabolism (e.g., Kamara et al., 1992), this influence on carbohydrate intake seems to be adaptive. Short-term intake appears to be mediated by type-2 corticosteroid receptors linked to carbohydrate intake (see Chapter 2), a quick but incomplete way of meeting energy needs. But glucocorticoids and their effects on insulin secretion and neuroeptide-Y expression influence long-term regulation of energy balance (Dallman et al., 1993).

There is also evidence that aldosterone, the other corticosteroid (or deoxycorticosterone, a biosynthetic precursor to aldosterone, see Chapter 2), increases body weight when given systemically in combination with corticosterone (Devenport, Torres, and Murray, 1983). Moreover, recent evidence suggests functions for the brain's mineralocorticoid receptor (type-1) sites other than maintaining the body's sodium balance (see Chapter 2). Type-1 corticosteroid-mediated functions have been linked to fat ingestion and to long-term regulation of energy balance (Tempel and Leibowitz, 1994).

Researchers have long known that the paraventricular nucleus of the hypothalamus is involved in food ingestion and that noradrenergic signals within this nucleus, as well as other hypothalamic nuclei, play roles in activating feeding behavior. More than 30 years ago it was shown that local injections of norepinephrine into the hypothalamus stimulated food intake (Groosman, 1960), a finding that led Neal Miller (1959–71) to speculate that scientists were uncovering the chemical messengers underlying the organization of behavior.

Interestingly, at the level of the paraventricular nucleus, noradrenergic-induced feeding depends on corticosterone. Adrenalectomized rats do not show adrenergic-induced feeding responses (Leibowitz et al., 1984). Of course, it could be that these rats are feeling stronger and that their increased carbohydrate intake reflects this. Nonetheless, the site-specific findings are intriguing. An implant of corticosterone directly into the paraventricular nucleus in an adrenalectomized rat will increase its intake of carbohydrates, as compared with rats implanted with cholesterol. This effect, mediated by type-2 corticosteroid receptors, varies with the time of day (Kumar, Papamichael, and Leibowitz, 1988). Corticosterone's effect appears to be exerted in part through norepinephrine in the paraventricular nucleus. Along with its anatomic circuitry (Sawchenko, 1993), the paraventricular nucleus, in addition to other forebrain sites (central nucleus of the amygdala), regulates behaviors involving food ingestion, energy metabolism, and adaptation to adversity.

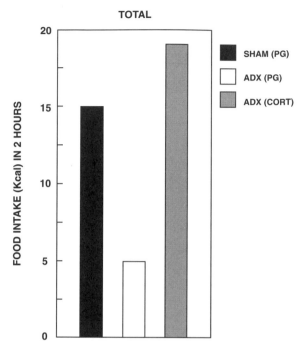

Figure 3.7. Effects of adrenalectomy (ADX) and glucocorticoid (CORT) replacement on food ingestion. PG is polyethylene glycol, a vehicle. (Adapted from Kumar et al., 1988.)

Corticosterone and Neuropeptide-Y-induced Food Ingestion

Like many neuropeptides, neuropeptide Y is produced in both the peripheral and central nervous systems. A pancreatic polypeptide, it is also synthesized in hypothalamic neurons, including those of the paraventricular nucleus and the arcuate nucleus (Olschowka et al., 1982). It is a 36-amino-acid peptide (Gray, O'Donohue, and Magnuson, 1986), and there are at least three different sets of receptors for the peptide.

In the brain, neuropeptide Y is synthesized in the paraventricular nucleus of the hypothalamus, in addition to the arcuate nucleus and central nucleus of the amygdala and the lateral bed nucleus of the stria terminalis (Gray et al., 1986). These neural sites project to brainstem sites such as the dorsal vagal complex and the parabrachial and solitary nuclei (Gray et al., 1986). Neuropeptide-Y efferents from the ventrolateral medulla project to the amygdala (Zardetto-Smith and Gray, 1995). These sites are among the brain regions that orchestrate central motive states. Experiments using c-fos as an anatomic marker for early gene expression have shown that many of these sites (paraventricular nucleus of the hypothalamus, central nucleus of the amygdala, parabrachial and solitary nuclei) are activated when neuropeptide Y is injected centrally (Wilding et al., 1993a, b).

Neuropeptide Y may well be the most potent stimulant of food intake (Clark et al., 1984, 1985), and it appears to integrate feeding responses with the regulation of the hypothalamic-pituitary-adrenal axis (Dallman et al., 1993, 1995).

Systemically administered neuropeptide Y does not elicit food intake. Only when neuropeptide Y is injected into the lateral or third ventricle or into specific regions of the brain does it stimulate food intake (Levine and Morley, 1984; Morley et al., 1987) or, perhaps more specifically, carbohydrate ingestion (cf. Tempel and Leibowitz, 1994; Levine and Billington, in press). This ingestive response has been demonstrated in sated rats. It is also elicited when neuropeptide Y is injected into the paraventricular nucleus of the hypothalamus or the central nucleus of the amygdala (Tempel and Leibowitz, 1994).

The concentration of neuropeptide Y in the paraventricular nucleus is elevated in anticipation of food reward and following food deprivation (Kalra et al., 1991). Neuropeptide Y elicits motivated behaviors to gain access to food sources (bar pressing for food) (Jewett et al., 1995). Interestingly, it does not elicit changes in intraoral intake (Seeley, Payne, and Woods, 1995), which is reminiscent of the findings in rats electrically stimulated in the lateral hypothalamus (Berridge and Valenstein, 1991).

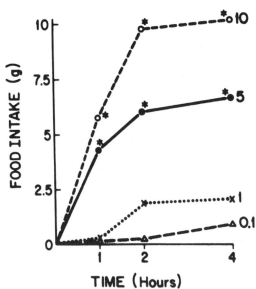

Figure 3.8. Effect of neuropeptide Y (μg) on food intake. (Adapted from Levine and Morley, 1984.)

But these ingestive responses also depend upon circulating corticosterone (Tempel and Leibowitz, 1994). Adrenalectomy abolishes neuropeptide-Y-induced feeding, and corticosterone replacement reinstates it (Stanley et al., 1989). Neuropeptide-Y injections promote corticosterone secretion, and stimulation of type-2 corticosteroid receptors increases neuropeptide-Y gene expression in the basomedial hypothalamic region or in the arcuate nucleus, whose terminal fields are in the paraventricular nucleus of the hypothalamus (Larson et al., 1994; Mercer, Lawrence, and Atkinson, 1996).

Functional studies have demonstrated that food-restricted and food-deprived rats show increases in expression of neuropeptide-Y mRNA in cell bodies in the arcuate nucleus and other hypothalamic sites (Brady et al., 1990), but decreases in other peptides (e.g., corticotropin-releasing hormone). Elevations of corticotropin-releasing hormone (CRH) decrease food intake in rats, and high concentrations of corticosterone, which also decrease food intake (Glowa et al., 1992; Dallman et al., 1995; Spina et al., 1996), will increase CRH in extrahypothalamic regions of the brain (e.g., Makino et al., 1994a, b). This interaction of neuropeptide Y and CRH appears to be fundamental in the regulation of food intake (Dallman et al., 1993; Schwartz, Dallman, and Woods, 1995). In contexts in which CRH is reduced by corticosteroids in the paraventricular nucleus and neuropeptide Y is increased by the steroid, rats increase their food ingestion (Heinrichs et al., 1992; Mercer et al., 1996).

Neuropeptide Y is regulated under conditions in which food access is compromised and rats are hungry. Intraventricular neuropeptide-Y infusions also increase insulin secretion and activate the paraventricular nucleus (Lambert et al., 1995; Kalra and Kalra, 1996). In hyperphagic rats, the peptide is upregulated and then reduced by neuropeptide-Y blockers (Leibowitz, 1995). Elevated concentrations of neuropeptide-Y may be linked to the enjoyment of rewards such as food intake (Heinrichs et al., 1992). In addition, it is known that manipulations of endogenous opiates can either increase or decrease food ingestion (Levine et al., 1995).

Concentrations of neuropeptide Y and levels of gene expression are elevated in rats by treatment with dexamethasone (Wilding et al., 1993a,b; Mercer et al., 1996), both in the paraventricular nucleus and in the arcuate nucleus of the hypothalamus. This same dexamethasone treatment further increases the food intake that normally results from centrally delivered neuropeptide Y (Heinrichs et al., 1992). For example, rats treated for several days with low doses of dexamethasone and then injected with neuropeptide Y directly into the paraventricular nucleus increased their food ingestion (Menzaghi et al., 1993). In other words, the steroid is facilitating the effects of the peptide on this ingestive behavior.

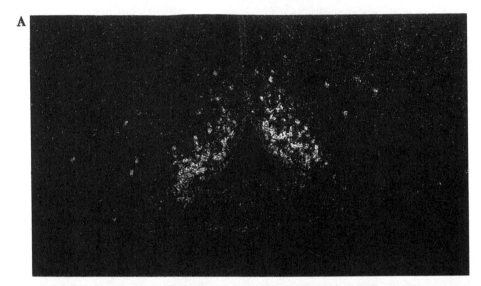

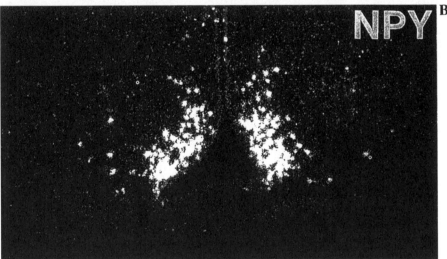

Figure 3.9. Darkfield photomicrographs of neuropeptide-Y mRNA in the arcuate nucleus in sated (A) and food-deprived (B) rats. (From L. S. Brady, M. A. Smith, P. W. Gold, and M. Herkenham, 1990, Altered expression of hypothalamic neuropeptide mRNAs in food-restricted and food-deprived rats, *Neuroendocrinology* 52:441–7. Reproduced with permission of S. Karger AG, Basel.)

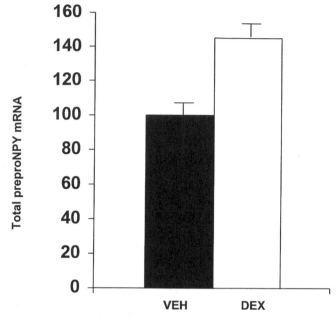

Figure 3.10. Total hybridization to pre-pro-neuropeptide Y (preproNPY) mRNA in the arcuate nucleus in vehicle-injected controls (VEH) and hamsters injected daily for 28 days with dexamethasone (DEX). (Adapted from Mercer et al., 1996.)

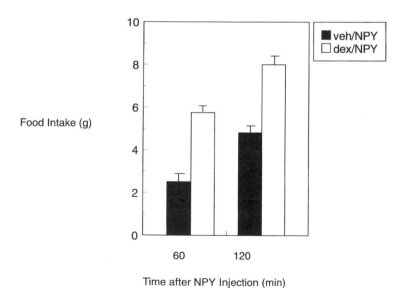

Figure 3.11. Increases in food intake for rats treated first with dexamethasone or vehicle and then given intracerebroventricular neuropeptide Y. (Adapted from Menzaghi et al., 1993.)

In the presence of dietary-induced obesity (Wilding et al., 1993a,b), neuropeptide Y is increased, and leptin is decreased (MacDougald et al., 1995). Leptin is synthesized in adipose tissue, and its production is altered in human obesity (Maffei et al., 1995). It is synthesized in regions of the hypothalamus and is influenced by both neuropeptide Y and insulin (Tartaglia et al., 1995).

In summary, corticosterone acts together with neuropeptide Y to regulate food intake, particularly carbohydrates. Specifically, the steroid corticosterone modulates or potentiates the behavioral effects of neuropeptide Y, and that increases the likelihood of greater food consumption. This is just one part of a much larger story in which both excitatory and inhibitory humoral and nonhumoral signals (Friedman and Stricker, 1976) regulate ingestive behaviors.

Insulin and Leptin: Functional Roles in the Regulation of Energy Balance

Insulin is produced in β cells within the pancreas. It is a peptide hormone that regulates glucose and fat metabolism (Woods et al., 1996). Circulating insulin derived from pancreatic tissue is thought to be a signal about adiposity in the long-term regulation of energy balance; the greater the adiposity, the greater the insulin secretion (Woods et al., 1996).

Leptin, a protein, also may signal adiposity in a manner analogous to that of insulin (Seeley and Schwartz, 1997). They both relay information to the central nervous system about peripheral metabolic balance, and both will reduce food intake when they are injected into the brain (Seeley et al., 1996). Insulin, like leptin, enters the brain through receptor-mediated transport systems and a transduction process that results in the cessation of ingestive behavior (Kaiyala, Woods, and Schwartz, 1995).

Corticosterone influences insulin's impact on food ingestion. Adrenalectomy, for example, will reduce insulin secretion, but corticosterone will restore or increase it (Dallman et al., 1993). Low concentrations of corticosterone will stimulate neuropeptide Y, whereas central infusions of insulin will decrease neuropeptide-Y synthesis (Woods et al., 1996).

There are insulin receptors in the brain, although, unlike the case for many other peptides, it remains unclear whether or not the hormone is produced there. The receptors are located in hypothalamic nuclei, such as the paraventricular nucleus and arcuate nucleus, in addition to other brain sites. Insulin injections into the brain will reduce food intake (Woods et al., 1979, 1996).

Intraventricular insulin will reduce food intake in a number of species,

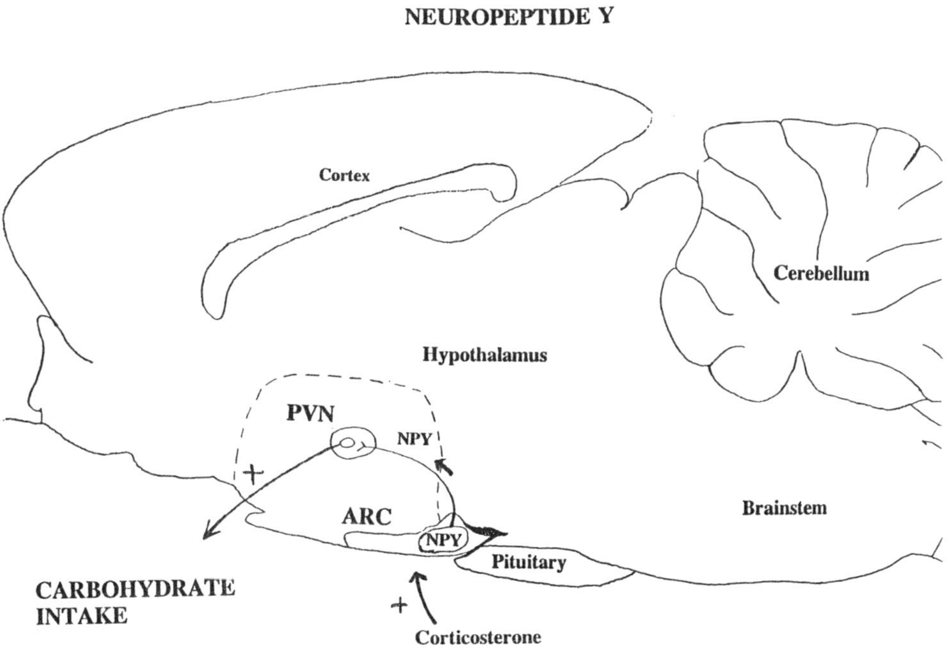

NEUROPEPTIDE Y

Figure 3.12. Side view of rat brain showing structures within the hypothalamus at the base of the brain that control appetite for carbohydrate. Corticosterone elevation in response to a decline in glucose metabolism activates cells in the arcuate nucleus (ARC) that synthesize neuropeptide Y (NPY). Activation of the NPY gene in the ARC increases NPY transport to another area of the hypothalamus, the paraventricular nucleus (PVN). The release of NPY in the PVN stimulates appetite for carbohydrate to restore the body's nutrient stores. (Courtesy of S. F. Leibowitz.)

perhaps by altering an animal's weight gain (Woods et al., 1979; Chavez et al., 1995), though the extent of the effect depends upon the diet that is offered (Friedman and Ramirez, 1987). Intraventricular implants in pancreatic tissue will also reduce body weight, and insulin-induced changes in food intake are not due to malaise (Chavez et al., 1995).

Leptin is expressed in regions of the brain such as the hypothalamus, and the gene for leptin is expressed in adipose tissue (Zhang et al., 1994). Leptin concentrations are correlated with adiposity (Schwartz et al., 1995). Its concentrations are reduced during food deprivation (Frederick et al., 1995), and peripheral infusions of leptin will reduce food ingestion (Camfield et al., 1995), as will central infusions (Seeley et al., 1996; Seeley and Schwartz, 1997).

Thus reduced concentrations of insulin or leptin will increase food intake, perhaps at the level of the hypothalamus by increasing neuropeptide-Y expression and decreasing CRH expression (Seeley et al., 1996). Insulin and

103

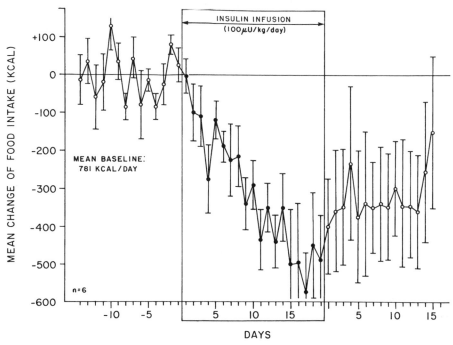

Figure 3.13. Effects of central insulin infusions on food intake by baboons. Points and bars represent means ± 1 SEM. Open symbols represent control intervals when synthetic cerebrospinal fluid was infused; solid symbols represent intervals when insulin was infused at one of three doses. (Adapted from Woods et al., 1979.)

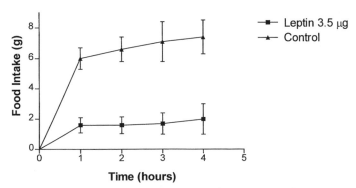

Figure 3.14. Effect of intracerebroventricular infusion of leptin into the third ventricle (3.5 μg of recombinant mouse leptin versus control). (From R. J. Seeley et al., Intraventricular leptin reduces food intake and body weight of lean rats but not obese Zucker rats, *Horm. Metab. Res.* 28 (1996):1–5, with permission from Georg Thieme Verlag – New York.)

perhaps leptin can send signals about the extent of adiposity that can influence ingestion.

Cholecystokinin-induced Satiation

Perhaps the clearest and best-documented humoral system involved in the cessation of food intake is that of cholecystokinin-induced satiety (Gibbs et al., 1973; Smith, 1997). Cholecystokinin is a polypeptide that is secreted by the intestines after a meal is ingested. It is produced in gastrointestinal sites and is widely distributed in the central nervous system.

Systemic injections of cholecystokinin will decrease food intake by both intact rats and gastric-fistulated rats (Gibbs, Kulkosky, and Smith, 1981). The effect is dose-dependent and varies with the sensory characteristics of food sources, and it depends upon type A cholecystokinin receptor sites (Moran et al., 1986). In this regard, cholecystokinin is analogous to an atrial peptide hormone (a hormone secreted by the heart, such as atrial natriuretic factor); the former is to food ingestion what the latter is to fluid ingestion (see Chapter 2).

Cholecystokinin antagonists can increase food intake (Brenner and Ritter, 1995). Thought to control short- but not long-term food intake (Smith et al., 1997), cholecystokinin decreases the incentive values of food sources, and it reduces the lever pressing that is associated with the acquisition of nutritious food (Balleine et al., 1995).

Cholecystokinin-induced suppression of food intake is not confined to mammals. The food intake of the white-crowned sparrow, for example, is reduced by cholecystokinin, while water ingestion remains intact (Richardson et al., 1993). Cholecystokinin studies in goldfish (*Carassius auratus*) have shown patterns of brain and gut distributions similar to those in other species, as well as similar reductions in food ingestion following cholecystokinin infusions (Himick and Peter, 1994).

This systematic effect of cholecystokinin-induced food reduction is determined via afferents from the vagal nerve that terminate in the solitary nucleus (Smith, Jerome, and Norgren, 1985; Norgren and Smith, 1988). Vagotomy abolishes the cessation of food ingestion that follows systemically delivered cholecystokinin. The hormone's effects work in conjunction with other gastrointestinal-related events that occur at the end of a meal (e.g., volume distension in the stomach, hepatic regulation) and are mediated by a subset of vagal afferents that integrate mechanical and peptide actions that result in the termination of a meal (Schwartz and Moran, 1996).

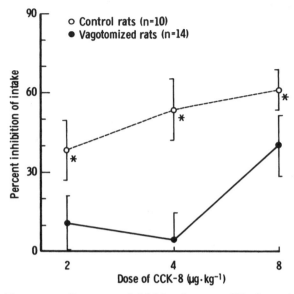

Figure 3.15. Percentage inhibition (mean ± SE) of 30-minute intake during sham feeding after three doses of cholecystokinin COOH-terminal octapeptide (CCK-8) compared with the intake after saline injection at 1 minute before the intake test (intake by sham-vagotomized rats after saline injection = 38.3 ± 2.3 ml/30 min; intake by vagotomized rats after saline injection = 42.6 ± 2.3 ml/30 min). (From Joyner et al., 1993, with permission.)

Food deprivation decreases cholecystokinin mRNA both peripherally and centrally, and food intake increases it. Cholecystokinin's neural origin led some investigators to inquire whether or not the hormone would change ingestive behavior if it were delivered directly to the brain. Experiments have shown that indeed it does. For example, cholecystokinin delivered to the third ventricle decreases the food ingestion of hungry rats. It also decreases motivated behavior (e.g., running speed) toward sucrose, and the decrease is dose-dependent (Stellar and Stellar, 1985). Despite these effects on feeding and motivated ingestive behavior, the hormone's impact is due largely to peripherally derived cholecystokinin acting on the central nervous system (Crawley et al., 1991).

Studies using c-fos as an anatomic marker have shown that cholecystokinin activation is expressed in both the brainstem and the forebrain. In the brainstem, the area postrema – a circumventricular region – is activated, as are regions of the solitary nucleus and the dorsal motor nucleus of the vagus nerve. In the forebrain, sites such as the paraventricular nucleus, central nucleus of the amygdala, and bed nucleus of the stria terminalis are activated (Rinaman et al., 1993).

Cholecystokinin also activates a number of other peptides. One of the

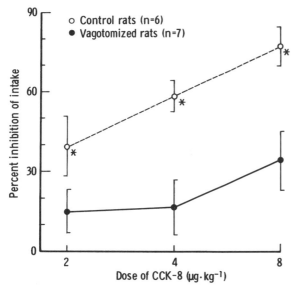

Figure 3.16. Percentage inhibition (mean ± SE) of 30-minute intake during real feeding. (From Joyner et al., 1993, with permission.)

most important is oxytocin in the paraventricular nucleus of the hypothalamus (Olson et al., 1991). Oxytocin appears to be a signal generally inhibiting ingestive behaviors (Verbalis et al., 1993). The paraventricular nucleus regulates gastric control of cholecystokinin and taste-aversion learning (Verbalis et al., 1993), perhaps by oxytocinergic mechanisms. Oxytocin (like CRH, etc.) can be functionally demarcated into two systems. One system is reflected in the regulation of the hypothalamic-pituitary-adrenal axis, and the other perhaps in the regulation of behavioral processes (e.g., taste aversion, inhibition of sodium appetite) (Stricker and Verbalis, 1987).

Neural Function and Hormone-induced Regulation of Food Ingestion

While considering the relationship between behaviors and levels of neural function, the social theorist and philosopher Herbert Spencer (1901) introduced the idea of evolution and dissolution in the brain. The idea is simple: Older, more rigid behavioral functions are controlled by phylogenetically ancient brain regions, whereas newer functions are controlled by more recently derived brain regions (James, 1952; Gallistel, 1980; but see Porges, 1996). This orientation is more like a working heuristic than a law in science.

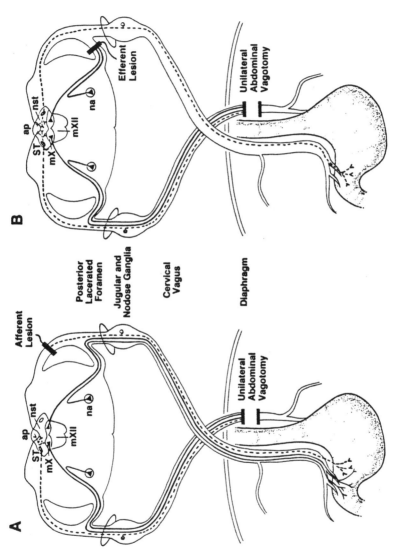

Figure 3.17. Schematic representation of the subdiaphragmatic vagus trunks and their branches from a ventral perspective in a supine rat in relation to the cervical vagus nerves and the medulla. Arrowheads indicate the approximate level at which nerves were severed and incubated in horseradish peroxidase. Abbreviations: ap, area postrema; nst, nucleus of the solitary tract; mX, dorsal motor vagal nucleus; mXII, hypoglossal nucleus. (From Smith et al., 1985, with permission.)

Using this level-of-neural-function approach, researchers have studied the relationships between the brain and several hormonally regulated behaviors, including food ingestion (Grill and Norgren, 1978a,b), water and sodium ingestion (Grill et al., 1986), sexual behavior (Beach, 1942), aggression (Flynn, 1969), rage (Bard, 1939), and fear (Rosen and Davis, 1988).

Hughlings Jackson (1958), a great nineteenth-century neurologist who still influences much of our thinking about the brain and its regulation of behavior (Rozin, 1976b), first demonstrated that the basic motor skills needed to ingest food remain intact even after significant cerebral damage. Specifically, he showed that brain-damaged humans who could not respond to verbal commands to stick out their tongues could do so when food was placed in front of them. Some years later, Miller and Sherrington (1915) demonstrated that a decerebrated dog that would not initiate food ingestion would eat if it was hungry and food was placed in its mouth. Sherrington thus illustrated the consummatory phase, but not the appetitive phase, of a central motive state.

In rats, in which the forebrain is disconnected from the brainstem, removing the cerebrum leaves intact the basic taste reactivity responses, as well as responses to a number of humoral and metabolic signals that underlie food ingestion (Grill, 1980; Flynn, Berridge, and Grill, 1986). Thus food intake is increased in decerebrates to appropriate metabolic signals. But by no means is food responsivity in the decerebrate animals normal – even when food is infused into the oral cavity (Kaplan, Seeley, and Grill, 1993). The animals are compromised in their ability to regulate nutrients, as well as in basic oral motor functions (Kaplan et al., 1993). They also do not demonstrate taste-aversion learning (Grill, 1980). Nonetheless, the animals do respond competently to a number of humoral signals. Insulin-induced feeding is expressed, for example, as is food-deprivation-induced caloric intake (the sucrose is infused into the oral cavity, as decerebrates are unable to ingest sucrose or other food freely available in the home cage). The insulin-induced feeding is dose-dependent, as is insulin secretion in response to infusions of glucose (Flynn et al., 1986).

Anencephalic human neonates retain the normal facial responses to the good taste of sucrose and the bad taste of quinine (Steiner, 1977). The same holds for decerebrate rats (Grill and Norgren, 1978a,b). But whereas the basic hedonic responses remain intact, brain-damaged subjects lose their ability to shift these responses when necessary – away from a food source following taste aversion, for example, or toward a salty food when the hormones of sodium hunger are elevated (see Chapter 2, this volume; Grill et al., 1986).

Experiments with decerebrates have helped researchers resolve an old debate over whether cholecystokinin-induced reduction in food intake is due to satiety or malaise. Of course, the two are not completely incompatible: As one becomes sated, a food loses its pleasant taste. But the issue for cholecystokinin is whether or not its impact is primarily due to the malaise. Humans given cholecystokinin report reductions in hunger, but they do not report feelings of malaise.

An experiment with decerebrate and control rats using taste reactivity addressed this question (Grill and Smith, 1988). First, researchers showed that control rats decreased their intake of intraorally infused sucrose following systemic cholecystokinin infusions. But they did not show the quinine-like reactions that are known to follow taste-aversion learning. That finding supported the idea that the cholecystokinin effect was not simply due to malaise. Second, the researchers showed that the cholecystokinin suppression of food intake could be mediated at the level of the brainstem (e.g., area postrema and the solitary nucleus or the parabrachial nucleus) (Spector, 1995). Changes in gustatory neurons at that level reflect changes in metabolic and mineral balances, in addition to changes in learned aversions (Scott and Mark, 1986).

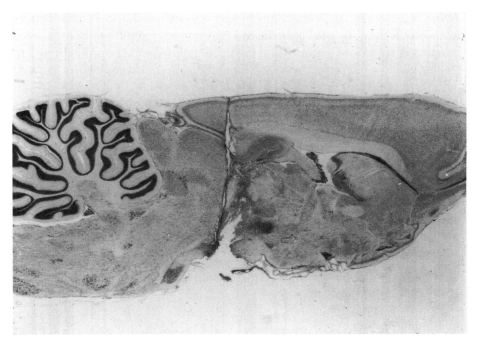

Figure 3.18. Representative sagittal section of the brain of a supracollicular chronic decerebrate rat. (Courtesy of H. Grill.)

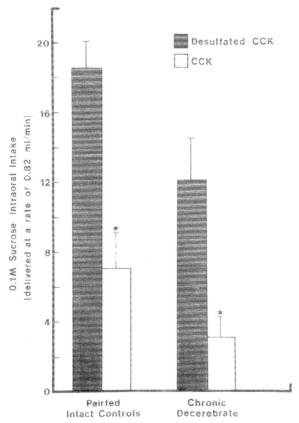

Figure 3.19. Effects of intraperitoneal peptide treatment on intraoral intake of a 0.1-M sucrose solution: differences in intake volumes between CCK-8 and desulfated CCK-8 conditions for decerebrate and pair-fed control groups. (From Grill and Smith, 1988, with permission.)

As stated earlier, increasing evidence indicates that central oxytocin projections to brainstem sites such as the area postrema, solitary nucleus, and dorsal motor nucleus of the vagus nerve influence ingestive behavior and may contribute to the reduction in food intake that follows administration of CRH (Olson et al., 1991) and more general inhibitory signals. In addition, it has been argued that both cholecystokinin- and lithium-chloride-induced sickness can activate oxytocinergic sites in the paraventricular nucleus of the hypothalamus that then project to the brainstem to decrease food intake by malaise (Olson et al., 1991). Again, malaise is not conceptually separate from satiety; as we satiate, we appreciate the food less – particularly its

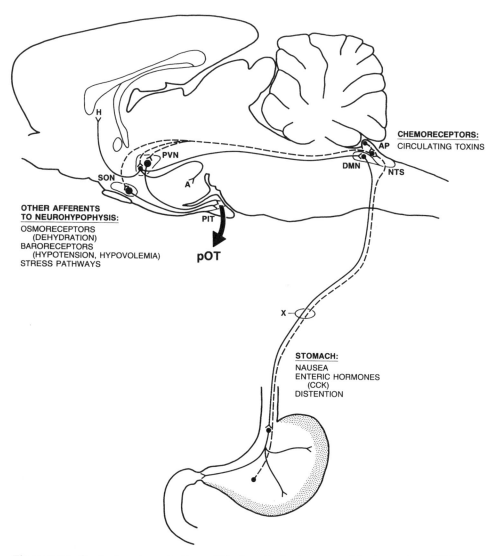

Figure 3.20. Oxytocinergic projections of the hypothalamic paraventricular nucleus (PVN). Oxytocin-containing neurons from magnocellular cells project to the posterior pituitary (PIT), from which oxytocin is secreted to raise plasma concentrations (pOT). Neurons from parvocellular cells in the PVN project throughout the brain, including the amygdala (A), the hippocampus (H), and the dorsal motor nucleus of the vagus nerve (DMN) and the nucleus tractus solitarius (NTS) in the brainstem. Oxytocin acting as a neurotransmitter appears to be involved in the central control of gastric function and ingestive behavior. In contrast, oxytocin-containing neurons from the hypothalamic supraoptic nucleus (SON) project only to the posterior pituitary. Stimuli for oxytocin secretion from both magnocellular and parvocellular PVN neurons are conducted through vagal afferent nerves (X) from the stomach and also arise from the area postrema (AP) and other afferents. (Courtesy of J. Verbalis.)

specific sensory characteristics (Rolls, 1986) – and can come to feel malaise. One model in which to envision oxytocin inhibition of ingestive behaviors is depicted in Figure 3.20.

Other peptides that reduce food intake also seem to operate at the level of the brainstem. Bombesin reduces food intake (Gibbs et al., 1981), but not water intake. Found in the peripheral visceral nervous system in addition to the central nervous system, bombesin is a gastric peptide hormone that is modulated by food deprivation and food ingestion (Flynn, 1992). It is concentrated in brainstem sites such as the solitary nucleus, area postrema, and parabrachial nucleus, in addition to hypothalamic nuclei (paraventricular and ventral medial hypothalamic nuclei) and the central nucleus of the amygdala.

Whereas bombesin is localized in peripheral sites, its inhibition of food intake is mediated in the brainstem. Bombesin injections into the solitary nucleus and area postrema will reduce food intake, and vagotomy will not abolish the effect. Bombesin infusions into the fourth ventricle, but not the third ventricle (in decerebrates and intact controls), will reduce both sucrose and NaCl intake (Flynn, 1992), indicating that the hormone's effects are not specific for food.

At other levels of the neural axis, what do we know about hormonal control of food intake? We know that if the whole of the cortex is removed, insulin-induced feeding is still expressed (Schulkin and Grill, 1980), as are other ingestive behavioral responses to humoral signals (Grill, 1980). The lateral and ventral medial areas of the hypothalamus are parts of the circuitry underlying food-motivated behaviors (Stellar, 1954), as are the catecholamines that traverse the hypothalamus and generally activate motivational states (Stricker and Zigmond, 1974) that are separable from hedonic judgments (Berridge, 1996). In conjunction with this brainstem activity is local activation of dopamine neurons in the nucleus accumbens (e.g., Wilson et al., 1995) during motivational states. It is the nucleus accumbens that Nauta (1961) envisioned as the site of the motor output in limbic functions (Mogenson, 1987).

At the level of the hypothalamus, galanin increases food intake, particularly when it is injected into the paraventricular nucleus (Leibowitz, 1995). (Galanin was originally isolated from the small intestine and is widespread throughout the brain.) Food intake also increases when galanin is injected in the central nucleus of the amygdala (Crawley et al., 1990; Tempel and Leibowitz, 1993, 1994). Unlike the case with neuropeptide Y, this peptide-induced food intake does not depend on circulating corticosterone. Galanin antagonists block galanin-induced food intake if they are implanted at either anatomic site (Corwin, Robinson, and Crawley, 1993). It is interesting

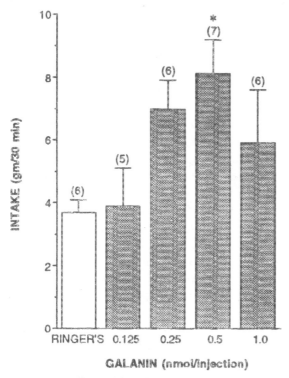

Figure 3.21. Effects of four doses of galanin, microinjected bilaterally into the central nucleus of the amygdala, on food intake in rats (mean ± SEM). Numbers in parentheses indicate the number of rats per treatment group. (From R. L. Corwin, J. K. Robinson, and J. N. Crawley, 1993, Galanin antagonists block galanin induced feeding in the hypothalamus and amygdala of the rat, *European Journal of Neuroscience* 5:1528–33. Reprinted by permission.)

that these two brain sites are also involved in humoral responses to stress (see Chapters 4 and 5). It is not yet clear whether or not galanin-stimulated animals will desire to gain access to food.

Summary

This chapter has outlined some of the humoral factors that underlie feeding behavior. Both steroid hormones and peptides can contribute to the processes of generating and terminating ingestion. Separately or together they can initiate both appetitive and consummatory behaviors (Stellar and Stellar, 1985). This theme of steroids and peptides interacting to regulate behavior through their actions in the brain will predominate throughout this book. Similarly, the neural circuits that underlie the desire for food, as well as other motive states, will reappear in many of the following chapters.

Feeding behavior (as well as other endocrine-controlled states) is not controlled by just one signal. Either neuropeptide Y or corticosterone, for example, can generate carbohydrate ingestion. Multiple signals in the regulation of behavior tend to be the rule, not the exception. In most cases there seems to be a confluence of both excitatory and inhibitory humoral signals that regulate behavior (Seeley and Schwartz, 1997). Corticosterone and neuropeptide Y stimulate food ingestion, whereas CRH, leptin, cholecystokinin, and insulin decrease it.

Like all behaviors, feeding strategies reflect evolutionary adaptations (Krebs and Davies, 1991) and vary with circumstances. For instance, the ability of birds to remember where food is located varies with the number of different kinds of foods they eat (Shettleworth and Krebs, 1982). Learning about food also requires social knowledge. In many species, conspecifics readily learn from others during development (Goodall, 1986; Galef, 1991). Rats, for example, learn to avoid foods that other rats have found to be aversive, and they learn to prefer foods that they have observed other rats ingest (Galef, 1986). Red-wing blackbirds that observe conspecifics eating out of colored containers will tend to do the same (Mason and Reidinger, 1982). But in addition, in a number of species, knowledge about food sources is built toward the avoidance of bitter sources that signal danger and poisoning (Janzen, 1977).

This chapter has also described several neural sites – traversing both the brainstem and forebrain – that regulate ingestive behavior. The degrees of competence that follow manipulation of different levels of the neural axis reflect the brain's evolution. A common circuit for hormone-induced behavioral responses has already been delineated. It includes brainstem sites such as the solitary nucleus and the parabrachial nucleus, and it also includes forebrain sites such as the paraventricular nucleus, lateral hypothalamus, ventral medial hypothalamus, and central nucleus of the amygdala. These sites will appear in connection with many of the hormone-induced behaviors described in subsequent chapters. Nature is economical, using the same neural circuits for a number of different behaviors.

CHAPTER FOUR

Hormones, Parental Care, and Attachment Behaviors

Introduction

Animals like ourselves seek attachment and security during development. A number of hormones underlie and help facilitate attachment behaviors and the distress behaviors that result from loss of attachment and security. Hormones facilitate the attachment behaviors of parents toward their young and the young toward their parents. Although this has been demonstrated for only a few species, most of us in the field think that eventually the phenomenon will be observed in most animal species in which there is parental investment and long-term bonding during development.

There are clear differences between the two major traditions of thought regarding endocrinology and the regulation of behavior. The scientific tradition in which I have been educated is that of Bernard, Cannon, and Richter, described at the end of Chapter 2. We are accustomed to thinking of bodily alterations and hormonal activation and regulation as being brought about by both physiological and behavioral means. The tradition that emanates from Bernard, Cannon, and Richter looks at the inside of the animal: that is, how changes in the internal milieu generate both physiological responses and then behavioral compensatory responses to provide stability.

But there is another rich tradition, as relevant as the Richter tradition. It emphasizes social facilitation in the regulation of the internal milieu (e.g., Levine, Chamoux, and Wiener, 1991), a view championed by Lehrman and Friedman (1968) and Hinde and Stevenson (1970). According to this view, social events are used to regulate the internal milieu: Suppressions and elevations of reproductive hormones, for example, as functions of the social environment, are commonplace phenomena across a wide range of species (Faulkes, Abbott, and Jarvis, 1991). It is the ability of some species to orchestrate their social surroundings that promotes secretion of the reproductive hormones that lead to long-term success for a species.

Consider a recent set of experiments on the social regulation of the internal milieu, involving cortisol concentrations, fear, and young children: A subset of young children who have high cortisol concentrations and are fearful use social contacts to reduce both their fear and their cortisol concentrations (Gunnar et al., 1987, 1989); that is, by reaching out to others (primary caretakers, or perhaps substitutes) they reduce their circulating cortisol concentrations and the consequent metabolic demands. Physical contact is essential for this to occur. Staying close to others and feeling secure, stable, and related to others are primary conditions for this effect. These are, of course, important items during normal development and throughout the course of one's life. We shall be talking more about cortisol and fear in this chapter and in Chapter 5. What is important to note here is that this demonstrates the importance of the social world for regulation of the internal milieu. The social attachments of children turn out to be critical in regulating behavior and cortisol physiology (Bowlby, 1973; Gunnar et al., 1987, 1989).

This chapter highlights several fundamental, hormone-mediated events in the organization of parental behavior, social bonding, and development. We shall begin with central peptides and the regulation of parental behavior in several species. We shall then turn to attachment and separation studies, with emphasis on the roles of cortisol and brain function. Finally, we shall briefly discuss the hormones that may underlie quiescence – arousal, attachment – separation, and fear.

Prolactin, Reproduction, and Parental Care

Prolactin is both a hormone produced by the pituitary gland and a neuropeptide expressed in a variety of brain sites. Prolactin has receptor sites in the periphery (e.g., ovary) and in the central nervous system. Researchers have found prolactin receptors in cells in the preoptic region, the ventral medial hypothalamus, the bed nucleus of the stria terminalis, the suprachiasmatic nucleus of the hypothalamus (SCN), and brainstem sites such as the central gray (Walsh, Slaby, and Posner, 1987; Buntin, Ruzycki, and Witebsky, 1993). One route by which peripheral prolactin gains access to the brain is through the choroid plexus; another is by way of the circumventricular organs (Walsh et al., 1987).

Central to the topic of this chapter, prolactin is also fundamental for brooding in birds and parental behaviors in birds and mammals. In order to get a foothold in the world, the offspring of a wide variety of animal species, including our own, require parental care and guidance. Underlying parental behaviors are numerous hormonal signals. It has been known for some time that high concentrations of estrogen, for example, are essential for the expression of maternal behavior (Beach, 1948; Rosenblatt, 1970). The behavior does not depend on the pituitary gland, but does vary with experience and circumstance, as reviewed by Numan (1994).

Several key researchers in psychobiology have sought to integrate a sense of central states with peripheral physiological changes when studying the organization of behavior. Danny Lehrman is one such figure. He outlined quite clearly that there is a central state of what one might characterize as parental care, nest building, that is induced and sustained by hormonal signals. One such signal that Lehrman (1955, 1961) and others (Riddle, Lahr, and Bates, 1935a,b; Silver, 1984) called attention to is that the concentration of prolactin plays a role in these behaviors.

Prolactin is directly involved in suckling as well as other care-giving behaviors. Suckling triggers prolactin secretion in both the mother and the offspring in a number of mammalian species at both the level of the pituitary and the level of the hypothalamus (Moyer et al., 1979). Decreases in attachment between a rat pup and its mother decrease prolactin and increase cortisol concentrations, and temperature regulation is essential in the symbiotic relationship between the pup and its mother (Adels et al., 1986). Prolactin, in addition to estrogen and progesterone, has been shown to play a role in nest building by female rabbits (e.g., Zarrow, Farooq, and Denenberg, 1963; Mariscal et al., 1996)

Prolactin is important in the sitting behavior (Lehrman, 1961) of some

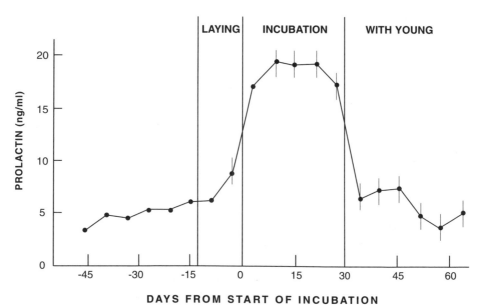

Figure 4.1. Mean plasma prolactin concentrations in female mallard ducks. (Adapted from Goldsmith et al., 1981.)

species of birds during reproduction (e.g., ringdove, *Streptopelia risoria*). Prolactin elicits brooding behavior in both sexes in several species of birds (Silver, 1984). Central infusions of prolactin into turkeys (*Meleagris gallopavo*), for example, facilitate brooding and nest building (Youngren et al., 1991). Electrical stimulation of the turkey's hypothalamic regions can also facilitate brooding behavior and increase prolactin concentrations, whereas hypothalamic lesions reduce them (Youngren et al., 1989, 1993). Interestingly, nest deprivation decreases prolactin concentrations. In zebra finches, high concentrations of prolactin are linked to parental behavior in both males and females.

It has been known for some time that prolactin facilitates the food regurgitation involved in parental care in several species of birds (Lehrman, 1955). Prolactin infusions in the medial preoptic region can facilitate food ingestion and regurgitation among young doves, and ibotenic acid lesions (cell-body lesions) of the medial preoptic region do the reverse (Buntin, 1992). Prolactin is fundamental in the regurgitation of food to neonates. Prolactin-induced regurgitation is increased in females with higher concentrations of estrogen.

Moreover, prolactin concentrations are elevated at different times in male and female ringdoves, reflecting the different times that male and female birds are responsible for feeding the offspring. In the beginning, it is the female that regurgitates more food to the young than does the male. Toward the end of the rearing period, the male takes on a greater share of this responsibility as the female begins to anticipate the next crop of eggs for the next offspring. These differences are reflected in the birds' prolactin concentrations (Silver, 1984).

Prolactin concentrations are also elevated in both female and male mice, in which both sexes participate in parental behaviors (Gubernick and Nelson, 1989), as well as in marmoset monkeys (*Callithrix jacchus*) (Dixson and George, 1982), cotton-top tamarins (*Saguinus oedipus*) (Ziegler, Wegner, and Snowdon, 1996), and the king penguin (*Aptenodytes patagonicus*) (Garcia, Jouventin, and Mauget, 1996). These are, of course, species in which males take care of the young.

The many observations of male parental care – accompanied by elevated prolactin – show that prolactin elicitation is not bound to estrogen. However, in ringdoves, males take their cues from the females, whose behavior is determined by the gonadal steroid hormones (Ramos and Silver, 1992).

In female rats, prolactin infused into the lateral ventricle – after a background of systemic estrogen treatment – rapidly elicits maternal behavior (e.g., pup retrieval, licking). In multiparous females, central infusions of

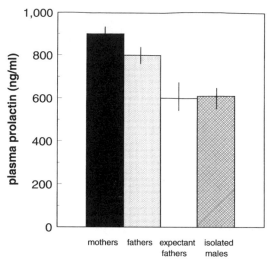

Figure 4.2. Plasma prolactin titers (means ± 1 SEM) in California mice. Blood samples were obtained from mothers (*N* = 11) and fathers (*N* = 11) at 2 days post partum, from expectant fathers within 10 days of parturition (*N* = 10), and from isolated, adult virgin males (*N* = 14). Fathers were in contact with pups prior to blood collection. (Adapted from Gubernick and Nelson, 1989.)

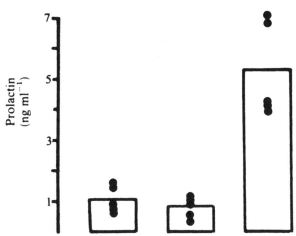

Figure 4.3. Plasma concentrations of prolactin in male marmosets in captive family groups containing, left to right, nonpregnant females, pregnant females, and females with infants aged 10–30 days. Each histogram shows the mean concentration for five males. Each point represents the mean plasma concentration for 10 samples from a single male, except for two of the males caged with pregnant females; six and seven samples were collected from these animals before their female partners gave birth. (Adapted from Dixson and George, 1982.)

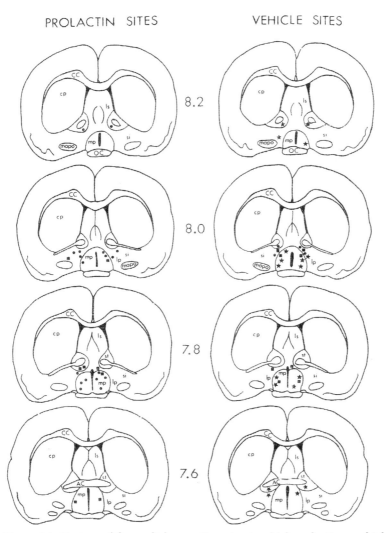

PROLACTIN SITES   VEHICLE SITES

Figure 4.4. Regions of the medial preoptic region in which prolactin or vehicle was directly delivered in estrogen-treated rats. (From R. S. Bridges et al., 1996, Endocrine communication between conceptus and mother: placental lactogen stimulation of maternal behavior, *Neuroendocrinology* 04:57–64. Reproduced with permission of S. Karger AG, Basel.)

prolactin alone can elicit the behaviors (Bridges et al., 1990; Bridges and Mann, 1994). Infusions of prolactin, specifically into the medial preoptic region, a region of the brain known for its involvement in maternal behavior (e.g., Rosenblatt, Hazelwood, and Poole, 1996), will elicit the behavior. Con-

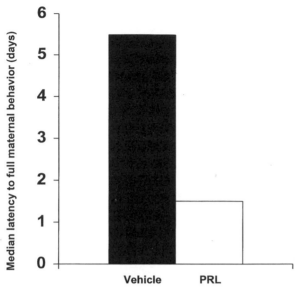

Figure 4.5. Effects of bilateral medial-preoptic-area infusions of prolactin (PRL) or vehicle on the induction of maternal behavior in steroid-primed, behaviorally inexperienced female rats. (Adapted from Bridges et al., 1996.)

versely, lesions of this brain region will impair maternal behavior (Bridges and Freemark, 1995).

Estrogen, by increasing prolactin expression in the brain, increases the likelihood of and onset of maternal behaviors in suitable environments. Pharmacologic blockade of these sites does the reverse (Bridges and Mann, 1994). It is believed that sensory input from the offspring activates brain regions such as the medial preoptic region and ventral bed nucleus of the stria terminalis (Numan and Numan, 1995), which then play important roles in maternal behavior.

Prolactin cells and receptors in the brain of the rat (e.g., Siaud et al., 1989; Bakowska and Morrell, 1997) are depicted in Figure 4.6, and researchers have outlined an elaborate anatomic circuit for maternal behavior in the brain of the rat (Numan, 1994). Important sites include the bed nucleus of the stria terminalis, the medial amygdala, and the preoptic region, in addition to brainstem sites such as the ventral tegmental area and the central gray.

In conclusion, prolactin appears to be linked to quiescence during the suckling period in mammals and the sitting behavior of birds. Both are elements of the reproductive behavioral responses.

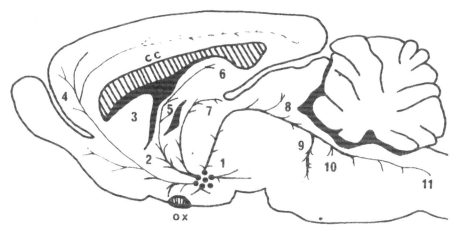

Figure 4.6. Prolactin system on a paramedian sagittal section of the rat brain: 1, perikarya and hypothalamic fibers; 2, bed nucleus of the stria terminalis and preoptic projections; 3, caudate nucleus almost devoid of fibers; 4, prefontal cortex; 5, septal nuclei; 6, dorsal hippocampus; 7, midline thalamic nuclei; 8, central gray; 9, raphe dorsalis; 10, locus ceruleus; 11, solitary-tract nucleus; ox, optic chiasm; cc, corpus callosum. (From L. Paul-Pagano, R. Roky, J.-L. Valat, K. Kitahama, and M. Jouvet, 1993, Anatomical distribution of prolactin-like immunoreactivity in the rat brain, *Neuroendocrinology* 58:682–95. Reproduced with permission of S. Karger AG, Basel.)

Oxytocin and Parental Behavior

Another important hormone, oxytocin, is linked to both the physiologic regulation of milk release from the nipple and the behavioral response of closeness. That is, hormones such as oxytocin, in humans, are elevated during periods in which infants are breast-feeding (Uvnas-Moberg, 1996). Although this role is controversial, oxytocin has been suggested as a fundamental hormone during the bonding period between mothers and their babies.

We have long known that maternal behavior in animals is under the control of gonadal and peptide hormones (Riddle et al., 1935a,b; Lehrman, 1955) and that both humoral and nonhumoral factors control postpartum attachments (Rosenblatt, 1994a,b).

In several mammalian species, central infusion of oxytocin also elicits maternal behavior (Pedersen and Prange, 1979; Fahrbach, Morrell, and Pfaff, 1984; Pederson ot al., 1994). A small (6-amino-acid) peptide, oxytocin appears to be a hormone linked to affiliation or attachment in general (Carter, 1992) and specifically to the onset of maternal behavior in several species (e.g., Levy et al., 1992). High oxytocin concentrations have been found to

123

Figure 4.7. Female and male prairie voles and parental care. (Courtesy of Sue Carter.)

facilitate maternal attachment, whereas low concentrations lead to its withdrawal. Moreover, central (but not peripheral) infusions of oxytocin will elicit partner preference in female prairie voles (Carter, 1992).

In addition, intracerebral injections of oxytocin will elicit parental behavior in several species studied in the laboratory (Pedersen et al., 1982; Carter, 1992; Kendrick et al., 1992; Levy et al., 1992). Oxytocin also elicits female aggression, which is linked to estrogen and to maternal behavior. Peripheral administration of oxytocin does not elicit these responses.

In mice, centrally infused oxytocin also elicits parental behavior in monogamous males that care for their young (Carter, 1992). Amazingly, in the California mouse (*Peromyscus californicus*), oxytocin is even elevated in expectant fathers (Gubernick et al., 1995).

Convincing support for oxytocin's role in parental behavior also comes from observations of animals in nature. As model systems, researchers have studied two species of voles, one of them monogamous – the prairie vole (*Microtus ochrogaster*) – and the other not – the mountain vole (*Microtus montanus*). In the monogamous species, males contribute to care of the offspring. Oxytocin-receptor densities reflect these differences in behavior. In several brain regions, including the bed nucleus of the stria terminalis (Insel, 1992), oxytocin receptors are more numerous in prairie voles than in mountain voles. The greater the amount of time they spend with pups, the

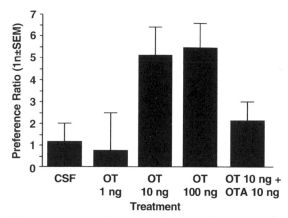

Figure 4.8. Oxytocin and attachment and partner preferences in female prairie voles following central infusions of oxytocin (OT) or oxytocin plus an antagonist (OTA). (Adapted from Williams et al., 1994.)

greater the number of oxytocin receptors. Interestingly, high corticosterone concentrations reduce attachment behavior in voles (Carter, 1992).

Oxytocin's activation of parental care is tied to the medial preoptic nucleus. Lesions of and stimulation to this region will abort and facilitate parental care, respectively. Oxytocin-producing cells are found in magnocellular and parvocellular regions of the paraventricular nucleus and supraoptic nuclei of the hypothalamus, as well as other brain sites (e.g., bed nucleus of the stria terminalis) (Insel, 1992). Moreover, estrogen is known to increase the number of oxytocin-producing cells in the bed nucleus of the stria terminalis in sheep.

The bed nucleus is known to be tied to affiliative behavior and the formation of partner preference (Insel, 1992). Oxytocin, as mentioned earlier, is correlated with a number of different affiliative and social-bonding behaviors. These include sexual arousal, pair bonding, and sexual satisfaction (Carter, 1992). Interestingly, oxytocin is also fundamentally involved with parturition and lactation. Always economical and opportunistic, nature uses the same hormonal systems to control a number of different behavioral responses.

The peptide hormone oxytocin plays several roles in successful reproduction. The first is its well-known role in physiologically regulating milk letdown in mothers. Second, depending on its site of action, oxytocin can elicit either sexual receptivity or maternal or paternal behavior (see Chapter 1) (e.g., Levy et al., 1992).

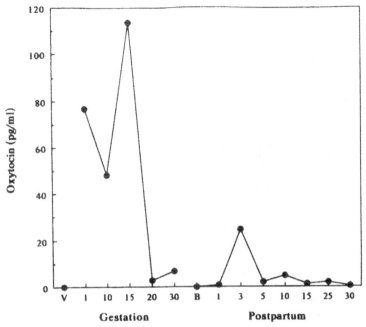

Figure 4.9. Median concentrations of plasma oxytocin in male mice across the reproductive cycle. Blood samples were collected from virgin males (V), expectant fathers throughout the gestational period, and fathers from the day of birth (B) of the litter to weaning. Expectant fathers had significantly higher concentrations of plasma oxytocin than did virgin males or fathers. (From D. J. Gubernick, J. T. Winslow, P. Jensen, L. Jeanotte, and J. Bowen, 1995, Oxytocin changes in males over the reproductive cycle in the monogamous, biparental California mouse, *Peromyscus californicus, Hormones and Behavior* 29:59–73. Reprinted by permission.)

Vasopressin, Testosterone, and Parental Behavior

Vasopressin, like prolactin and oxytocin, is both a pituitary hormone and a neuropeptide. In Chapter 1 we considered some of the effects of vasopressin on aggression and scent-marking behaviors (Ferris, 1992). But an additional role for central vasopressin has been demonstrated in at least one species: Intraventricular central infusions of vasopressin will increase parental behavior in prairie voles.

Central vasopressin may underlie pair bonding in this monogamous vole: Vasopressin antagonists infused into the lateral ventricle will reduce pair bonding, whereas vasopressin potentiates it (Winslow et al., 1993; Carter, 1994).

Recall that vasopressin's cellular levels of action in the medial nucleus of

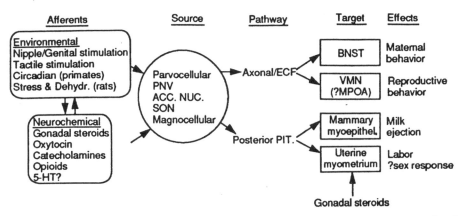

Figure 4.10. A model of oxytocin (OT) physiology posits parallel processing in central and peripheral pathways. Several environmental inputs, acting via a range of neurochemical pathways, are believed to induce OT synthesis in the hypothalamus. Parvocellular neurons, either by direct axonal projections or through exocrine release into the extracellular fluid (ECF), affect cells in the bed nucleus of the stria terminalis (BNST) and the ventral medial nucleus (VMN) to influence the initiation of maternal (BNST) and reproductive (VMN) behaviors. Concurrent neurohypophyseal release induces milk ejection and uterine contraction. In this model, the key determinant of OT's functional effects is the induction of receptors at the target organ by circulating gonadal steroids. The regional specificity of gonadal-steroid-receptor distribution (high in the BNST and VMN) permits the selective induction of receptors. The physiologic variations in estradiol and progesterone may determine whether maternal or reproductive behaviors are expressed. These parallel central and peripheral pathways can be dissociated, resulting in adequate labor and nursing in the absence of normal maternal care. (Adapted from Insel, 1992.)

the amygdala and medial bed nucleus of the stria terminalis are dependent upon testosterone (DeVries, 1995). Concentrations of testosterone will influence vasopressin receptors in the brain. Eliminating testosterone and lowering central vasopressin concentrations will reduce male parental behavior in voles (DeVries, 1995).

Central vasopressin infusions facilitate parental behavior in male voles, which is further augmented by background levels of testosterone (Albers, Liou, and Ferris, 1988; Albers and Cooper, 1995) (Figure 4.11 shows the same effect in male hamsters). Testosterone sustains the fiber projections where vasopressin acts, for example, in the medial amygdala and medial bed nucleus of the stria terminalis that are necessary for this behavior (Wang and DeVries, 1993). Moreover, castration reduces both parental behavior in this species and vasopressinergic expression in the brain (DeVries, 1995; see Chapter 1, this volume). Figure 4.13 depicts vasopressin expression in the brain.

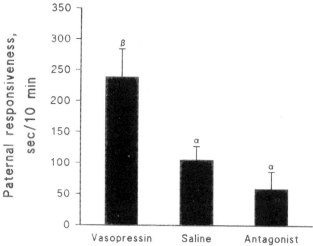

Figure 4.11. Differences in paternal responsiveness among hamsters injected with arginine vasopressin (AVP), saline, or the V1 antagonist. Hamsters injected with vasopressin (β) showed more paternal responsiveness than did the other two groups. (From Z. Wang, C. F. Ferris, and G. J. DeVries, 1994, Role of septal vasopressin innervation in paternal behavior in prairie voles (*Microtus ochrogaster*), *Proceedings of the National Academy of Sciences U.S.A.* 91:400–4. Copyright 1994 National Academy of Sciences, U.S.A. Reprinted by permission.)

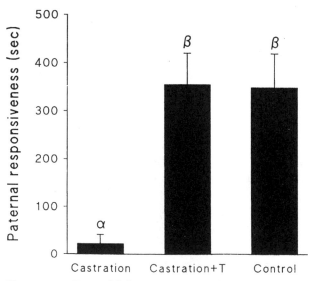

Figure 4.12. Parental behavior in castrated, castrated and testosterone-treated, and control voles. (From DeVries, 1995, with permission.)

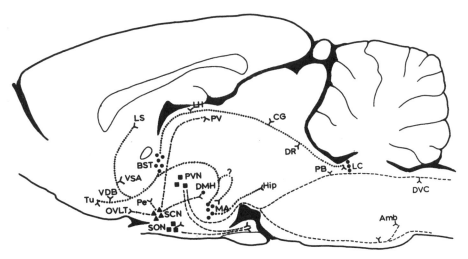

Figure 4.13. Scheme of the most prominent vasopressin pathways. Dotted path: BNST and medial anterior projections to the lateral septum (LS), ventral septal area (VSA), perimeter of the diagonal band of Broca (VDB), olfactory tubercle (Tu), locus ceruleus (LC), and ventral hippocampus (Hip); suprachiasmatic nucleus (SCN) projections to the perimeter of the organum vasculosum laminae terminalis (OVLT), the periventricular (Pe) and dorsomedial nucleus of the hypothalamus (DMH). Dash-line path: paraventricular nucleus (PVN) projections to the parabrachial nucleus (PB), dorsal vagal complex (DVC), and ambiguous nucleus (Amb). Dot-and-dash path: PVN and supraoptic nucleus (SON) projections to the neurohypophysis. (From G. J. DeVries, R. M. Bujis, F. W. VanLeeuwen, A. R. Caffe, and D. F. Swaab, 1985, The vasopressinergic innervation of the brain in normal and castrated rats, *The Journal of Comparative Neurology* 233:236–54. Copyright© 1985 Wiley-Liss, Inc. Reprinted by permission of Wiley-Liss, Inc., a subsidiary of John Wiley & Sons, Inc.)

Separation, Attachment Behaviors, and Cortisol

Having discussed the hormonal signals for attachment behaviors from the parental side, let us look a bit at the development of attachment and the loss of the attachment object from the perspective of the offspring. There are many species differences, most of which we shall not be taking into account in this discussion.

In a number of species, early developmental experiences can either strengthen or weaken future attachments to others (Uvnas-Moberg, 1996). The best-known example is that of imprinting – the phenomenon, described by Lorenz (1981), in which many different kinds of birds follow the first moving object they see after hatching. Because that first object is typically the mother, this behavior is a logical adaptation.

Attachment to objects is one way animals get a foothold in the world and learn to solve problems (Mahler, Pine, and Bergman, 1975). During the process of attachment, the concentrations of various hormones will rise and fall, including those of corticosterone, prolactin, and growth hormone. High concentrations of glucocorticoids will decrease prolactin expression and perhaps the sense of attachment and comfort.

In many species, including our own, an animal's attachment to its parents is a fundamental determinant of success at survival in the world. Parents bring us into the world, and they provide us with our first orientations. Providing offspring a sense of closeness is one of the most important things care-givers do for their young (Ainsworth, 1969; Bowlby, 1973). There is a sense of social symbiosis (Bowlby, 1973; Hofer, 1994). How close we felt to our parents as we gained a foothold in the world – including how often and how we were touched – has helped determine our sense of who we are (Mahler et al., 1975).

Hormonal Influences on Parenting: Getting a Foothold in the World

There is a considerable literature on the neuroendocrine aspects of social separation and loss, and we now know something about the neural substrates of social attachment. From a behavioral perspective, one might presume that social attachment is simply the inverse of social separation. Indeed, many of the neuroendocrine systems activated by separation are inhibited by social contact (Levine, 1993).

Prominent among the ways of becoming attached are the suckling behavior of mammals and the regurgitation of food to their young by birds. The search for and selection of appropriate foods take up major portions of the lives of many animals. Initiation of this behavior represents the transition from infancy to adulthood. Hormones play fundamental roles in this process as they elicit innate preferences, learning, and (in a number of species) perhaps the development of social knowledge (Wingfield, 1994). For example, studies in rats have shown that when young rats watch conspecifics ingest food sources, that ensures and facilitates their ingestion of foods that are appropriate. This was demonstrated in the laboratory when young rats learned to eat unpalatable foods and learned to identify food sources because they watched conspecifics do so (Galef, 1986).

Young mammals follow an elaborate procedure in the search for and attachment to the mother's nipple in order to acquire basic nutrients (Hall, 1990). The behavior can be broken down into an appetitive phase (the search for the mother) and a consummatory phase (the actual suckling bouts).

Among mammals, the nursing period is not a continuous prelude to adult ingestive behavior (Hall, 1990). Infant rats, for example, are largely controlled by the mother while they suckle. If the mother is anesthetized, pups will continue to ingest milk from the nipple until they burst. The mother controls the milk supply. The normal adult vagal and hormonal inputs do not signal satiety in the pups (Lorenz, 1994). Moreover, cholecystokinin, which inhibits food intake in adults, is less potent in suckling rat pups (Lorenz, 1994). There thus appear to be two independent systems in the development of ingestive behavior – one controlled by the mother, and the other controlled by the offspring. Although independent regulatory signals can be demonstrated in a pup separated from its mother, it is normally the mother that controls ingestive behavior, and the pup concedes this control (Hall, 1990). If a rat pup is tested away from its mother, on the other hand, many of the hormonal signals discussed in Chapter 3 can elicit or terminate food intake (e.g., Hall, 1990).

As we turn to the topics of separation and attachment, note that both sodium depletion and maternal separation can have long-term consequences on sodium ingestion when the neonate becomes an adult (Leshem, Maroun, and DelCanho, 1995); in both cases, sodium ingestion will be higher than in control rats from the same litter. We have discussed why and how angiotensin and aldosterone generate a lifelong avidity for sodium (see Chapter 2). Perhaps maternal separation, which is known to activate the stress axis (ACTH, corticosterone), is also sufficient for a lifelong increase in sodium ingestion by the adult.

The Sense of Being Connected or Separated Has Hormonal Consequences for the Offspring

Researchers have demonstrated that during breast feeding, women are less responsive to activation of the hypothalamic-pituitary-adrenal axis (Altemus et al., 1995; Carter and Altemus, 1997). Moreover, lactation is a period in which cortisol concentrations are independent of hypothalamic-pituitary activation (Fischer et al., 1995a). Moreover, cortisol concentrations are related to maternal behavior in rats and people; they can enhance the responsiveness to infant cues (e.g., odor) (Fleming, Corter, and Steiner, 1995). Corticosterone can facilitate attachment behaviors in the male prairie vole (Carter, DeVries, and Getz, 1995),

For mammals, breast feeding serves a number of functional requirements, of which nutrient transfer is but one. The symbiotic homeostatic relationships between rat pups and the mother have been carefully observed for many years by Myron Hofer (1994). Hormones such as oxytocin can increase the social bonding of the neonate with the mother (Panksepp et al., 1988; Nelson and Panksepp, 1996) and decrease the distress signals when they begin to separate during developmental periods.

Secreted by the offspring during contact with its mother, prolactin and growth hormone are essential for normal growth to proceed. Infants separated from their mothers exhibit decreased secretions of growth hormone and prolactin, as well as elevated cortisol concentrations (Kuhn et al., 1991).

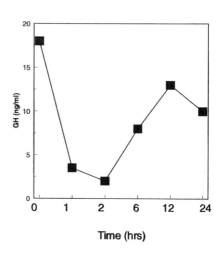

Growth Hormone

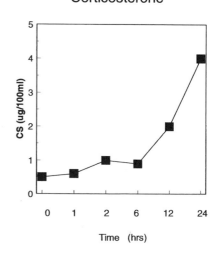

Corticosterone

Figure 4.14. Serum growth hormone and corticosterone in 10-day-old rat pups with increasing durations of maternal deprivation. Pups were left with the dam (controls) or were placed in an incubator for the indicated time; they were then killed, and serum growth hormone and corticosterone were determined. (From C. M. Kuhn, J. Pauk, and S. M. Schanberg, 1990, Endocrine responses to mother-infant separation in developing rats, *Developmental Psychobiology*, 23:395–410. Copyright © 1990 John W. Wiley & Sons, Inc. Reprinted by permission of John Wiley & Sons, Inc.)

Figure 4.15. Picture of my wife, April Oliver, and my daughter Danielle: strong and comforting attachments.

They also lose body weight. Conversely, active parenting promotes growth-hormone secretion and reduces the production of stress hormones by the young (Schanberg and Field, 1988).

Even heavy stroking – that is, active contact – is enough to increase growth hormone in pups and decrease corticosterone secretion (Pauk et al., 1986). In other words, hormones linked to arousal, such as cortisol, are elevated under duress, and the hormones linked to growth and well-being (growth hormone and prolactin) are elevated during rest and physical contact.

Freud (1960) recognized that a sense of security is essential for normal development. He coined a term for the later aberration that results from inappropriate separation during development: "separation anxiety." Separation studies in monkeys have shown that untimely separation patterns can facilitate detrimental behavioral adaptations. The distress that is experienced is represented in the brain, perhaps by the activation of peptides such as corticotropin-releasing hormone. The result is a vulnerability to fearfulness.

Watching my own daughter, I am constantly reminded that her orientation comes from the secure base provided her. Her strengths and vulnerabilities, both inherited and acquired, are linked to what her parents do. As a parent, one tries to provide a secure, stable base. Biology intercedes with our great ability to adapt, to cope, to overcome adversity. These features are at the heart of natural selection.

Temperament, Attachment, and Cortisol

I have already described some of the work done with shy children and their social adaptation, of reaching out to others to ameliorate their internal states (reduced cortisol, etc.). Now we shall consider the roles of temperament, social interaction, and cortisol. As indicated earlier, cortisol is elevated in a subset of temperamentally fearful and inhibited children (Kagan et al., 1988). They also display heightened, potentiated startle responses, fear of novel objects, and poor eye contact with strangers (Schmidt et al., 1997). Maternal ratings of shyness map onto behavioral inhibition and excessive shyness, as well as cortisol concentrations.

What is striking is that even at the age of several months, behavioral expressions of shyness can appear in this group of infants, continuing at least up to 7 years of age, and probably throughout their lifetimes (J. Kagan, personal communication).

As indicated in the Introduction, one mammalian response is to adapt, reducing fear via the social strategy of being close to others. This is a piece of adaptation that in this case serves the child well (Gunnar et al., 1987, 1989). Loss of attachment increases cortisol concentrations, creating a greater sense of vulnerability; this is particularly true in toddlers who feel less secure in their attachments with their mothers.

Perhaps these attachments are even more crucial when they are linked with temperament (Gunnar et al., 1987, 1989). Those infants with insecure attachments or disorganized attachment behaviors demonstrate greater cortisol responses to unfamiliar situations (Hertsgaard et al., 1995).

Anxiety disorders in adulthood have been linked to behavioral inhibition during childhood, and behavioral inhibition in children may be linked to parents with panic disorder (Rosenbaum et al., 1988) and parents who tend to be introverted and subject to greater avoidance behaviors (Rickman and Davidson, 1994).

It has been known for some time that institutionalized children are vulnerable to a number of psychological abnormalities. Such children often have higher concentrations of cortisol, which can be reduced by physical

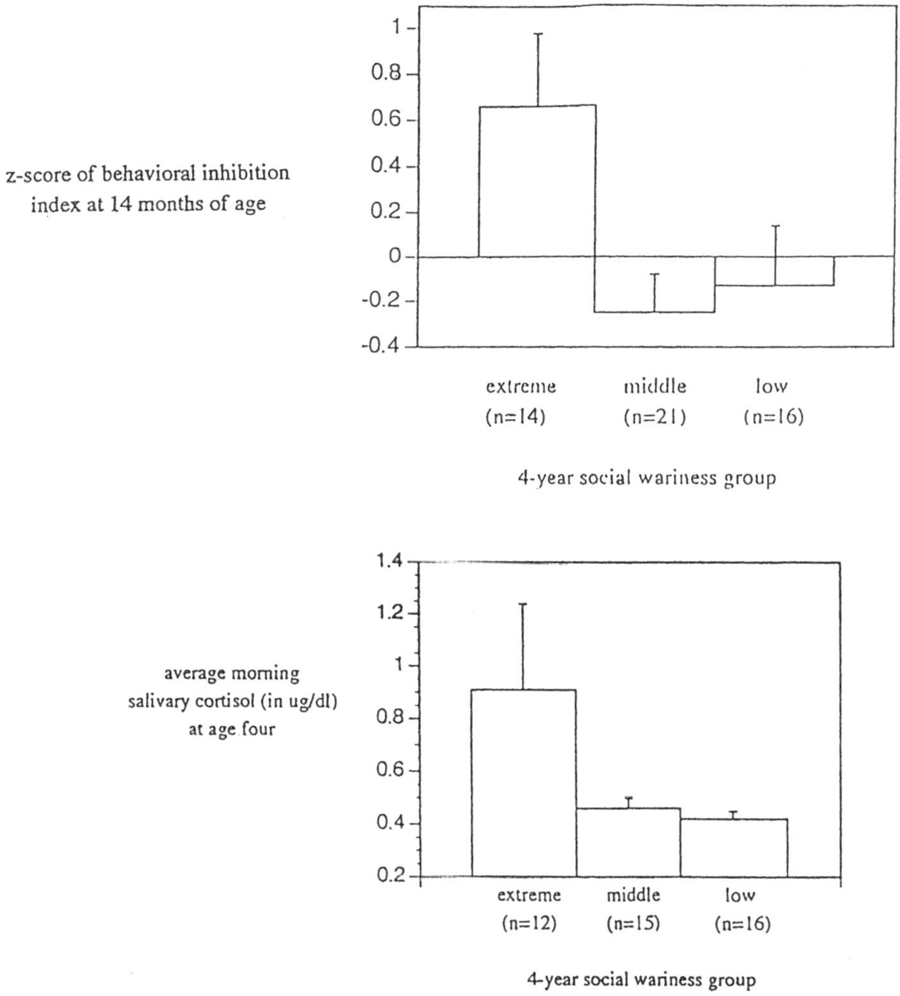

z-score of behavioral inhibition
index at 14 months of age

extreme   middle   low
(n=14)   (n=21)   (n=16)

4-year social wariness group

average morning
salivary cortisol (in ug/dl)
at age four

extreme   middle   low
(n=12)   (n=15)   (n=16)

4-year social wariness group

Figure 4.16. Behavioral inhibition or shyness and salivary cortisol concentrations in 4-year-olds. (From L.A. Schmidt, N. A. Fox, K. H. Rubin, E. M. Sternberg, P. W. Gold, C. C. Smith, and J. Schulkin, 1997, Behavioral and neuroendocrine responses in shy children, *Developmental Psychobiology*, 30.127–40. Copyright © 1997 John Wiley & Sons, Inc. Reprinted by permission of John Wiley & Sons, Inc.)

contact and by fostering a sense of closeness to others (e.g., Gunnar et al., 1989).

Some questions can never be totally resolved: Who can overcome this adversity and develop normally? Who is vulnerable? There are two senses of vulnerability here: The first is the vulnerability due to genetic heritabil-

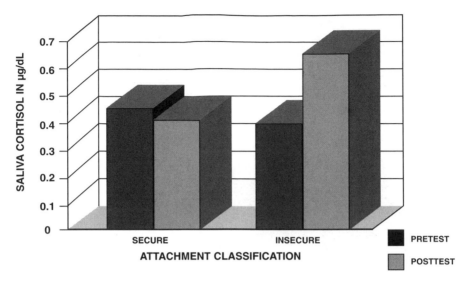

Figure 4.17. Salivary cortisol concentrations in 18-month-old infants who feel securely attached to their mothers, versus insecure infants. (Redrawn from Nachmias et al., 1996, with permission.)

ity. The second is the vulnerability linked to experiences during critical stages in development, a period in which hormones can have greater effects than at any other time in life. The fact that the brain is more malleable during critical stages is no accident. We are born with more brain cells than we can use, and we begin losing neurons at birth. Neurons, in fact, compete with one another, forming circuits through programs provided by their genes.

One might envision both types of vulnerability interacting with and facilitating one another. As Kagan (1989) has speculated, fearful and inhibited children may possess "a more excitable circuit from the amygdala" (p. 172) that is sustained by high cortisol concentrations and environmentally induced duress during critical periods in development. Perhaps a high cortisol concentration facilitates attention toward aversive events and memory of them in these temperamental children. Transgenic models in which corticotropin-releasing hormone is expressed to a greater degree in the brain, and in which there are greater fear-related behaviors, may reflect this state (Stenzel-Poore et al., 1994).

Glucocorticoid Effects on the Hippocampus During Development: Hypothalamic-Pituitary-Adrenal Axis and Behavior

During both gestation and postnatal development, glucocorticoid concentrations can have major impacts on brain organization (and subsequent behaviors) in a variety of different animals. If a mother experiences severe stress during pregnancy – an event that will raise her cortisol concentration – her offspring may be affected both physiologically and behaviorally. In humans and several other mammals these effects range from decreased motor performance to lower body weight, lower production of growth hormone (Schneider, 1992a,b), and sometimes greater fear-related responses to unfamiliar events.

In rats, researchers have looked at three brain regions that are affected when mothers are exposed to stressful events: In their offspring, the central nucleus of the amygdala, the lateral bed nucleus of the stria terminalis, and the paraventricular nucleus of the hypothalamus all show higher concentrations of corticotropin-releasing hormone (CRH) than in control rats whose mothers did not experience stress during pregnancy (Cratty et al., 1995; see Chapter 5, this volume).

After an animal is born, stress hormones still have an impact on future behavior. In rats, there is a period during development when the hypothalamic-pituitary-adrenal axis is relatively refractory (Sapolsky, 1992; Meaney et al., 1991) and is regulated by the mother (Levine et al., 1991). Between 4 and 14 days, the rat's adrenal cortical responses are less sensitive to stressors (Sapolsky, Krey, and McEwen, 1986; Levine et al., 1991; Levine, 1993). However, negative feedback by corticosterone on CRH neurons in the paraventricular nucleus of the hypothalamus can be demonstrated during this period. It is not that the system is not responsive at all, but it is less so than when the rat matures. This so-called hyporesponsive period varies with different species.

The developmental calendar for corticosterone secretion in rats is well documented (Levine, 1994). Corticosterone concentrations in rats decline following birth and do not reach fetal values until day 35, past the time the offspring are weaned. During fetal development, glucocorticoid hormones are essential for normal CRH gene expression.

Maternal separation results in activation of the hypothalamic-pituitary adrenal axis. This has been demonstrated in a number of species (Coe et al., 1985; Levine et al., 1991; Levine, 1994). One such demonstration is illustrated in Figure 4.18.

Maternal deprivation of neonates during development, which will raise

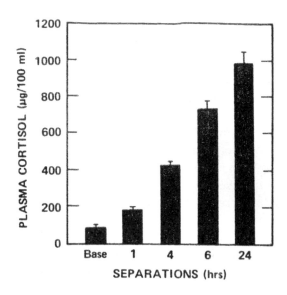

Figure 4.18. Plasma cortisol responses of infant squirrel monkeys to maternal separation of various durations. (From Coe et al., 1985, with permission.)

corticosterone concentrations, will have an impact on CRH expression in the amygdala when they become adults. For example, neonatal rats that are separated from their mothers for several hours each day will, when they are tested under adverse conditions as adults, synthesize greater amounts of CRH than will controls (P. M. Plotsky, unpublished observations; Ladd, Owens, and Nemeroff, 1996). Such maternally deprived offspring also have a greater tendency to become listless and hopeless in difficult situations. They show altered levels of CRH-receptor activity in the amygdala, prefrontal cortex, and locus ceruleus, in addition to CRH activation of the hypothalamic-pituitary-adrenal axis (P. M. Plotsky, unpublished observations). Like prenatal events, these postnatal events have long-term consequences for brain and behavior.

A great many species secrete cortisol (corticosterone) in threatening situations that are associated with greater fear-related behaviors (e.g., freezing) (Kalin and Shelton, 1989). Interestingly, basal cortisol concentrations in rhesus monkeys are correlated with "freezing" behavior observed in both mother and her offspring; the greater the circulating concentration of cortisol, the greater the manifestations of freezing behavior (Kalin and Shelton, 1989; Kalin et al., 1998). There is evidence in other nonhuman primates of increases in CRH as a function of early experiences. For example, ma-

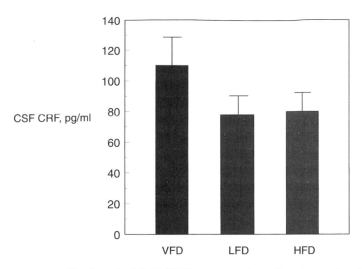

Figure 4.19. Cerebrospinal fluid (CSF) concentrations of corticotropin-releasing factor (CRF) in differentially reared adult bonnet macaques whose mothers were exposed to low, high, and variable foraging demands as infants. (Adapted from Coplan et al., 1996.)

caques that were exposed to unpredictable foraging outcomes as youngsters tended to have higher concentrations of CRH circulating in the cerebrospinal fluid as adults (Coplan et al., 1996).

There are critical periods during which corticosterone must be elevated for normal expression of behaviors to counteract fear (L. K. Takahashi, 1995). In rats, for example, adrenalectomy at 10 days (but not after 14 days) impairs the normal expression of freezing behavior and perhaps other fear-related behaviors. There seems to be a critical window in which corticosterone acts in the brain to organize the expression of behavioral inhibition (Takahashi and Rubin, 1993).

Intracranial or systemic injections of corticosterone in neonatal rat pups facilitates the expression of behavioral inhibition (Takahashi and Kim, 1994), perhaps by activating the dentate gyrus (L. K. Takahashi, 1995). In fact, there are data showing that intracranial injections to 9-day-old rats facilitate fear-induced freezing when faced with anesthetized strange males, and lesions in the hippocampus abolish this behavior (L. K. Takahashi, 1995).

Whereas the hippocampus and its link to the hypothalamic-pituitary-adrenal axis are major components in the regulation of systemic physiologic processes during duress (Herman et al., 1989) and during context-dependent fear conditioning (LeDoux, 1995), it is the ventral hippocampus that appears to regulate CRH production in the paraventricular nucleus in the rat (Herman et al., 1995). The amygdala/hippocampal transition zone is within this vicinity, thus linking amygdala function to stress-related responses.

One view, which will be made clearer in the next chapter, is that there are two CRH-regulated systems: One is linked to systemic physiology, and the other to behavior. Restraint of CRH production curtails the exhaustion of bodily events, while activating sites in the brain that regulate vigilance (amygdala) (see Chapter 5).

Social Control of Cortisol Effects on the Hippocampus: Hypothalamic-Pituitary-Adrenal Axis and Behavior

Handling rats during critical stages in their development can influence the concentrations of corticosterone (Levine, 1993) and the numbers of glucocorticoid receptors in the hippocampus when they become adults (Meaney et al., 1991). In fact, there may be a critical window during which glucocorticoid-receptor activation is particularly vulnerable to the long-term consequences of glucocorticoid reduction (Meaney et al., 1991).

Interestingly, neonatally handled rats have lower ACTH-corticosterone concentrations and perhaps demonstrate less fearfulness as adults than do nonhandled rats when placed in adverse situations (Levine and Mullins, 1968; Meaney et al., 1994). These events are linked to the development of inhibitory control; there is greater negative-feedback control in the handled rats, as well as lower activation of the hypothalamic-pituitary-adrenal axis (Meaney et al., 1994; Liu et al., 1997b).

It logically follows that the handling of rats during critical stages in development will have consequences for behaviors mediated by hormone concentrations. Nonhandled rats, for example, show decreased performances in cognitive tasks, presumably because of the effects of corticosterone-induced vulnerability and degeneration in the hippocampus (Meaney et al., 1994).

Given the studies that have been described, corticosterone or cortisol appears to be linked to critical stages in development in which the brain and subsequent behaviors can be altered by this hormone's actions during a sensitive period. As indicated earlier, Meaney et al. (1991, 1994) have demonstrated that hippocampal receptors for glucocorticoids may be particularly vulnerable during development. In addition, there are differences in CRH expression in the brain between handled and nonhandled neonates (Meaney et al., 1994).

With adaptation and application of broad social intelligence there are ways in which to reduce cortisol concentrations and reduce fear by gaining

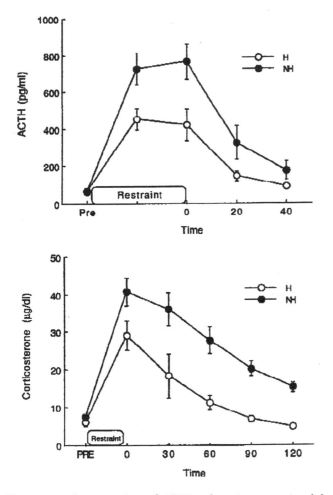

Figure 4.20. Concentrations of ACTH and corticosterone in adult rats who were handled or not handled as neonates. (From Meaney et al., 1994, with permission.)

social contact, which sustains feelings of security. There are numerous examples in which social effects can modify the activation of the hypothalamic-pituitary-adrenal axis (Lyons and Levine, 1994).

Opponent Processes: Arousal and Quiescence, Separation and Attachment

Hormones such as cortisol are essential to the organization of behavior in most species. For species that rest during the dark phase of the diurnal

141

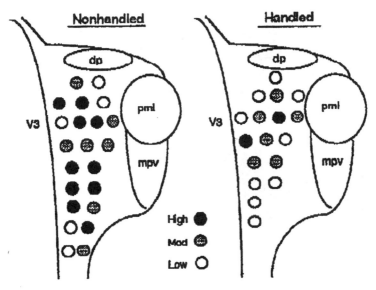

Figure 4.21. Schematic representation of the results of in situ hybridization studies of CRH mRNA expression in the medial portion of the paraventricular nucleus of rats who were or were not handled as neonates. Areas are designated as high, moderate, or low in density of neurons with high levels of CRH mRNA expression. (From Meaney et al., 1994, with permission.)

cycle, hormones such as prolactin, melatonin (see Chapter 6), and growth hormone are needed and are linked to quiescence or rest. Corticosterone and CRH increase fear (Kalin and Shelton, 1989; Levine et al., 1989; see Chapter 5, this volume), and neuropeptide Y decreases it (Koob et al., 1994). Prolactin also may be linked to alleviation of fear (Wehr et al., 1993).

Perhaps prolactin can decrease fear-related responses, the opposite of CRH-treated rats increasing their fear-related responses (P. Holmes and J. Schulkin, unpublished observations). The hormone may act to decrease fearfulness by facilitating a sense of quiescence (Wehr et al., 1993). In depressed humans there is evidence that drugs such as fluoxetine (a serotonin-uptake blocker) can decrease cortisol and CRH concentrations and increase prolactin concentrations (Upadhyaya et al., 1991), thus reducing anxiety and promoting a greater sense of relief and quiescence.

Researchers have found that prolactin concentrations rise in humans while they are meditating (Jevning, Wilson, and VanderLaan, 1978). Other metabolic and humoral processes are also affected by quiet attentiveness or meditation (Davidson, 1976), such as elevated concentrations of vasopressin and decreased cortisol concentrations (Jevning et al., 1978). In other words, the relaxed state of meditating releases prolactin, which perhaps then feeds back to reinforce the relaxed state (Wehr et al., 1993).

The metabolic story is instructive. One consequence of having the hormones of energy balance elevated is that there will be decreases in other hormones that also have metabolic consequences; when the hormones of stress are elevated, the hormones of reproduction are decreased (Sapolsky, 1992; Moore et al., 1994).

In the case of prolactin, there may be two roles: One is to promote quiescence; the second is to combat stress. Secreted in response to a variety of stressors, prolactin may act to protect against stress in general (Drago, 1989, 1990). In humans attempting to solve mathematical problems, for example, both cortisol and prolactin are elevated, just as they are in a number of other stressful situations. Prolactin is also elevated during defensive fighting, but not necessarily during offensive fighting. The hormone may work by promoting stillness, which in turns provides an antidote to stress. Prolactin infusions reduce the impact of stress (as measured by reduced gastric abnormalities) (Drago, 1989). In other words, prolactin, in some contexts, may be one of the body's own tranquilizers (T. A. Wehr, unpublished observations).

Abnormal prolactin concentrations have been reported in depressed pa-

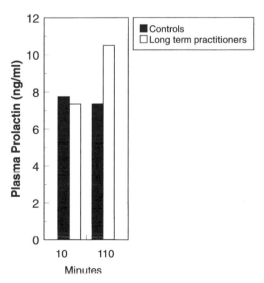

Figure 4.22. Plasma prolactin concentrations in short-term practitioners (restudied controls) and long-term practitioners of meditation. (Adapted from Jevning et al., 1978.)

tients (Linkowski et al., 1989), particularly those who are suicidal. The hormone is elevated in patients with unipolar depression, in which, as one psychiatrist says, patients become so quiet that they are just "blobs" that do not move (T. A. Wehr, personal communication). Though the evidence is conflicting, there have been studies suggesting that increases in prolactin follow the use of neuroleptics in schizophrenic patients, in whom the hormone is sometimes low. Manipulations that return prolactin concentrations to normal or increase them, while decreasing cortisol, have had significant therapeutic effects in many depressed patients. High CRH concentrations are reduced (and prolactin levels are increased) by pharmacologic agents such as Zoloft, a selective serotonin reuptake inhibitor (Upadhyaya et al., 1991).

Conclusion

In this chapter we have discussed several model systems, emphasizing two traditions: one that concentrates on how hormones activate the brain to influence behavior to correct internal instability, and the other the social regulation of the internal milieu. It is the social facilitation that is so essential in the parenting relationship in animals like ourselves and in the offspring's regulation of the relationship and its internal states. We have considered several instances in which steroids and peptides interact in influencing behaviors (e.g., estrogen and central oxytocin; prolactin, testosterone, and vasopressin).

CHAPTER FIVE

Hormonal Regulation of Fear and Stress

Introduction

A natural part of our psychobiology, the central motive state of fear is controlled at least in part by hormones. Fear is not uniform. Fear of the unfamiliar, for example, differs from conditioned fear and perhaps from those forms due to innate releasers of fear (Kagan and Schulkin, 1995). Hebb (1946) and others pointed this out some 50 years ago, noting that there is more than one sense of fear, a fact that has biological significance. Hebb emphasized fear of the unfamiliar and fear associated with innate perceptual sensibility. Both humans and chimpanzees (*Pan troglodytes*), for instance, are afraid of headless animals. This chapter will explain how humoral signals – specifically glucocorticoids and corticotropin-releasing hormone (CRH) – help to sustain some fear-like states.

Many thinkers in this century have construed fear either as an intervening variable (e.g., Miller, 1959–71; Davis et al., 1993) or as a tool with which to predict behavior, "but nothing inside in which to frame a science" (e.g., Skinner, 1938). That is not my view. Fear is a property of the nervous system, as well as a product of our evolutionary history.

Although emotions are often construed as passive (e.g., by Spinoza and Freud) (Parrott and Schulkin, 1993), fear is an active response, rather than a passive response. As a product of our evolution, fear generates adaptive, problem-solving behaviors. When emotions such as fear are placed in this functional light, they come closer to what Charles Darwin (1965), William James (1952), and John Dewey (1894, 1895), had in mind – that fear is a biological phenomenon linked to the perception of danger and the behavioral reactions to avoid it.

Fear – like thirst, sodium hunger, and sexual appetite – is a central motive state that potentiates readiness to behave in specific ways (Frijda, 1986). The central state prepares the animal for action, which in some cases may involve freezing (LeDoux, 1987, 1995), and in others may involve fleeing or avoiding the source of the danger.

Certain states of fear are linked to the biology of avoid-or-approach responses (Schnierla, 1959; Davidson et al., 1990; Lang, 1995). Some of the
most elegant experiments on fear were performed in contexts of conflict and
choice by Neal Miller and his group at Yale University during the 1940s
through the 1960s. Fear can create inner conflict when one simultaneously
receives signals to approach and other signals to avoid something. This
central motive state is expressed in appetitive and consummatory behaviors: behaviors to avoid the fearful event and consummatory behaviors of
relief.

Hormonal changes go hand in hand with fear. Specifically, glucocorticoids, which are produced by the adrenal gland, are essential for perceiving
and appropriately reacting to fear, and perhaps also for sustaining the fear
response. They do this in two important ways: by restraining physiologic
events (perhaps so that the animal will not be overwhelmed) and by activating regions of the brain (amygdala) that respond to environmental signals
of danger. Without the glucocorticoid cortisol, animals, including humans,
are much less able to cope with extreme adversity (Richter, 1976; Pugh et
al., 1997).

This chapter provides a framework in which to integrate the many studies
of the regulation of fear in animals. This chapter builds on the preceding
chapter, in which hormones such as cortisol (in humans and monkeys),
corticosterone (in rats), and central CRH were discussed. They will be discussed in more detail here. Moreover, this chapter brings together a cluster
of psychological symptoms in humans – including fear, anxiety, agitated
depression, panic disorder, and posttraumatic stress disorder – that, while
distinct and possibly operating through different mechanisms, all result in
a fixation on anticipated negative events.

Finally, this chapter describes a neurohormonal model system that may
underlie fear states. I argue that brain regions such as the amygdala (and its
associated structures) may, under normal circumstances, help regulate the
body's internal milieu (by responding to internal signals such as CRH and
the glucocorticoids) and integrate it with external events.

Hormones and Helplessness

Curt Richter at the Phipps Clinic at Johns Hopkins University described an
animal model for what he called "hopelessness" (Richter, 1976): He held
rats until they stopped struggling or were immobile; then he placed them in
water. Whereas rats typically can swim for long periods of time, those that
had been held drowned within minutes. Richter suggested that the animals
had experienced hopelessness – they "expected failure" – and as a conse-

quence they just gave up. In all of the rats that died, corticosterone concentrations were elevated.

Over the past 10 years, the terminology for this phenomenon has changed from "hopelessness" to "learned helplessness." The most widely used animal model of helplessness is induced by uncontrollable shock. A variety of species will become what is called "helpless" when they are placed in situations marked by uncontrollable and unpredictable aversive events such as shocks. After learning that their attempts at adaptive behaviors will not ameliorate the situation, the animals eventually give up. Even if at a later time they have the capacity to do something to change the situation (e.g., avoid it by performing an operant), they still will give up, even though they seem anxious and fearful as well (Seligman, 1975; Maier et al., 1993). Helpless animals not only lose the incentive to perform a task that would save them in a situation in which previously they had lost control; they also become apathetic and express decreased hunger and sex drives. They lose incentive and disengage. Disengagement may also be an adaptive response, reducing behavioral and physiologic activities that would place the animal in danger.

One of the most important determinants of animal learning is prediction and perceived control. As an important psychobiological adaptation, animals track events and draw causal relationships among them (Dickinson, 1980; Gallistel, 1990). The ability to detect causal relationships arises specifically when expectations are thwarted; the discrepancy between what is expected and what actually happens can induce learning (Rescorla and Wagner, 1972; Dewey, 1989) and enhanced attention (Mackintosh, 1975).

In the presence of unpredictability, however, learning declines and attention dissipates (Miller, 1959–71). Avoidance behavior takes precedence over approach behavior. This is particularly true under stress. In the laboratory, a shock preceded by a signal may cause less bodily harm than an unsignaled shock (e.g., gastric abnormalities, central catecholamine depletion, decrease in appetite). This means that the unsignaled shock is particularly anxiety-provoking. Many models of helplessness in the laboratory are based on the loss of predictability and perceived control and their impact on anxiety and fear (Seligman, 1975; Maier et al., 1993). Creating a perpetual state of anticipation, such loss may affect brain sites involved in maintaining vigilance toward events that are expected over both short and long periods of time.

In humans (Brier, 1989) and rats (Kant et al., 1992), the perception of control can affect glucocorticoid concentrations. In several laboratory species it has been found that when a footshock is predictable, and thus anticipated by the test animals, glucocorticoid concentrations are lower than in

147

yoked control animals who are unable to predict the shock (Levine et al., 1991; cf. Maier and Ryan, 1986). Perceived control apparently results in lower concentrations of the hormone (Mason et al., 1961). Though the data are inconsistent, it appears that without perceived control, the elevations of glucocorticoids in rats and monkeys persist for longer periods of time than in control animals. Predictability can reduce the circulating concentrations of glucocorticoids (Kant et al., 1992). Conversely, the greater the sense of vulnerability to an aversive event, the higher the circulating concentrations of glucocorticoids (Levine et al., 1991; Sapolsky, 1992).

Glucocorticoids are also secreted when rats and monkeys are placed in novel environments and when rats experience acute anxiety (e.g., Mason et al., 1961; Levine et al., 1991). Interestingly, rats handled extensively during neonatal development have lower circulating concentrations of glucocorticoids than do nonhandled controls. In addition, they have a greater tendency to explore novel environments (Meaney et al., 1991; see Chapter 4, this volume).

Facilitation of Fear Through Glucocorticoid Action

In a wide variety of species (e.g., reptiles, birds, and mammals, including humans), glucocorticoid concentrations are high when animals are fearful. Research has shown that monkeys that are afraid have higher systemic concentrations of glucocorticoids than do those that are not afraid (e.g., Champoux et al., 1989; Sapolsky, 1992). In cats (*Felis catus*), cortisol production is most pronounced during periods of fear and restlessness (Kojima et al., 1995). Monkeys that are more isolated than those living in social groups also have higher concentrations of cortisol (Lyons and Levine, 1994), as do inhibited human children who are shy and fearful (Kagan et al., 1988; see Chapter 4, this volume). Depressed adults, perhaps particularly those with agitated depression, have elevated cortisol concentrations (Sachar et al., 1970; Brown et al., 1988). Attempted suicide, in fact, has been correlated with elevated cortisol in depressed and panic-disorder patients (Fawcett, 1992). Hospitalized suicidal patients not only have high concentrations of cortisol but also exhibit fewer receptor binding sites for CRH in the frontal cortex than do controls (Nemeroff et al., 1988).

Among reptiles, several species of lizards secrete glucocorticoids when they are exposed to fearful situations, including being held in collecting bags (Moore, Thompson, and Marleer, 1991), or experiencing defeat in struggles for male dominance. These events all seem to determine glucocorticoid concentrations (Knapp and Moore, 1995).

The sizes of adrenal glands, which produce glucocorticoids, also correlate

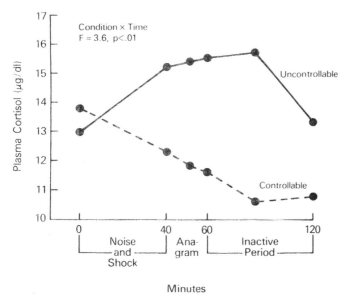

Figure 5.1. Plasma cortisol concentrations in uncontrollable and controllable stress paradigms in healthy human volunteers. (Reprinted with permission from *Biological Psychiatry* 26, A. Brier, Experimental approaches to human stress research: assessment of neurobiological mechanisms of stress in volunteers and psychiatric patients, pp. 438–62, 1989, Elsevier Science Inc.)

with dominance. The more dominant a male tree lizard is, the higher its testosterone and the lower its glucocorticoid (Greenberg, Chen, and Crews, 1984; Tokarz, 1987). The converse holds for the subordinate animal. Elevated glucocorticoid concentrations have been found in submissive rats, in addition to mice, pigs, and birds, as reviewed by Sapolsky (1992). This phenomenon also occurs in the green iguana (Daugherty and Callard, 1972) and the American alligator (*Alligator mississippiensis*) (Lance and Elsey, 1986). In some of these species, skin pigmentation is associated with social status and glucocorticoid concentrations (Greenberg et al., 1984); these reptiles also have much lower glucocorticoid concentrations when they are exposed to females. In the former context, testosterone concentrations are low, whereas in the latter they are high.

It has been known for some time that dominant male monkeys secrete less cortisol under basal conditions than do subordinates (e.g., Sapolsky, 1992). The social stress of being lower in the pecking order results in lower testosterone, but higher glucocorticoids. This increase in glucocorticoids and decrease in testosterone in turn reduce the subordinate animals' prospects for successful reproduction. Elevated glucocorticoid concentrations may reflect loss of control, although there most likely are other differences – such as different degrees of fear and anxiety – between dominant and subordinate males. Interestingly, dominant monkeys that are subjected to uncontrollable situations actually hypersecrete cortisol, indicating that they are shocked by the loss of control.

Anatomic and Functional Studies of the Amygdala and Bed Nucleus of the Stria Terminalis

The amygdala is functionally involved in many of the same events as the hypothalamus, including appetite, drive, and enjoyment (Fonberg, 1974). The amygdala is also involved in the anticipation of fearful and anxiety-producing events and perhaps chronic arousal. Activation of the amygdala is known to lead to anticipation of fearful events (LeDoux, 1987, 1996).

Regions of the amygdala can be demarcated in a number of ways. For our purposes, we shall distinguish between the basal-lateral and central-medial

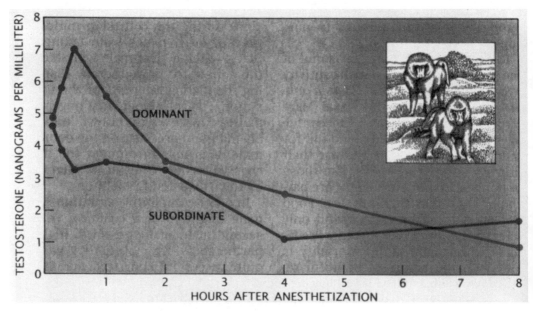

Figure 5.2. Average testosterone concentrations in dominant and subordinate male baboons are essentially equal when the animals are at rest, but typically diverge strikingly when the animals are exposed to a stressor – in this case anesthesia. The concentration in the subordinate male plummets immediately, whereas that in the dominant male rises sharply at first and remains elevated for approximately an hour. (From Sapolsky, 1990, with permission. Figure by Gabor Kiss and Patricia J. Wynne.)

regions. The basal-lateral is a funnel for all sensory information from the neocortex (Aggleton and Mishkin, 1986; LeDoux, 1995), and the central-medial for the activation of behavioral responses. Both regions contain nuclei that receive and process distinct sensory information (LeDoux, 1995). The central-medial is subcortical, and the basal-lateral is considered more cortical, presumably because of input from, or connectivity with, neocortical sites (Krettek and Price, 1978; Alheid, de Olmos, and Beltramino, 1996).

The central nucleus of the amygdala (CEA) receives the largest number of visceral projections from the solitary and parabrachial nuclei in the lower brainstem; it projects directly to these regions, as well as to other areas of the brainstem that are involved in arousal. It also receives input from neocortical sites and from the lateral hypothalamus (e.g., Schwaber et al., 1982; LeDoux et al., 1990).

The CEA receives sensory information from the lateral amygdala (LeDoux et al., 1990), a critical site for fear conditioning (perhaps because of sensory projections to this region) (Kapp et al., 1979). The CEA is also continuous with the bed nucleus of the stria terminalis (Johnston, 1923; Alheid and Heimer, 1988; Alheid et al., 1996), a site that is derived embryologically from CEA tissue during neuronal development and thus can be considered "extended amygdala" (Johnston, 1923). The lateral bed nucleus is linked to the CEA, and the medial bed nucleus is linked to the medial nucleus of the amygdala (Alheid and Heimer, 1988).

These anatomic linkages suggest that the bed nucleus of the stria terminalis may regulate many of the same behaviors as regions of the amygdala (Herman and Cullinan, 1997). In fact, anatomically the CEA and bed nucleus are tied to the same sites in the brain (e.g., Gray, Carney, and Magnuson, 1989) – both projecting to the parabrachial nucleus and solitary nucleus, for example (Schwaber et al., 1982; Gray et al., 1989).

Early studies in humans indicated that stimulation of the amygdala would result in activation of the hypothalamic-pituitary-adrenal axis (LeDoux, 1987). Functional studies in rats have demonstrated that direct electrical or chemical stimulation of the CEA and bed nucleus will provoke release of hypothalamic CRH and increase the systemic circulation of ACTH and glucocorticoids (Gray et al., 1989; Roozendaal, Koolhaas, and Bohus, 1992). Such stimulation decreases corticotropin-releasing factor (CRF) in the paraventricular nucleus and in the median eminence.

Lesions of the CEA can interfere with the release of pituitary hormones in immobilization-induced stress (e.g., Beaulieu et al., 1987). CEA lesions also decrease the glucocorticoid elevation that results from short-term immobilization-induced stress (Van de Kar et al., 1991; Herman and Cullinan, 1997) – specifically, elevation of glucocorticoids in conditioned stress

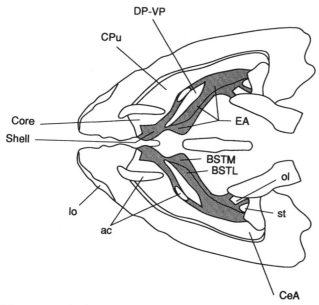

Figure 5.3. The bed nucleus of the stria terminalis (BSTM, BSTL) and the central nucleus of the amygdala (CeA). Abbreviations: ac, anterior commissure; CPu, caudate putamen; DP-VP, dorsal and ventral pallidum; EA, extended amygdala (e.g., BNST); lo, lateral olfactory tract; st, stria terminalis. (Courtesy of G. Alheid. Reprinted from *Neuroscience and Biobehavioral Reviews* 18, J. Schulkin, B. S. McEwen, and P. W. Gold, Allostasis, amygdala and anticipatory angst, pp. 385–96, copyright 1994, with kind permission from Elsevier Science Ltd, The Boulevard, Langford Lane, Kidlington OX5 1GB, UK.)

(Roozendaal et al., 1992). Lateral lesions within the amygdala do not produce this effect. Moreover, lesions of the lateral bed nucleus (the extended amygdala that contains CRH neurons and glucocorticoid receptors) produce the same decrease in stress-induced glucocorticoid secretion as do lesions of the CEA (Gray et al., 1989; Herman, Cullinan, and Watson, 1994). Electrophysiologic evidence indicates that the bed nucleus of the stria terminalis is activated under aversive or stressful conditions (Henke, 1982).

Other functional studies have shown that the CEA, in particular, plays a role in conditioned arousal as well as in attention and learning (Kapp et al., 1979; Gallagher and Holland, 1994) and anticipatory anxiety (Schulkin et al., 1994a; Rosen et al., 1996). CEA damage abolishes conditioning to a fearful stimulus, without disrupting the unconditioned response (Davis et al., 1993). In addition, CEA lesions reduce the freezing behavior associated with inescapable shock or hopelessness (Grahn et al., 1992). Whereas lesions of the CEA (and, to some extent, the lateral nucleus of the amygdala) interfere with the hypoalgesic response to conditioned fearful stimuli, they do not interfere with the response to unconditioned stimuli (Helmstetter, 1992). These results extend earlier findings that damage to this region of the amygdala reduces fear and anxiety (Kapp et al., 1979), whereas its activation increases these feelings. CEA activation sets the stage for the anticipation of adversity.

CRH and Glucocorticoid Brain Regulation

The hypothalamus is the principal site in the brain that maintains endocrine homeostasis and, in many instances, behavioral (vegetative) homeostasis. Under negative-feedback regulation and set-point control, hypothalamic-releasing hormones strongly influence pituitary regulation of the thyroid, gonadal, and adrenal axes (Vale et al., 1981). CRH activates corticotropin (ACTH), which in turn activates glucocorticoids from the adrenal glands (e.g., Dallman et al., 1992).

CRH is located in both the paraventricular nucleus and extrahypothalamic sites. Just as they are known to restrain other physiologic processes (Munck et al., 1984), glucocorticoids restrain CRH elevation at the level of the paraventricular nucleus of the hypothalamus (e.g., Swanson and Simmons, 1989). For example, adrenalectomy will augment noradrenergic activation of the paraventricular nucleus, and glucocorticoid will reduce or return it to normal (Pacak et al., 1993). Neither effect is expressed during behavioral paradigms of adversity (immobilization-induced stress). Negative feedback is one of the body's primary mechanisms for self-control. In other words, regulation of CRH – restraining it or restraining catecholamines or vasopressin (Munck et al., 1984; McEwen and Stellar, 1993) – is under negative-feedback control at the level of the hypothalamus. Interestingly, during stress, there is a subset of vasopressin neurons that become CRH sites (Whitnall, 1993), and vasopressin and CRH are synergistic in regulating the hypothalamic-pituitary-adrenal axis, and perhaps even behavioral functions (startle) (Davis et al., 1993).

During threatening situations, organisms experience disturbances in hypothalamic regulation of growth, reproduction, hunger, and principal neuroendocrine systems, namely, the hypothalamic-pituitary-adrenal axis (e.g., Gold, Goodwin, and Chrousos, 1988). Several researchers have suggested that inhibition of vegetative functions and hyperarousal under conditions of stress are mediated through the action of CRH. Support for this idea comes from the fact that intracerebroventricular administration of CRH activates a coordinated series of behavioral and physiologic responses, including enhanced arousal, fear, and inhibition of vegetative functions (e.g., Owens et al., 1990). This phenomenon has been demonstrated in rats and rhesus monkeys (e.g., Kalin and Shelton, 1989; Koob et al., 1993).

In addition to their well-known ability to restrain physiologic processes (e.g., inflammation) and regulate and maintain a variety of neuropeptides and neurotransmitters, such as enkephalin mRNA (Chao and McEwen, 1990), GABA receptors (Majewska, Bisserbem, and Eskay, 1985), and 5-hydroxtryptamine (5-HT) receptor mRNA (Chalmers et al., 1993), glucocorticoids can induce positive effects. These include upregulation of vasopressin receptors or cell bodies (Colson et al., 1992; see Chapter 2, this volume), increasing neurotensin in the CEA and the bed nucleus (Watts and Sanchez-Watts, 1995a,b), and increasing proenkephalin gene expression in the caudate putamen and the hippocampus (Chao and McEwen, 1990).

More germane in the present context, glucocorticoids can facilitate amphetamine or even morphine self-administration (Piazza et al., 1991). That

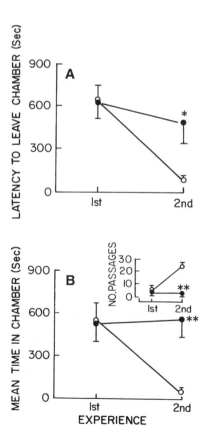

Figure 5.4. Facilitation of fear response by intraventricular infusions of CRH (filled circles) or vehicle 20 minutes prior to the second experience in the test environment. Panel A shows the latency to leave the chamber. Panel B shows the mean time spent in the chamber per entry. The number of passages made between the chamber and the open field is shown in the inset. (From L. K. Takahashi, N. H. Kalin, J. A. Vanden Burgt, and J. E. Sherman, 1989, Corticotropin-releasing factor modulates defensive-withdrawal and exploratory behavior in rats, *Behavioral Neuroscience* 103:648–54. Copyright 1989 by the American Psychological Association. Reprinted with permission.)

is, high concentrations of glucocorticoids can result in enhanced self-administration of amphetamine, in addition to cocaine-induced kindling (Kling et al., 1993). Chronic release of corticosterone (7 days) at stress levels, in contrast to acute release (1 day), can increase peripheral levels of sympathetic adrenal responses to immobilization-induced stress (Kvetnansky et al., 1993). Glucocorticoids can also raise the rate of dopamine synthesis in the nucleus accumbens and increase norepinephrine synthesis in the prefrontal cortex. Some of the psychomotor effects should be due to this increase in catecholamine synthesis.

Some studies have drawn a clear distinction between the effects of glucocorticoids on systemic physiologic regulation and the effects of centrally administered CRH (Britton et al., 1986). Centrally infused CRH also facilitates kindling, specifically amygdala-induced seizures (Weiss et al., 1991). The hormone also potentiates cocaine-induced kindling, a phenomenon that depends (at least in studies with infant rats) on the amygdala (Kling et al., 1993). Glucocorticoids facilitate CRH-induced kindling (Rosen et al., 1994): Whereas injections of CRH into the third ventricle do not normally facilitate kindling, the same doses will do so when rats have been treated for several days with corticosterone. The same doses will increase CRH mRNA in cell bodies in the CEA. These findings contradict the notion that glucocorticoids restrain seizures by restraining the hypothalamic-pituitary axis. Perhaps the glucocorticoids' effects on kindling are due to activation of extrahypothalamic CRH.

The foregoing findings suggest that whereas glucocorticoids restrain CRH in the paraventricular nucleus to regulate homeostasis, they may facilitate

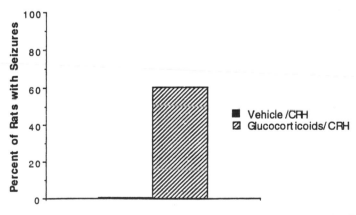

Figure 5.5. The effect of intraventricular infusions of CRH on kindling is facilitated by glucocorticoids. (Adapted from Rosen et al., 1994.)

CRH expression in the amygdala to potentiate conditioned fear, anxiety, and cautious avoidance. The different roles played by hypothalamic and extra-hypothalamic CRH to mediate glucocorticoids constitute one of the theoretical underpinnings of this chapter (Schulkin et al., 1994a).

There are CRH cells in the central nucleus of the amygdala, bed nucleus of the stria terminalis, paraventricular nucleus of the hypothalamus, and parabrachial and solitary nuclei of the lower brainstem (e.g., Gray, 1990; Sawchenko, 1993), as well as in the dorsal motor nucleus, the central gray, and Barrington's nucleus adjacent to the locus ceruleus (Swanson et al., 1983). These regions constitute a part of the limbic circuit that is involved in generating motivated behaviors. They are reciprocally connected with one another, many containing both type-1 and type-2 corticosteroid receptor sites (De Kloet, 1991; McEwen and Stellar, 1993).

In addition, there is CRH mRNA in brainstem sensory areas and regions of the central auditory system that may be linked to fear conditioning (LeDoux et al., 1990). Moreover, there is colocalization of CRH mRNA and glucocorticoid receptors in several of these regions (e.g., CEA, bed nucleus, parabrachial nucleus), which means they contain CRH cells and corticosteroid receptor sites.

Whereas CRH receptor sites are largely found outside of the areas where CRH-producing cells are located (e.g., Potter et al., 1994), CRH cells and fibers are reciprocally connected with one another (e.g., Swanson et al., 1983; Gray, 1990) and are well constituted to underlie autonomic and affective responses. Experiments utilizing c-fos have shown that intracerebral injections of CRH activate early gene transcriptions in the CEA and bed nucleus of the stria terminalis (Andrease and Herbert, 1993; Herbert, 1997), and c-fos experiments with adrenalectomized rats treated with glucocorticoid have shown that the hormone activates parvocellular paraventricular CRH (Dallman et al., 1993).

Dual Roles of Glucocorticoids in Regulating CRH

We now know that glucocorticoids, to reduce the stress response, not only decrease CRH or vasopressin in the brain but also increase CRH mRNA (Swanson and Simmons, 1989). Whereas glucocorticoid hormones increase CRH in the placenta and magnocellular neurons of the paraventricular nucleus of the hypothalamus, for example, they decrease CRH or vasopressin production in parvocellular neurons in the paraventricular nucleus of the hypothalamus (Swanson and Simmons, 1989). The hormones also reduce

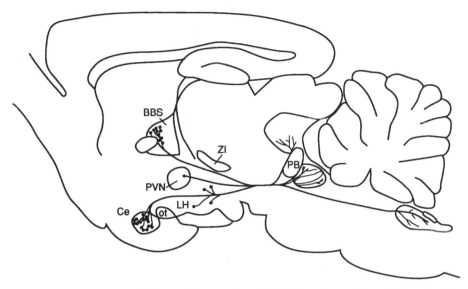

Figure 5.6. CRH sites in the brain. Abbreviations: BBS, bed nucleus of the stria terminalis; Ce, central nucleus of the amygdala; LH, lateral hypothalamus; ot, optic tract; PB, parabrachial nucleus; PVN, paraventricular nucleus of the hypothalamus; ZI, zona incerta. (Reprinted with permission from T. S. Gray, 1990, The organization and possible function of amygdaloid corticotropin-releasing factor pathways, in *Corticotropin-releasing Factor: Basic and Clinic Studies of a Neuropeptide,* ed. E. B. DeSousa and C. B. Nemeroff. Boca Raton: CRC Press. Copyright CRC Press, Boca Raton, Florida.)

cyclic GMP and intracellular calcium proteins (Strauss, Schulkin, and Jacobowitz, 1995), in the amygdala and other central sites (McEwen and Stellar, 1993).

Thus, whereas vegetative or homeostatic functions such as food intake are inhibited by elevated CRH (e.g., Glowa et al. 1992), other functions such as hyperarousal are stimulated. This phenomenon may reflect the fact that one system (CRH mRNA) regulates the pituitary-adrenal axis and control of vegetative functions, and another system (CEA) regulates behavioral functions such as fear and anxiety. Whereas adrenalectomy elevates CRH mRNA in the paraventricular nucleus, it reduces CRH mRNA in the CEA; this effect is reversed by corticosterone (M. Palkovits, unpublished observations; S. Makino, P. W. Gold, and J. Schulkin, unpublished observations). That finding suggests differential regulation by glucocorticoids of CRH mRNA in different neural sites (Albeck et al., 1997). Moreover, evidence indicates that there are two different types of CRH receptors (Lovenberg et al., 1995), which may be linked to this functional distinction.

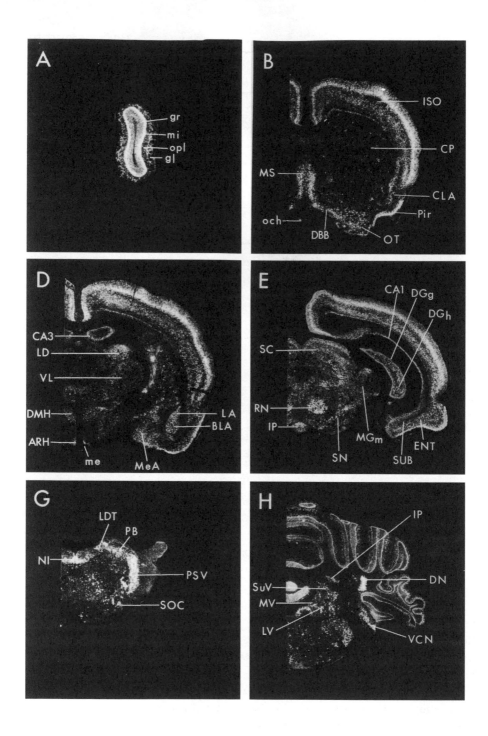

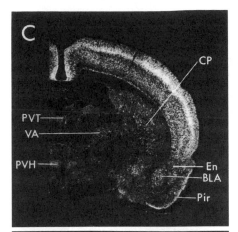

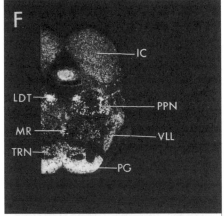

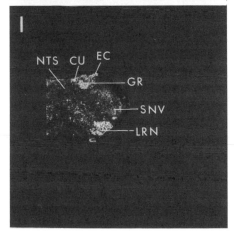

Figure 5.7. Hybridization histochemical localization of CRF-R mRNA in the rat brain. A series of darkfield photomicrographs is arranged from rostral (A) to caudal (I) to show the regional distribution of neurons positively hybridized with an antisense probe generated from full-length (1.3-kb) cDNA encoding the CRF-R mRNA (all × 1.5). Abbreviations: ARH, arcuate nucleus (n.); BLA, basolateral n. (amygdala); CA1, field CA1 (Ammon's horn); CA3, field CA3 (Ammon's horn); CLA, claustrom; CP, caudoputamen; CU, cuneate n.; DBB, n. of diagonal band of Broca; DGg, granule-cell layer (dentate gyrus); DGh, hilar region (dentate gyrus); DMH, dorsomedial n. (hypothalamus); DN, dentate n.; EC, external cuneate n.; En, endopiriform n.; ENT, entorhinal cortex; gl, glomerular layer (olfactory bulb); gr, granule-cell layer (olfactory bulb); GR, gracile n.; IC, inferior colliculus; IP, interpeduncular n.; ISO, isocortex; LA, lateral n. (amygdala); LD, laterodorsal n. (thalamus); LDT, laterodorsal tegmental n.; LRN, lateral reticular n.; LV, lateral vestibular n.; me, median eminence; MeA, medial n. (amygdala); mi, mitral-cell layer (olfactory bulb); MGm, medial geniculate body (medial division); MR, median raphe n.; MS, medial septal n.; MV, medial vestibular n.; NI, n. incertus; NTS, n. of the solitary tract; och, optic chasm; opl, outer plexiform layer (olfactory bulb); OT, n. of the optic tract; PB, parabrachial n.; PG, pontine gray; Pir, piriform cortex; PPN, pedunculopontine n.; PSV, principal sensory n. of the trigeminal nerve; PVH, paraventricular n. (hypothalamus); PVT, paraventricular n. (thalamus); RN, red nucleus; SC, superior colliculus; SN, substantia nigra; SNV, spinal trigeminal n.; SOC, superior olivary complex; SUB, subiculum; SuV, superior vestibutal n.; TRN, tegmental reticular n.; VA, ventral anterior n. (thalamus); VCN, ventral cochlear n.; VL, ventral lateral n. (thalamus); VLL, ventral n. of the lateral lemniscus. (From E. Potter, S. Sutton, C. Donaldson, R. Chen, M. Perrin, K. Lewis, P. E. Sawchenko, and W. Vale, 1994, Distribution of corticotropin-releasing factor receptor mRNA expression in the rat brain and pituitary. *Proceedings of the National Academy of Sciences U.S.A.* 91:8777–81. Copyright 1994 National Academy of Sciences, U.S.A. Reprinted by permission.)

In experiments with tree shrews, submissive males that had high concentrations of glucocorticoid demonstrated differential binding of CRH receptors in the hippocampus and amygdala (Fuchs and Flugge, 1995). That finding and other evidence support the idea of differential regulation on target tissues by glucocorticoids, as well as possible mechanisms that may be responsible for this. On the other hand, we have found that high concentrations of corticosterone that will increase CRH mRNA in cells in the central nucleus will decrease CRH mRNA receptors in the basal-lateral amygdala (Makino et al., 1995). What this means we do not know.

The greater the degree of glucocorticoid activation of the CEA, the greater the amount of CRH mRNA synthesized (Swanson and Simmons, 1989; Makino et al., 1994a; Watts and Sanchez-Watts, 1995a,b; Schulkin, Gold, and McEwen, in press; see chapter 4, this volume). This occurs in adrenalectomized rats treated with corticosterone (Swanson and Simmons, 1989) and in adrenally intact rats treated with high concentrations of corticosterone (Makino et al., 1994a). Given the fact that fear is correlated with high glucocorticoid concentrations (see Chapter 4), it is possible that glucocorticoids are magnifying CRH in the CEA, and one result may be expectations of adversity (e.g., fear or anxiety). The hypothesis suggests that activation of CEA CRH may set the context for expecting adverse events.

These effects are not confined to the CEA. They also occur in the dorsal lateral bed nucleus, an anterior extension of the CEA that increases its CRH production in response to sustained high concentrations of glucocorticoids (Makino et al., 1994b; Watts and Sanchez-Watts, 1995a,b; Watts, 1996). This is the same region of the brain that is anatomically tied to brainstem sites (Schwaber et al., 1982). Because the CEA massively projects to the bed nucleus, and is only barely connected to the paraventricular nucleus, it may influence the hypothalamic-pituitary-adrenal axis through the bed nucleus (Gray et al., 1989; Herman et al., 1995). This view would place the bed nucleus as the head ganglion of the hypothalamic-pituitary-adrenal axis (Schulkin, McEwen, and Gold, 1994).

In studies of immobilization-induced stress, in situ hybridization and northern-blot techniques have demonstrated increased CRH expression (e.g., Kalin, Takahashi, and Chen, 1994). Koob's laboratory has revealed another important relationship between the CEA and CRH: As measured by microdialysis, CRH is increased in the CEA during immobilization-induced stress (Pich et al., 1993a,b) as is norepinephrine (Pacak et al., 1993). This is a condition in which glucocorticoids are elevated and may be activating CRH and other neuropeptides and neurotransmitters in the CEA, as well as perhaps other sites in the CRH circuit (bed nucleus of the stria terminalis,

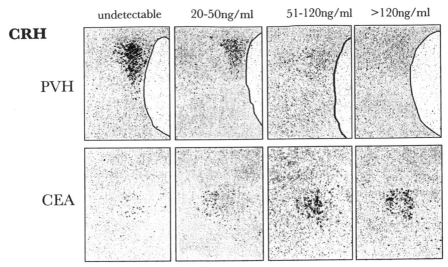

Figure 5.8. Darkfield photomicrographs of CRH mRNA hybridization in the paraventricular nucleus of the hypothalamus (PVN) and central nucleus of the amygdala (CEA) after increasing plasma corticosterone concentrations for 5 days. (Adapted from Watts and Sanchez-Watts, 1995b.)

parabrachial and solitary nuclei in the lower brainstem) (Swanson and Simmons, 1989). The result is to arouse anxiety and fear and the expectation of adversity (Lee and Davis, 1997).

Whereas the focus here is on the hormones of stress, antisense experiments have suggested that reducing the neuropeptide-Y concentration in the brain will increase anxiety in rats. Interestingly, this response is linked to the CEA and is ameliorated by using antisense techniques and reducing CRH-producing cells (Koob et al., 1994).

CRH and Fear

It is also known that CRH injected into the lateral ventricles reduces exploration and increases fear (e.g., Koob and Bloom, 1985). Moreover, CRH injections enhance fear and defense reactions to novel environments (Takahashi et al., 1989; Koob et al., 1993). Interestingly, in familiar environments, CRH intracerebral infusions reduce freezing responses and increase movements, and the converse in unfamiliar environments (Takahashi et al., 1989). These effects are centrally mediated, and interference with CRH within the amygdala reduces fear (Koob et al., 1993). In averse-conditioning experiments, chronic CRH injections increase the extent of freezing re-

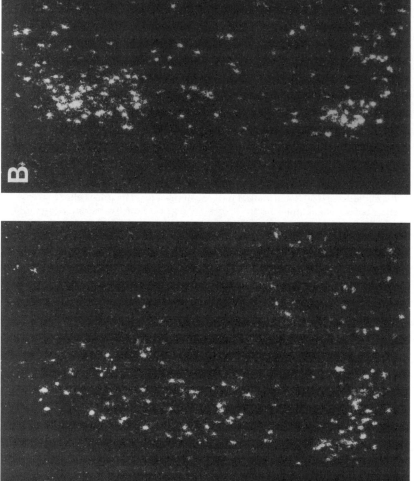

Figure 5.9. Photomicrographs of CRH mRNA in neurons of the lateral bed nucleus of the stria terminalis (A) before corticosterone treatment and (B) following corticosterone treatment. (Reprinted from *Brain Research* 657, S. Makino, P. W. Gold, and J. Schulkin, Effects of corticosterone on CRH mRNA and content in the bed nucleus of the stria terminalis; comparison with the effects in the central nucleus of the amygdala and the paraventricular nucleus of the hypothalamus, pp. 141–9, copyright 1994, with kind permission from Elsevier Science.)

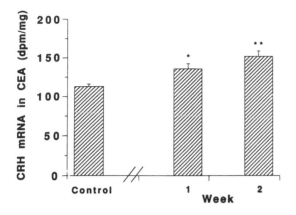

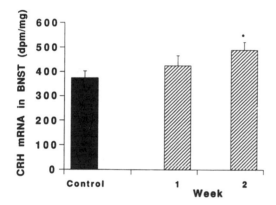

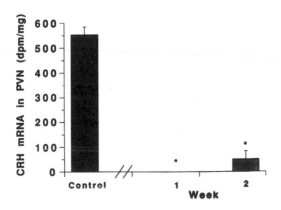

Figure 5.10. CRH mRNA in three sites after no treatment, after 1 week of corticosterone treatment, and after 2 weeks of treatment. (Reprinted from *Brain Research* 640, S. Makino, P. W. Gold, and J. Schulkin, Corticosterone effects on CRH mRNA in the central nucleus of the amygdala and the paraventricular nucleus of the hypothalamus, pp. 105–12, and from *Brain Research* 657, S. Makino, P. W. Gold, and J. Schulkin, Effects of corticosterone on CRH mRNA and content in the bed nucleus of the stria terminalis; comparison with the effects in the central nucleus of the amygdala and the paraventricular nucleus of the hypothalamus, pp. 141–9, both copyright 1994, with kind permission from Elsevier Science.)

sponses compared with acute injections (5 days versus 1 day). Additionally, whereas CRH antagonists ameliorate stress-induced reduction of exploration, CRH injections into the CEA restore this effect. CRH antagonists reduce defensive fear (Korte et al., 1994) and anxiety (Lee and Davis, 1997), in addition to generally reducing stress-related responses (Heinrichs et al., 1992). An essential link between glucocorticoids and the activation of CRH and the amygdala is apparent (Gray and Bingaman, 1996).

Corticosterone and Conditioned Freezing

Glucocorticoids can also enhance freezing responses to conditioned aversive stimuli (e.g., Jones, Beuring, and Blokhuis, 1988). We have shown this phenomenon using a Pavlovian conditioning paradigm in rats that had been trained to associate a shock with an acoustic signal. Treatment with high doses of corticosterone potentiated their freezing responses (Corodimas et al., 1994).

Other evidence suggests that corticosterone is linked to the development of fear-induced freezing (Takahashi and Rubin, 1993; Takahashi, 1995; see Chapter 4, this volume). In rhesus monkeys, high concentrations of cortisol have been linked to exaggerated freezing responses (Kalin and Shelton, 1989).

But the effects of systemically given corticosterone are different from

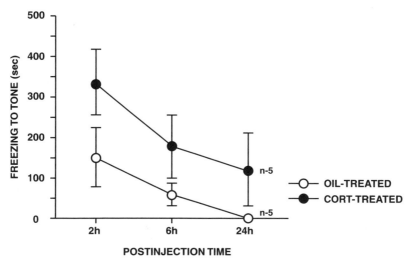

Figure 5.11. Amounts of time animals froze when faced with a novel context and a 600-second tone (corticosterone-treated group versus placebo group). (Adapted from Corodimas et al., 1994.)

164

those of centrally infused CRH. Whereas CRH almost always induces enhanced freezing – and comes much closer to being strongly linked to the sense of adversity (P. Holmes and J. Schulkin, unpublished observations) – glucocorticoids behave in a more complicated fashion. Why? The molecule of energy homeostasis, corticosterone is elevated when organisms are expending energy to do many things. It takes energy to fear and avoid predators, for example, and corticosterone's actions on the amygdala may increase fear. But glucocorticoids, as indicated earlier, can also induce dopamine synthesis in the nucleus accumbens, which may induce movement.

There are two possible reasons why it is difficult (though not impossible) to potentiate freezing reactions with corticosterone. First, though they are linked to this system and important in stress maintenance, glucocorticoids are not exclusively hormones of fear. Second, the hormones may induce competing responses, namely, one to move and another to freeze.

CRH, Corticosterone, and the Startle Response

Researchers use the startle reflex as a model of anxiety and a measure of anticipation; CRH also facilitates this response (Davis, Walker, and Lee, 1997; Lee and Davis, 1997). Both melancholic depressed patients and anxious individuals exhibit enhanced startle responses to certain stimuli (Grillon et al., 1993). Moreover, fear-potentiated startle is reduced by antidepressants. In rats, CRH infused into the cerebral ventricles potentiates the acoustic startle reflex, which is increased by CEA stimulation. The CEA pathways that mediate startle have been anatomically determined by tract tracing and electrophysiologic studies (Rosen and Davis, 1988; Rosen et al., 1991).

Lesions of the CEA (but not the paraventricular region of the hypothalamus) abolish the startle response (Liang et al., 1992). CRH antagonists infused directly into the CEA also impair startle (Britton et al., 1986). These results indicate that the anticipatory control of behavior (as measured by startle) to central CRH resides in the amygdala and not the hypothalamus. Specifically, cutting the ventral amygdalafugal pathway interferes with the potentiation of CRH-induced startle (Davis et al., 1993); this pathway is rich in CRH fibers (Gray, 1990; Gray and Bingaman, 1996) and is known to be involved in other steroid- and peptide-induced regulatory behaviors (see Chapters 1, 2, and 4).

Corticosterone pretreatment potentiates CRH-induced startle responses in rats (Lee, Schulkin, and Davis, 1994); that is, by pretreating the animals with corticosterone doses known to induce CRH in the CEA or the bed nucleus

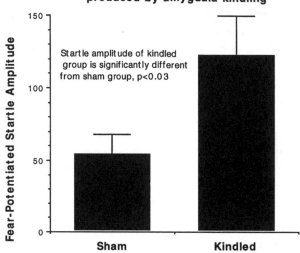

Exaggerated fear-potentiated startle produced by amygdala kindling

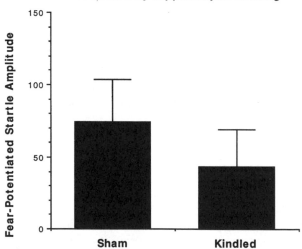

No Exaggerated Fear-Potentiated Startle Produced by Hippocampus Kindling

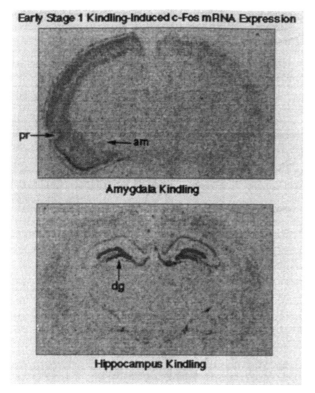

Figure 5.12. Facing page: Effects of partial kindling on fear-potentiated startle. Rats were conditioned to be fearful of a light by pairing the light with footshock. The next two days they received either sham stimulation or bilateral kindling stimulation (400 μA of a 1-sec, 100-Hz train of 1-msec biphasic square-wave pulses) in the amygdala or hippocampus. The following day they were tested for fear-potentiated startle by presenting an acoustic startle stimulus in the presence or absence of the light (fear-potentiated startle amplitude was derived by subtracting startle amplitude in the absence of the light from the amplitude in the presence of the light). Top: Both sham and amygdala-kindled rats displayed fear-potentiated startle, but amygdala-kindled rats displayed significantly greater fear-potentiated startle. Bottom: Both sham and hippocampus-kindled rats displayed fear-potentiated startle, but hippocampus kindling did not produce exaggerated fear-potentiated startle. The data suggest that exaggerated fear-potentiated startle is specific to amygdala kindling. This page: Neuronal activation by partial unilateral amygdala (top) and hippocampus (bottom) kindling as visualized by ³⁵S in situ hybridization of c-fos mRNA expression 15 minutes after kindling stimulation. Amygdala kindling induced widespread ipsilateral cortical c-fos mRNA expression that included parts of the fear circuit [amygdala (am) and perirhinal cortex (pr)]. Hippocampus kindling induced bilateral c-fos mRNA expression that was confined to the hippocampus and dentate gyrus (dg). The results suggest that amygdala kindling activated structures important for the exaggerated fear potentiated startle response seen at the top of the facing page, whereas hippocampus kindling did not. (From Rosen et al., 1996.)

of the stria terminalis, CRH-potentiated startle is elevated. The effect is specific: Aldosterone neither facilitates this response nor induces CRH (Watts and Sanchez-Watts, 1995a,b; Y. Lee, J. Schulkin, and M. Davis, unpublished observations). In the Introduction, as well as in Chapter 4, I have described evidence that young children who are fearful and inhibited, and who also have higher cortisol concentrations, have greater startle responses than do children who are not (Schmidt et al., 1997).

In rats, experiments have shown that stress resulting in increased corticosterone also facilitates the expression of conditioned startle later (Rosen et al., 1996). Other results suggest that amygdaloid-kindled rats, a process that also elevates corticosterone, have enhanced startle responses when the animals are tested at a later time; no such effect is observed following dorsal hippocampal kindling (Rosen et al., 1996). It is as if the amygdala, once activated, becomes more sensitized and prepared to react to aversive events (Rosen et al., 1996; Rosen and Schulkin, 1998).

CRH, Corticosterone, and Adversity

In studies with rats, elevated CRH is linked to adverse events (e.g., Koob and Bloom, 1985; Albeck et al., 1997). Moreover, transgenic mice who overproduce CRH (Stenzel-Poore et al., 1994) or have decreased numbers of glucocorticoid receptors show greater anxiogenic behaviors (enhanced startle) (Beaulieu et al., 1994). Whereas it is known that CRH infused into the lateral ventricles facilitates behavioral arousal (Koob and Bloom, 1985) and can interfere with learning in animals, it is not known if CRH remains elevated for longer periods under conditions of perceived lack of control than under conditions of perceived control. What we do know is that (1) CRH is involved in responsiveness to aversive events and that antagonists injected into the CEA inhibit this response, (2) stress does induce CRH (measured by in vivo dialysis), and (3) glucocorticoids can increase CRH mRNA in brain regions involved in anxiety and fear (e.g., CEA, bed nucleus of the stria terminalis) (Makino et al., 1994a,b) or decrease it (e.g., cells in the parvocellular region of the paraventricular region of the hypothalamus) (e.g., Swanson and Simmons, 1989).

The literature on animals has thus demonstrated the importance of the amygdala during periods of stress. The response is influenced by corticosterone or cortisol and CRH-producing neurons. This idea does not negate the importance of the hypothalamus, but only suggests that the CEA and perhaps the bed nucleus of the stria terminalis or extended amygdala are ideally suited to underlie the chronic arousal of expectations of adversity. Their importance in arousal is suggested by the following facts: (1) These forebrain regions receive the densest projections of visceral information from the brainstem and send bilateral projections to them; (2) these two regions exhibit some control over hypothalamic sites; and (3) these receive projections from the locus ceruleus (which contains CRH) (e.g., Gray, 1990; Gray and Bingaman, 1996).

When one places the anatomy of the amygdala and its circuits in the context of hormonal and psychological factors, it seems plausible that they are the sites that regulate arousal abnormalities. In humans, the amygdala and the hippocampus are not separated so drastically as they are in other mammals. Moreover, regions of the amygdala that merge into the hippocampus contain corticosteroid receptor sites, project to the paraventricular hypothalamus, and receive catecholaminergic input (e.g., Swanson and Simmons, 1989). The locus ceruleus, which projects to the amygdala, is activated by CRH (Valentino, Foote, and Jones, 1983; Valentino and Curtis, 1991). Importantly, CEA lesions reduce CRH content in the locus ceruleus (Owens and Nemeroff, 1991). CRH antagonists injected directly into the locus interfere with shock-induced freezing behavior (Swiergiel, Takahashi, and Kalin, 1992) and increase anxiety responses, just as they do in the CEA (Weiss et al., 1994). The regulation of the locus ceruleus by glucocorticoids on CRH expression is different from that in the paraventricular nucleus (Pavcovich and Valentino, 1997).

Glucocorticoid–CRH Immune Interactions

Researchers have uncovered significant interactions between humoral and immune functions. It is now known that repeated or prolonged fear leads to a breakdown of immune regulation in rats and people (Maier et al., 1993) Similarly, psychological stress is known, in some contexts, to increase susceptibility to the common cold. As described at the beginning of this chapter, the induction of helplessness – leading to hormonal changes – has been linked to immune impairment, as has prolonged chronic depression (Maier et al., 1993) and the induction of cytokines (e.g., Minami et al., 1991) and immunologic messengers (Chrousos, 1995).

Chronic CRH infusions compromise immune function in rats (Chrousos, 1995). Glucocorticoid–CRH interactions with cytokines in the bed nucleus (and perhaps the CEA) suggest that cytokine or interleukin receptors are regulated by corticosterone and that stress or interleukins can activate CRH. In fact, dexamethasone (a synthetic glucocorticoid), at least at the level of the pituitary, can actually increase the activation of a subset of interleukins (Chrousos, 1995).

Glucocorticoid hormones play roles in regulating immune cells during

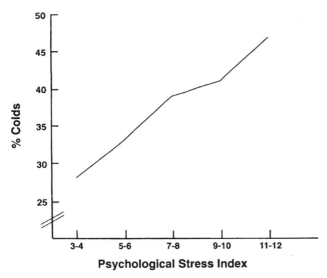

Figure 5.13. Susceptibility to the common cold is directly proportional to the intensity of psychological stress. Healthy volunteers were exposed to one of five viruses known to cause colds. The probability of getting sick increased with the severity of the psychological stress the subjects reported experiencing. (Courtesy of S. Cohen.)

infection and certain pathologic conditions (autoimmune disease). Figure 5.14 depicts the range of functions that glucocorticoids have within the immune system (McEwen and Stellar, 1993; Chrousos, 1995). Glucocorticoids can, in different situations, either enhance or suppress immune expression in the brain and periphery. The hormones may exert some of their effects on immune function through regulation of guanine nucleotide (Chrousos, 1995).

Interleukin given systemically or injected centrally into the ventral medial hypothalamus can decrease food intake (Weingarten, 1996), and probably all appetitive behaviors, because they make the animal sick. They also induce nerve growth factor in the hippocampus, perhaps to counteract the elevations in corticosterone and CRH. Interleukin-1, a cytokine produced by macrophage cells, induces sickness when injected in the brain (Van Dam et al., 1992).

Interleukins or cytokines can also activate the circumventricular organs (e.g., OVLT; see Chapter 2). They are known to excite neurons in the bed nucleus of the stria terminalis, suggesting interactions among glucocorticoids, CRH, and interleukins – all of which are activated under stress and can lead to pathologic conditions (Dunn, 1993). Interleukin-1b and other

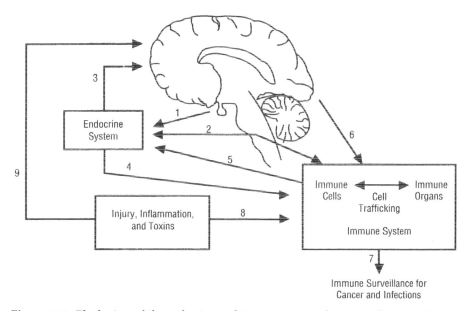

Figure 5.14. The brain and the endocrine and immune systems have complex interrelationships in response to pathogens and tumor cells: (1) Neuroendocrine products control endocrine function. (2) Neural activity also regulates endocrine function, in addition to affecting the immune system. (3) Hormones affect the brain and pituitary. (4) Hormones also affect immune cells and organs. (5) Immune-system messengers (e.g., cytokines, thymic hormones) affect endocrine function. (6) Immune-system messengers have direct and indirect effects on the pituitary gland and brain. (7) Immune function, which includes the movement of "trafficking" of immune cells to various tissues and organs through the circulation and lymphatic system, performs surveillance against tumor cells and pathogens as well as foreign cells and substances. (8) Injury, inflammation, and toxins from infection stimulate immunity as well as cytokine production. (9) Injury, inflammation, and toxins also signal the brain and pituitary gland. (From B. S. McEwen and E. Stellar, 1993, Stress and the individual: mechanism leading to disease, *Archives of Internal Medicine* 153:2093–101. Copyright 1993, American Medical Association. Reprinted by permission.)

mild stressors induce early gene expression in the paraventricular nucleus, which reflects activation of CRH neurons (Chan et al., 1993).

Both type-I and type-II interleukin receptors are located in the medial preoptic area, ventral medial hypothalamus, hippocampus, internal capsule, and piriform cortex (Schobitz, Voorhuis, and DeKloet, 1992). They are also found in the basal lateral amygdala, arcuate nucleus, and medial thalamic nuclei. Whereas interleukin-1 is known to activate the CEA, as well as the bed nucleus, paraventricular nucleus, solitary nucleus, and area postrema (in addition to CRH neurons in both the CEA and paraventricular nucleus), the thyroid is not activated, though the hypothalamic-pituitary-adrenal axis is (Brady et al., 1994). One likely site for peripherally derived cytokines is the CVO (median preoptic nucleus). Interleukin-1 also activates mineralocorticoid and glucocorticoid expression, and interleukin-6 activates vasopressin in humans (Chrousos, 1995).

Chronic activation of the immune signaling system during duress compromises bodily function. High concentrations of cortisol and CRH that remain elevated for long periods of time can compromise immune function, and chronic fear and anxiety are conditions that can lead to breakdown of normal immune/hormonal functioning. (McEwen, 1998).

Central Sites and Fear

This chapter has provided evidence that the hormones of stress – CRH and glucocorticoids – arouse an organism's sense of adversity, perhaps by their actions on the amygdala. Prolonged activation of these hormones can produce pathologic arousal (Schulkin et al., 1994a).

Support for the amygdala's role in integrating external events with internal humoral signals comes from (1) its known involvement in recognizing and remembering the significance of biological and emotional events and (2) the fact that it receives information from all neocortical sites, in addition to visceral input from the brainstem.

Both glucocorticoid receptors and CRH are found in the CEA and in other sites along the visceral neural axis, including the parabrachial nucleus (Kainu et al., 1993). The findings that CRH induction in this nucleus follows stress (Kalin et al., 1994) and that increased concentrations of CRH mRNA in the CEA follow glucocorticoid elevation (Makino et al., 1994a,b) provide further evidence that both hormones participate in the expectation of adversity in this extrahypothalamic site. More evidence is needed to test the hypothesis and place it in the context of other neuropeptides and neurotransmitters that may be altered. Still, the results of animal experiments are persuasive: The CEA is importantly involved in responding to the expectation of stress and chronic arousal and in regulating internal and external events. The amygdala responds to humoral signals induced by the organism's relationship to the outside world, regulating internal and external events over the long term (Gloor, 1978).

Whereas the literature in clinical endocrinology has emphasized the hypothalamic-pituitary-adrenal axis (specifically, CRH activation causing the secretion of ACTH and subsequently glucocorticoids), basic neuroscience has emphasized the negative feedback (and the loss of it) that results from hippocampal damage following prolonged glucocorticoid activation of receptor sites. The result of losing negative feedback is a further elevation of CRH due to the loss of inhibitory signals (Jacobson and Sapolsky, 1991).

Hippocampal damage induced by glucocorticoids, stress, age, or experimental manipulations can in fact result in increased CRH in the paraventricular nucleus (e.g., Sapolsky, 1992).

In rats, the septal-hippocampal region is known to be involved in anxiety (Gray, 1987), and stress impairs hippocampal function (McEwen and Sapolsky, 1995) and alters a number of neuropeptides in this region (e.g., nerve growth factor.) (Smith et al., 1995a–c). It may not be surprising, therefore, that regions within the hippocampus seem to have inhibitory control on paraventricular CRH and vasopressin. Disconnecting pathways from the hippocampus via the lateral fornix will lead to upregulation of both peptides (Herman et al., 1992). Both peptides also regulate the pituitary-adrenal axis and are regulated by glucocorticoids (e.g., Whitnall, 1993). But in rats it is the ventral hippocampus, just behind the amygdala, that regulates the hypothalamic-pituitary-adrenal axis (Herman et al., 1995).

Interestingly, amygdala stimulation can elicit CRH in hippocampus neurons that normally do not produce the hormone (Smith et al., 1991). These same sites project to the amygdala and to glucocorticoid receptors and CRH-containing neurons in the CEA (Honkaniemi et al., 1992). They also facilitate memory encoding of aversive events (Lee et al., 1993) and are differentially regulated (McEwen, 1992). The circuit for fear therefore includes sites in the hippocampus that mediate context learning of fear as well as sites in the amygdala that mediate conditioned fear (but also cortical sites) (LeDoux, 1995). This view is compatible with the role of the hippocampus in inhibiting hypothalamic CRH and possibly regulating homeostasis (see Chapter 4). The positive induction of CRH in the amygdala (and perhaps in other sites such as the bed nucleus of the stria terminalis) by glucocorticoids may further contribute to arousing the experience of adversity.

Putting together the experimental results to date, it seems that CRH and glucocorticoids together regulate the expectation of adversity, as well as perhaps a variety of psychopathologic states associated with increased fear, anxiety, and the sense of hopelessness. The larger circuit would include the solitary nucleus, the parabrachial nucleus, and other regions of the limbic circuit, in addition to the specific circuits for fear-related behavioral responses (LeDoux, 1995; Rosen et al., 1996; Lee and Davis, 1997).

The Human Condition

Researchers have identified elevated central concentrations of CRH in the cerebrospinal fluid of depressed (Nemeroff et al., 1984) and obsessive-compulsive humans (Altemus et al., 1992). Such elevations of CRH may

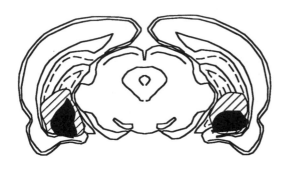

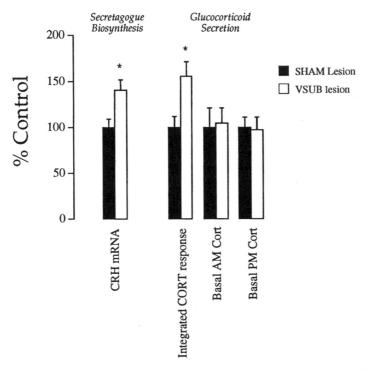

Figure 5.15. Upper panel: Typical large (crosshatched fill) and small (black fill) lesions of the ventral subiculum region. Note that both large and small lesions occupy areas containing substantial proportions of the ventral subiculum, while sparing the dorsal subiculum, CA1–4, and dentate gyrus. Effective lesions did not include the amygdalohippocampal transition zone, entorhinal cortex, or posterior cortical amygdala. Lower panel: Effects of ventral subiculum (VSUB) lesions on the hypothalamic-pituitary axis (HPA). VSUB animals exhibited a significant increase in basal CRH mRNA expression in the paraventricular nucleus, indicative of increased CRH biosynthesis. CRH changes were accompanied by an increase in stress-induced corticosterone (CORT) secretion, as estimated by plasma CORT concentrations integrated over 2 hours following induction of restraint. No changes were seen in basal CORT secretion in the morning or evening, indicating that VSUB effects were specific for stress-induced HPA activation. (From J. P. Herman, W. E. Cullinan. M. I. K. Morano, H. Akil, and S. J. Watson, 1995, Contribution of the ventral subiculum to inhibitory regulation of the hypothalamo-pituitary-adrenocortical axis, *Journal of Neuroendocrinology* 7:475–82. Reprinted by permission of Blackwell Science Ltd.)

also appear during posttraumatic stress syndrome, and though cortisol is low basally, these same individuals hypersecrete cortisol in the presence of adverse, fearful conditions (Yehuda et al., 1994; Yehuda, 1997). Agitated and melancholic depressions are condition in which central CRH and systemic glucocorticoids are linked and elevated (cf. Carroll et al., 1976; Nemeroff et al., 1984; Gold et al., 1984). Interestingly, whereas patients with Cushing's disease, as well as those with melancholic depression, have elevated cortisol, only depressed patients show elevated CRH in the cerebrospinal fluid (Gold et al., 1988; Kling et al., 1991). These findings suggest that glucocorticoid regulation may be impaired in depression (Young et al., 1991; Akil et al., 1993).

People who are actively trying to cope with anxiety and fear exhibit several well-known syndromes, including a sense of impending doom. Melancholia, for example, is an organized state of anxiety in which anticipation of negative events predominates, and no clear homeostatic set point is being regulated. Moreover, when depressed individuals are placed in contexts in which stress is uncontrollable, they secrete greater amounts of cortisol than nondepressed individuals, but secrete the same amount when the stress is controlled (Brier, 1989).

Positron-emission tomography (PET) measurements of regional blood flow have demonstrated increased amygdala activation in depressed pa-

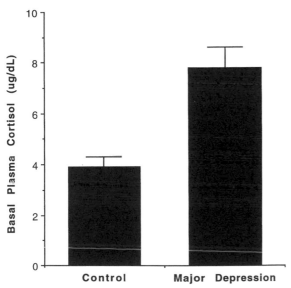

Figure 5.16. Cortisol concentrations in patients with major depression and in controls. (Adapted from Kling et al., 1991.)

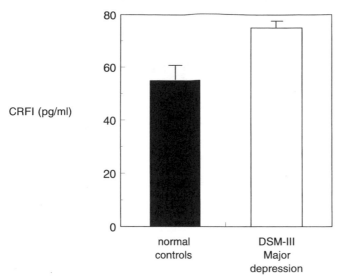

Figure 5.17. Concentrations of CRF in cerebrospinal fluid in normal controls (5 males and 5 females) and patients with DSM-III major depression. (Adapted from Nemeroff et al., 1984.)

tients (Drevets et al., 1992). More pertinent, lidocaine (which in rat brain induces cortisol release and CRH activation) also activates the amygdala in humans (Ketter et al., 1994).

Darwin understood the importance of facial expressions as indicative of the emotions (Darwin, 1965; Ekman, 1972). One such depiction from his book is shown in Figure 5.18. What it depicts is a person in terror, which is the extreme of a fearful state.

A role for the amygdala has long been known in the human experience of fear. Recent evidence, for example, indicates that lesions of the amygdala impair the recognition of faces that are fearful to others, in addition to impairing fear (Adolphs et al., 1995). Moreover, brain-imaging studies have revealed that the amygdala is activated in normal subjects who are shown fearful as opposed to happy faces (Morris et al., 1996).

Summary

Diverse thinkers who have reasoned about the brain, including Hughlings Jackson (1958) and William James (1952), have envisioned it in terms of evolution. Following this view, levels of neural function reflect evolutionary ascent, with lower levels of the neural axis serving more primitive func-

Figure 5.18. Face of terror. (From Darwin, 1965.)

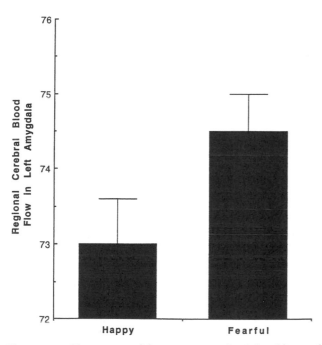

Figure 5.19. Human amygdala activation to fearful and happy faces. (Adapted from Morris et al., 1996.)

tions. Through natural selection, the brain has acquired an increasingly greater range of responsivenes.

A colleague of mine, now deceased, argued that only mammals possessing a central limbic system, including an evolved neocortex, are capable of experiencing fear or, for that matter, expressing motivated behavior (Epstein, 1982a). He argued that the motivated behavior of fear requires instrumental control, and he suggested that perhaps reptiles might not have that. I argued with him that his thesis was misguided, though it would be difficult to know whether or not an iguana felt fear. What we do know is that reptiles express endocrine events that are linked to fear.

As noted at the beginning of this chapter, fear is a central motive state that induces both appetitive and consummatory behaviors. Functionally tied to avoiding danger, it is much like the other central states described throughout this book (Lashley, 1938; Beach, 1942). Fear itself is a functional

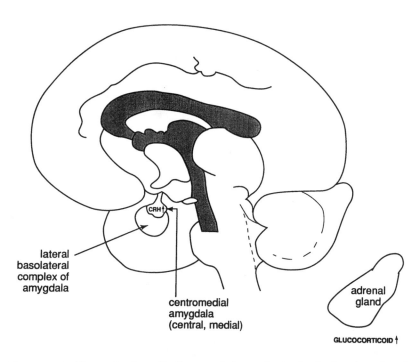

Figure 5.20. Human brain, with the amygdala emphasized. The basolateral and centromedial complexes are indicated; also shown is the fact that CRH is elevated in the central region, along with elevated glucocorticoids. (Reprinted from *Neuroscience and Biobehavioral Reviews* 18, J. Schulkin, B. S. McEwen, and P. W. Gold, Allostasis, amygdala and anticipatory angst, pp. 385–96, copyright 1994, with kind permission from Elsevier Science Ltd, The Boulevard, Langford Lane, Kidlington Ox5 1GB, UK.)

property of the nervous system, which in species such as mammals is linked to the amygdala (LeDoux, 1995). The amygdala plays a prominent role in regulating fear, particularly conditioned fear and fear that is linked to elevated glucocorticoids and CRH.

Glucocorticoids and CRH – known to be the hormones of stress – are activated during duress and may be closely linked with amygdala function and the central state of fear. As mentioned earlier, there is not just one state of fear: This chapter has focused largely on the fear and stress that result from losing predictability. Such loss of predictability and the fear that it generates occur, for example, during animals' social relationships. Endocrine changes are expressions of coping with or adjusting to such events. Prolonged exposure to stress hormones, however, can lead to physiologic problems such as tissue and mental deterioration.

CHAPTER SIX

Hormones, Behavior, and Biological Clocks

Introduction

One of evolution's most successful adaptations is the self-generated rhythmicity (endogenous rhythmicity) with which organisms adjust to periodic influences in their environments, including diurnal, lunar, seasonal, and tidal cycles. By measuring the passage of time, endogenous biological clocks regulate both behavior and physiology, with one complementing the other. There are several kinds of biological clocks, and their anatomic locations and functions vary across species.

In most species, a behavior alternates between an active phase and a restful phase. Recall that during active periods the concentrations of hormones such as cortisol are elevated. The active phase reflects energy use, and the rest phase energy conservation. However, hormones such as melatonin, which is secreted at night, can be elevated during the active phase or the rest phase, depending upon the presence of light/darkness. In rodents, two oscillators have been hypothesized: One oscillator is synchronized to dusk, and the other to dawn, and the two are entrained to one another (e.g., Pittendrigh and Daan, 1976).

In most animal species, sleep is regulated both by circadian (daily) mechanisms and by mechanisms that maintain homeostasis. One mechanism monitors time, while the other monitors the need for sleep. In some cases the circadian clock may facilitate wakefulness against the opposing homeostatic need to remain asleep (Edgar et al., 1992). Sleep deprivation acts on the homeostatic mechanisms for sleep; within limits, the more sleep an animal loses, the more it will need to make up. Sleep homeostasis can also vary significantly among individuals, some needing much more, or longer bouts of sleep, than others (Wehr, 1989).

Light influences the patterns of hormone secretion. Under constant lighting conditions in the laboratory, researchers can induce the phenomenon of splitting. In studies with rodents under constant-light conditions, the dusk/dawn oscillators are out of phase with one another, resulting in behaviors and physiology that are out of phase (e.g., running activity, secretion of luteinizing hormone) (Turek et al., 1987). Light intensity plays a fundamental role in inducing such events (Pickard, Turek, and Sollars, 1993).

Non-photic sensory stimulation also influences clock-controlled behavioral and hormonal events. We now know that social stimulation and other kinds of arousal-inducing activity can influence phase-shifting activities in laboratory animals (e.g., free running activity in hamsters) (Mrosovsky, 1995). Such behaviors can in turn influence hormone-regulated behaviors. For example, running activity by hamsters can inhibit the reproductive quiescence induced by melatonin (Pieper et al., 1988).

This chapter begins by discussing the differences between predictive and reactive homeostasis. The former is tied closely to anticipatory control over behavior, and the latter is linked to immediate alterations of the internal milieu. The chapter also describes the idea of multiple clocks and some of the relationships of these clocks to endocrine events. Focusing on several humoral systems (e.g., melatonin, cortisol, prolactin), the chapter also discusses hormonal effects on behavior, states of activity and rest, and mental health.

Predictive versus Reactive Homeostasis

Moore-Ede (1986) has made an important distinction between predictive and reactive homeostasis. Linked to predicting when events will occur in the environment, predictive homeostasis is anticipatory (Moore-Ede, 1986). An animal uses predictive homeostatic mechanisms, for example, when it forages for food and anticipates when a food source will become available (i.e., time of day). It is an important evolutionary adaptation (Gallistel, 1990). Anticipatory behavior oriented toward food sources has been demonstrated in insects, fish, birds, and a number of mammals (Gallistel, 1990). These animals' circadian clocks organize their behavior around when objects can be expected to appear; an animal uses its clock to anticipate rewards (Rosenwasser and Adler, 1986).

Laboratory rats are nocturnal animals. Therefore, during the light period they rest, and during the dark period they are active, as reflected in their ingestive patterns. Nonetheless, in the laboratory they can use their circadian clocks to anticipate food sources before those sources appear, whether it be during the dark period or the light period. Their anticipation is reflected in behaviors such as running on a wheel or beginning to bar-press several hours before food is scheduled to become available (Mistlberger, 1994). These are all laboratory analogues of wild animals moving toward edible objects in the world in which they are foraging. One of the most important roles played by circadian clocks is to facilitate such predictive homeostasis by predicting the time of day when food rewards will become available. One clear example of the circadian clock is the anticipatory behaviors before dusk or dawn that a number of animals express. They predict the time of day and head back to safety – heading home before it gets dark. Anticipatory behaviors are active, not passive.

Hormones such as insulin (required for energy balance) and vasopressin, angiotensin, and aldosterone (necessary for the body's fluid balance) are involved in both predictive and reactive homeostasis (Fitzsimons, 1979; Schulkin, 1991a). In the first case, hormones prepare the animal for what is to come (e.g., insulin secretion and the utilization of nutrients), for what might be needed. In the second, they represent a response to an alteration in homeostatic balance, and then they mediate the restoration of the balance.

The Circadian Clock

In animals such as ourselves, circadian clocks provide a certain stability in the regulation of biological systems, including behavior. They internalize

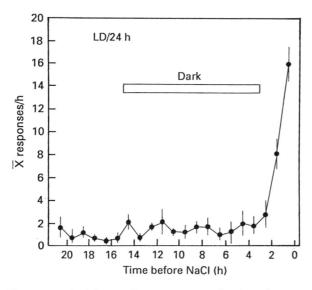

Figure 6.1. Anticipatory lever responses of sodium-hungry rats to the availability of NaCl, linked to 24-hour cycles. (From A. M. Rosenwasser, J. Schulkin, and A. N. Adler, 1988, The behavior of salt hungry rats to limited periods of salt, *Animal Behavior*, 16:324–9. Reprinted by permission of Academic Press Ltd.)

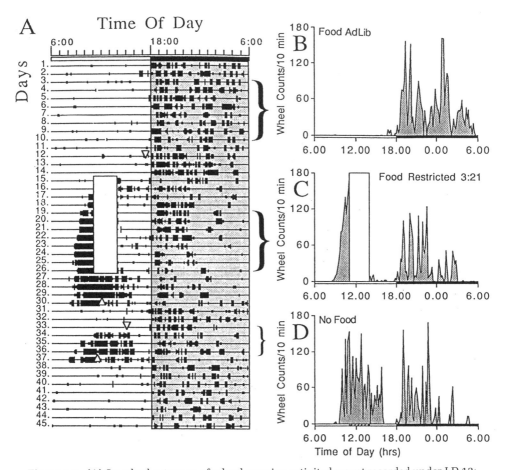

Figure 6.2. (A) Standard actogram of wheel-running activity by a rat recorded under LD 12: 12 (lights on from 0600 to 1800 hours) during ad libitum and restricted food access. Each line represents 24 hours with time in 10-minute bins plotted left to right. Time bins during which wheel revolutions were registered are represented by vertical deflections. The hollow vertical bar represents mealtime during the restricted feeding schedule (3 hours food, 21 hours no food). Triangles represent the beginning (pointing down) or end (pointing up) of total food deprivation. The rat exhibits little diurnal activity when food-deprived prior to the 3:21 feeding schedule. During the feeding schedule, the rat exhibits a robust food-anticipatory wheel-running rhythm. This rhythm persists during total food deprivation. (B, C, D) Waveforms of wheel running during ad libitum food access (7-day average), food restriction (7-day average), and total food deprivation (3-day average), respectively; unpublished data. (Reprinted from *Neuroscience and Biobehavioral Reviews* 18, R. E. Mistlberger, Circadian food-anticipatory activity: formal models and physiological mechanisms, pp. 171–95, copyright 1994, with kind permission from Elsevier Science, Ltd, The Boulevard, Langford Lane, Kidlington Ox5 1GB, UK.)

183

the natural cycles of the external world. Being able to predict when objects (e.g., food sources, opportunities for social or sexual contact) will appear is obviously of great adaptive value for all species trying to cope and survive (Gallistel, 1990).

Found in both invertebrates and vertebrates, as well as many unicellular organisms and plants, the endogenous circadian clock is ancient (Turek and Van Cauter, 1994). Though dependent upon certain sites in the hypothalamus in many mammalian species, circadian rhythmicity is found in a number of animals that have no hypothalamus at all, indicating that the clock can be embodied in diverse tissues. Although it can be embodied in different tissues, the circadian clock is part of the cognitive machinery that allows an animal to measure time and solve problems. These clocks have been shown to time everything from running activity by rats (e.g., Richter, 1922) to defecation by humans (Aschoff, 1994). As a measurement of the 24-hour day (Richter, 1965), the circadian clock evolved in response to natural cycles of darkness and light, eventually becoming incorporated into the hardware of the brain in which light regulates gene expression (J. S. Takahashi, 1995).

Anatomy of the Circadian Clock

To determine the circadian clock's location in the body, Richter removed various glands from rats to see if that would have an impact on the clock's regulation of predictive and reactive homeostatic behaviors. He found that removing the pituitary, adrenals, thyroid, and pineal gland did not disrupt circadian rhythmicity (Richter, 1965). Richter noted, however, that humans with tumors of the anterior hypothalamus had sleep disturbances and that lesions of the anterior hypothalamus disrupted the circadian rhythmicity that is expressed in a number of behaviors (e.g., wheel-running activity) (Richter, 1965). Later, researchers discovered that it was damage specifically to the suprachiasmatic nucleus of the hypothalamus (SCN) that altered circadian rhythmicity (e.g., Moore and Eichler, 1972). Damage to this region results in widespread loss of rhythmic activity.

Retinal projections to this region of the hypothalamus transmit information about light received through the retina. Other projections to the SCN are from the raphe nucleus in the midbrain and from the ventral (intergeniculate leaflet) lateral geniculate nucleus (Moore, 1993). These other projections, particularly from the raphe nucleus and its serotonergic pathways, may convey the arousal of circadian rhythmicity (Mrosovsky, 1995).

The anatomic sites underlying the circadian clock include the SCN, par-

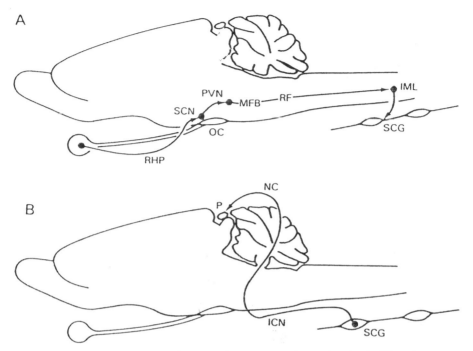

Figure 6.3. Proposed circuits in the suprachiasmatic-nucleus–pineal axis. There is a monosynaptic projection from cells in the retina (retina–hypothalamic projection, RHP) to the suprachiasmatic nucleus (SCN). SCN cells project to the paraventricular nucleus (PVN) of the hypothalamus, which in turn sends efferent fibers via the medial forebrain bundle (MFB) to the spinal cord. This projection terminates on cells of the intermediolateral cell column (IML); processes of these neurons then make synaptic connections in the superior cervical ganglion (SCG) of the sympathetic chain. Postganglionic noradrenergic fibers proceed in the inferior carotid nerve (ICN) and, later, in the nervil coronaril (NC) to innervate the pineal gland (P). (Reprinted from *Brain Research Bulletin* 10, D. C. Klein, Lesions of the PVN disrupt suprachiasmatic–spinal cord circuit in the melatonin rhythm generating system, pp. 647–52, copyright 1983, with kind permission from Elsevier Science.)

aventricular nucleus, pineal gland, and retina. Recent experiments in the rat have shown that changes in the light–dark cycle activate many of these brain regions (Moore, 1993). Specifically, changes in lightness and darkness activate SCN early gene expression, which is linked to behavioral changes (J. S. Takahashi, 1995; Wisor and Takahashi, 1997).

In mammals, the SCN is the pacemaker that is fundamental in the regulation of circadian rhythmicity. In some birds and reptiles it is the pineal gland that serves as the primary circadian clock. There is some evidence, however, to suggest that SCN damage can disrupt circadian rhythmicity in some birds (Cassone, 1990).

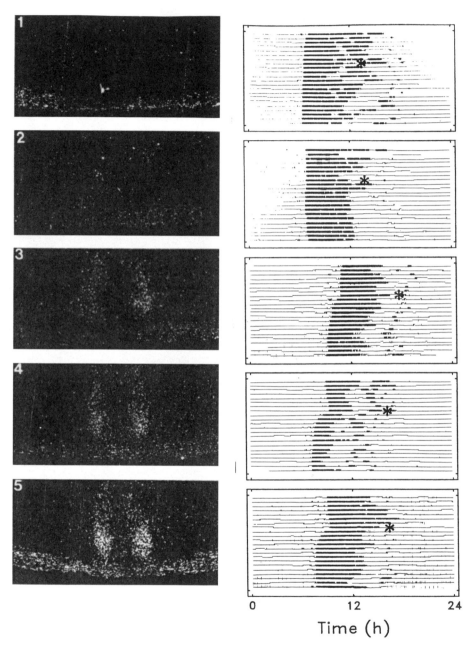

Time (h)

Figure 6.4. Effects of pulses of light of varying irradiance on c-fos expression in the SCN in hamsters maintained in constant darkness (left); representation of running activity (right). (From J. S. Takahashi, 1995; with permission, from the *Annual Review of Neuroscience*, vol. 18, © 1995, by Annual Reviews Inc.)

Melatonin and Biological Clocks

Secreted in anticipation of sleep by humans, who have consolidated long-term nocturnal sleep patterns (as opposed to rats, which wake and sleep many times), and secreted in reaction to stress, melatonin is involved in both anticipatory and reactive homeostasis. The hormone's secretion is a classic example of a photoperiod-dependent phenomenon. Pineal secretion of melatonin is elevated in the dark and suppressed in the light. Its daily secretion pattern is depicted in Figure 6.5 (Reiter, 1993).

Melatonin is a 400-amino-acid hormone relaying critical information about light and darkness (e.g., Reiter, 1993; Wehr, 1991), and the amount of melatonin secreted is determined by the levels of light and darkness. A common secretion pattern for melatonin in several different species has been well documented [e.g., humans, quail (Wehr, 1991), fungi]. In a variety of species, temperature, in addition to light, influences melatonin secretion (Reiter, 1993). The number of melatonin-binding sites is highest in the optic areas, a phenomenon that has been preserved through phylogeny (Iigo et al., 1994).

Like serotonin, melatonin is an indole amine. A metabolic conversion process transforms serotonin to melatonin. The structure of melatonin and its known physiologic effects are depicted in Figure 6.6.

Melatonin receptors, in some but not all animals (Cassone, 1990; Cassone, Brooks, and Kelm, 1995), appear in a wide array of anatomic sites, including the pituitary gland, regions of the hypothalamus and hippocampus, and the striatum and midbrain regions (Weaver et al., 1993). Although there is some species variation, melatonin receptors have been highly conserved across evolution in mammals, including humans, and in birds, with receptor sites in the SCN that in many species are linked to light and reproduction (Weaver, Keohan, and Reppert, 1987; Weaver et al., 1993).

The brain seems to have at least three different subtypes of melatonin receptors linked to a G-protein receptor family (Reppert et al., 1995). Subtypes of the melatonin receptors have now been cloned (Reppert et al., 1995; Roca et al., 1996). What awaits discussion is the functional link between the molecular biology and melatonin's behavioral and physiologic effects. Systemic injections of melatonin are known to influence the firing rate of neurons in the SCN in rats, and circadian rhythms are synchronized by this treatment (e.g., Lu and Cassone, 1993).

Importantly, melatonin injections also facilitate the onset of sleep in humans (Dollins et al., 1994). Melatonin, in humans, is linked to sleep enhancement, and melatonin infusions potentiate sleepiness in humans (Dol-

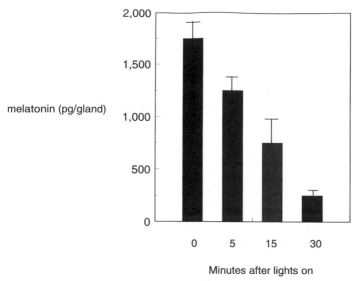

Figure 6.5. Rapid drop in pineal melatonin concentration as a consequence of acute light exposure. (Adapted from Reiter, 1993.)

lins et al., 1994; Haimov and Lavie, 1996). Amazingly, the melatonin receptors that develop when the fetus is an embryo (in a rat study) and the circadian pattern of secretion are linked to the mother's secretion of melatonin (Weaver et al., 1987).

Melatonin concentrations are importantly linked to reproductive behaviors. Melatonin has differential effects on reproductive hormones and behavior, depending upon whether the animal is a long-day breeder or short-day breeder. Elevated concentrations of melatonin decrease reproductive hormones (e.g., testosterone, estrogen), as well as the behaviors linked to seasonal reproduction (Turek, Desjardins, and Menaker, 1975; Tubbiola and Bittman, 1995; Powers et al., 1997), in long-day breeders. More recent evidence suggests that melatonin infusions directly into the SCN, paraventricular nucleus, and nucleus reuniens will decrease reproductive responses in Siberian hamsters (Badura and Goldman, 1992a).

Rhythmicity and Prolactin

A hormone linked to quiescence, darkness, sleep, reproduction, and homeostatic responses to stress (see Chapter 4), prolactin plays several roles that are critical to maintaining both anticipatory and reactive homeostasis. Prolactin is produced in the pituitary gland, and its concentrations are deter-

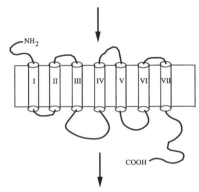

Melatonin
(N-acetyl-5-methoxytryptamine)

Physiological Effects

Seasonal reproduction

Circadian rhythms

Retinal physiology

Hypnotic effects

Figure 6.6. Melatonin acts through high-affinity G-protein-coupled receptors to elicit physiologic effects. (From S. M. Reppert and D. R. Weaver, 1995, Melatonin madness, *Cell* 83:1059–62. Copyright 1995 Cell Press. Reprinted by permission.)

mined by light intensity in a number of species (e.g., Wehr et al., 1993; Turek and Van Cauter, 1994). In rats, there is a distinct expression of prolactin secretion that is linked to midday and is under the influence of the circadian clock; SCN lesions abolish the hormone's midday secretion, as does ovariectomy. In rats, artificial light can alter natural prolactin and melatonin secretions from the pituitary, an effect that can vary with the seasons (Ravault, Reinberg, and Mechkouri, 1987).

In a greater number of species there is a surge in prolactin secretion that begins as day turns to night (Wehr et al., 1993). The longer the night, the greater the prolactin concentrations in many species (Borer, Ketch, and Corley, 1982). During the day, high prolactin concentrations have been linked to naps and shifts in daytime–nighttime activities (Turek and Van Cauter, 1994).

Prolactin concentrations also vary with seasonal photoperiods and are influenced by melatonin secretion (Lincoln, 1979). Prolactin is linked to an animal's breeding season, and in many species prolactin secretion increases

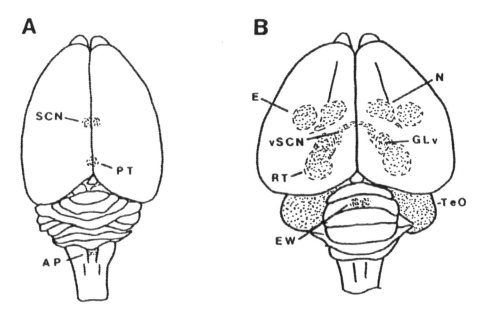

Figure 6.7. Schematic distributions of melatonin binding in rodent brain (A) and avian brain (B). (Reprinted from *Trends in Neuroscience* 13, V. M. Cassone, Effects of melatonin on vertebrate circadian systems, pp. 457–64, copyright 1990, with kind permission from Elsevier Science Ltd, The Boulevard, Langford Lane, Kidlington Ox5 1GB, UK.)

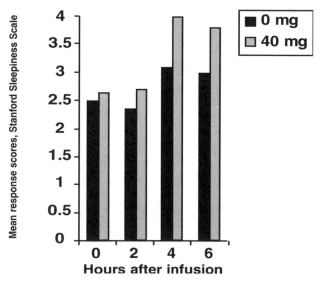

Figure 6.8. Infusions of melatonin to normal subjects and the sense of being sleepy during the daytime. (Adapted from Dollins et al., 1993.)

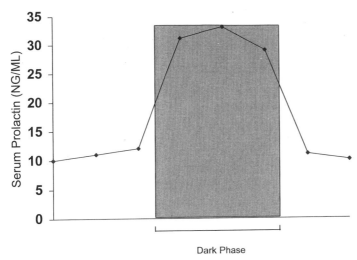

Figure 6.9. Mean plasma prolactin concentrations. (Adapted from Brainard et al., 1981.)

during the summer months, when it plays an important role in female re-productive functions. In hamsters, long photoperiods increase both prolac-tin concentrations and the fur coating, and short photoperiods do the con-verse (Duncan and Goldman, 1985).

Prolactin is known to facilitate the sleep process in humans, and sleep deprivation interferes with prolactin secretion (Spiegel et al., 1994). Quiet rest before sleep correlates with elevated prolactin in humans (Wehr et al., 1993). Prolactin concentrations rise during other quiet times as well. One paradigmatic time is during breast-feeding (Moyer et al., 1979), when both peripheral prolactin and central prolactin are elevated. Prolactin secretion also rises when birds nest or when they are sitting on eggs (e.g., Goldsmith et al., 1981) (see Chapter 4) – the reason Danny Lehrman (1961) called it the hormone of "sitting behavior."

Prolactin's Activity and Rapid-Eye-Movement (REM) Sleep

As noted in the preceding section, prolactin influences the sleep process. Infused peripherally or centrally into either the hypothalamus (Obal et al., 1994; Roky et al, 1994) or amygdala in rats, the hormone can influence REM sleep (Sanford et al., 1997), and prolactin blockers will inhibit the expres-sion of REM sleep.

Vasoactive intestinal peptide (VIP) also increases prolactin secretion and

191

REM sleep, whereas VIP inhibitors block those processes (Obal et al., 1992), as well as possibly blocking photoinduction of c-fos in the SCN (Martinet et al., 1995). That is, intracerebral VIP elicits increases in hypothalamic pro- lactin mRNA, which should also increase REM sleep. The effects of prolac- tin do not depend on the pituitary gland, because rats without pituitary glands show the same decreases in REM sleep in response to prolactin in- fusions to the brain (Roky et al., 1994). The phenomenon may be linked to hypothalamic and amygdala prolactin.

Glucocorticoids, Neuropeptides, Rhythms, and Ingestive Behavior

Glucocorticoids are regulated by the circadian clock. In humans, for exam- ple, cortisol secretion in response to a meal is greater in the morning than in the evening (Dallman et al., 1993). This makes sense, because cortisol, the "wake-up" hormone, is elevated higher in the morning than at other times of the day (Leibowitz, 1995).

Corticosterone is normally elevated during the active phase of the light– dark cycle, the time when glucose metabolism and energy utilization are high. For the rat, which is nocturnal, this means that corticosterone is ele- vated during the night, its activity phase (Dallman et al., 1993).

Food ingestion, like water and sodium ingestion, occurs typically during the active phase. Hormones such as corticosterone influence food consump- tion at various times during the light–dark cycle (Tempel and Leibowitz, 1994). Interestingly, corticosterone may be more flexible than some other hormones that are rigidly linked to a particular time of day. The glucocor- ticoids are important for feeding behavior and follow the patterns of access to food resources. Corticosterone, which can facilitate food intake, may do so by increasing neuropeptide-Y synthesis in the brain (see Chapter 3). Rats offered a more desirable diet during the day than at night, for example, will alter their pattern of corticosterone secretion as they become daytime eaters instead of nighttime eaters (Dallman et al., 1993). This elevated response, however, is decreased in SCN-damaged rats.

Corticosterone's negative-feedback control over the paraventricular nu- cleus of the hypothalamus is circadian-dependent (Dallman et al., 1993). In addition, receptor sites are differentially regulated under circadian rhythm- icity. For instance, type-1 (but not type-2) corticosteroid receptors are under circadian control (De Kloet, 1991).

Interestingly, neuropeptide-Y-induced food intake is linked to the circa- dian regulation of corticosterone concentrations, which is reflected in noc- turnal elevations of neuropeptide-Y concentrations in the arcuate nucleus and paraventricular nucleus (see Chapter 3). Neuropeptide-Y concentrations are reported to rise several hours before the onset of the activity period, and neuropeptide Y is known to influence circadian rhythmicity (Huhman and Albers, 1993). Again, corticosterone, in turn increases neuropeptide-Y concentrations, resulting in the search for and ingestion of food (see Chapter 3).

Several neuropeptides are under circadian control, including corticotropin-releasing hormone (CRH), vasopressin, and neuropeptide Y, independent of corticosterone (e.g., Watts and Swanson, 1989). There appears to be an inverse relationship between CRH secretion in the cerebrospinal fluid and systemic concentrations of cortisol in humans (Kling et al., 1991). What this means, for example, is that the concentration of CRH in brain does not strictly depend upon cortisol secretion (see Chapters 4 and 5). The brain is generating a central state by activating neuropeptides that facilitate behavioral expressions. In this case, the neuropeptides are endogenously generated (Koob et al., 1994). Elevated concentrations of neuropeptide Y may indicate activation of the search for food (see Chapter 3).

A 4-Day Clock in Rodents: The Estrous Cycle

The estrous cycle for a hamster is depicted in Figure 6.10. Though varying widely among different mammals, the length of the cycle in this species is 4 days. The hormonal changes that occur during estrus affect a number of behaviors. We shall consider several examples.

Estrogen, Running, and Behavior

More than 70 years ago, Richter (1922) used a running wheel to measure the running activity of rats. This is depicted in Figure 6.11. As mentioned earlier, running can influence hormonal secretion. In hamsters, for instance, running reduces the quieting effects of melatonin, as well as the hormone's impact on reproductive functions (Borer et al., 1982).

Manipulation of light levels can influence estrus, as can lesions of brain regions known to be involved in circadian rhythmicity. SCN lesions, for example, can lead to constant estrus, as can constant light (Zucker, 1988). In the golden hamster, melatonin injections can also induce acyclicity. Moreover, the SCN projects to neurons that contain both estrogen receptors and GRH (de la Iglesia, Blaustein, and Bittman, 1995).

What Richter highlighted, however, was that running activity varied depending upon the hormonal state of the animal. Estrogen can alter the daily pattern of running activity by female hamsters (shorten the period, depend-

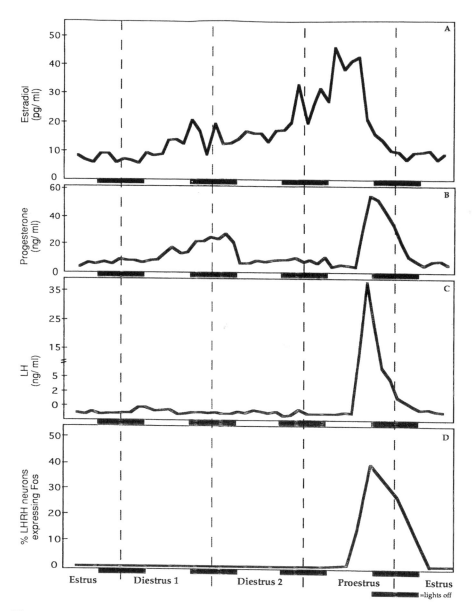

Figure 6.10. Concentrations of hormones circulating in the blood of hamsters during the estrous cycle. The abscissa shows the approximate time of ovulation for animals housed in a light–dark cycle of 14:10 (lights off at 2100 hours). The intermittent black bars indicate the hours of darkness. Day 4 corresponds to "proestrus." (From Lee et al., 1990b, with permission.)

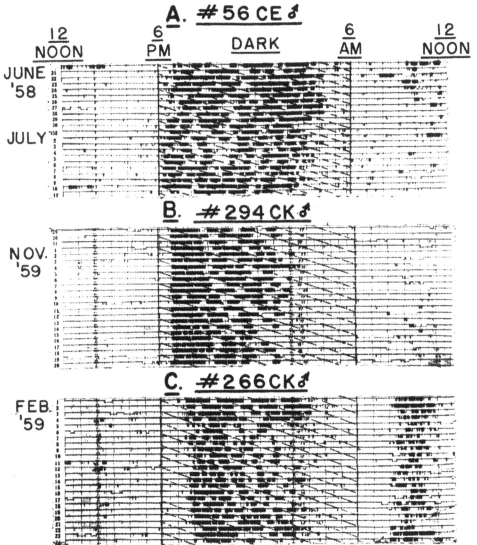

Figure 6.11. Running activity by rats during the light–dark cycle. (From C. P. Richter, *Biological Clocks in Medicine and Psychiatry,* 1965. Courtesy of Charles C Thomas, Publisher, Ltd., Springfield, Illinois.)

ing on the hormone's concentration) (Morin, Fitzgerald, and Zucker, 1977; Richter, 1976; Albers, 1981; Albers, Gerall, and Axelson, 1981).

Further studies have demonstrated that estrogen works to preserve the interactions of endogenous oscillators (e.g., reduce the occurrences of splitting under constant light) (Morin, 1980; Morin and Cummings, 1982). Sex differences in running activity and the effects of estrogen on running activity are dependent upon neonatal hormonal organizational effects (Zucker, Fitzgerald, and Morin, 1980).

Estrogen's effect on running behavior may be through its action on the SCN. Estrogen is known to influence neuropeptide-Y expression, and infusions of neuropeptide Y directly into the SCN will influence phase advances in running activity (Huhman, Babagbemi, and Albers, 1995).

Estradiol affects not only the estrous cycle and the circadian clock but also circannual rhythms. Ground squirrels implanted with estradiol lose weight, and the period of their circannual rhythm can be reduced by up to 57 days (Lee and Zucker, 1988; Zucker, 1988).

Estrogen, Cyclic Variation, Brain Function, and Behavior

In the rat brain, changes in estrogen concentrations during the estrous cycle can modify the dendritic spinal density in the ventral medial hypothalamus (Frankfurt et al., 1990) or dorsal hippocampus (Woolley et al., 1990, 1997). Spinal density is highest during estrus and proestrus, when estrogen concentrations are highest, and it is lowest during diestrus, when estrogen levels are lowest. Interestingly, ovariectomized rats show a decreased spinal density that can be reversed with estrogen treatment to the hypothalamus (Frankfurt et al., 1990) and hippocampus (Woolley and McEwen, 1993).

Decreases in spatial performance in voles and humans also stem from these hormonal changes (Hampson, 1990, 1995). For example, in female deer mice (*Peromyscus maniculates*), spatial performance decreased when estrogen concentrations were elevated; when estrogen was not elevated, they performed on a spatial task as males did (Galea et al., 1995). This decrease in spatial performance in the female may be seasonally mediated; that is, during the breeding season, spatial performance is decreased in both meadow voles (*Microtus pennsylvanicus*) and deer mice (Galea, Kavaliers, and Ossenkopp, 1996).

Food intake and some food preferences (preference for sweets) decline during estrous cycles (Tarttelin and Gorski, 1971). Estrogen concentrations acting in the brain affect food ingestion (Wade, Huang, and Crews, 1993). Many events, including the availability of metabolic fuels, can affect the

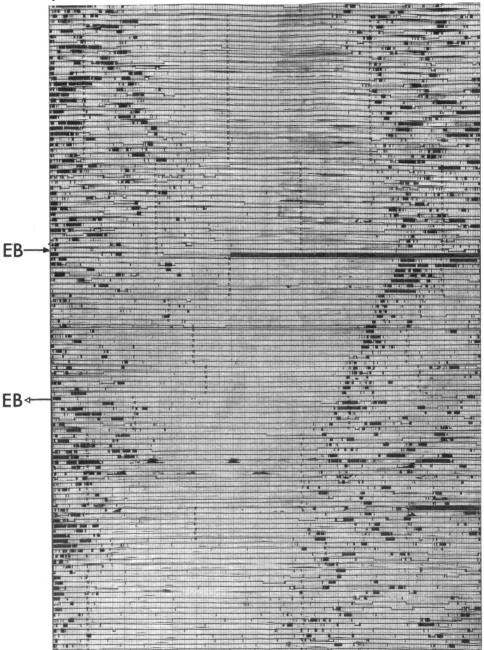

Figure 6.12. Continuous record of wheel-running activity by an ovariectomized and blind hamster in which a capsule containing estradiol benzoate (EB) was implanted and subsequently removed. The record has been photographically duplicated and double-plotted on a 48-hour time base to aid visual inspection of the free-running rhythm. (Courtesy of L. P. Morin.)

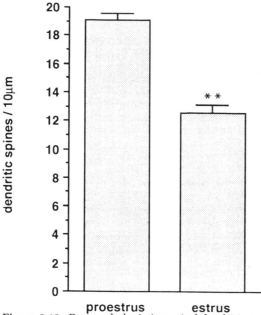

Figure 6.13. Bar graph depicting apical dendritic spine density (mean ± SEM) in the dorsal hippocampus in intact rats in the proestrous phase and estrous phase of the estrous cycle. Note that spine density is decreased in the estrous phase compared with the proestrous phase of the cycle. (From C. S. Woolley, E. Gould, M. Frankfurt, and B. S. McEwen, 1990, Naturally occurring fluctuation in dendritic spine density on adult hippocampal pyramidal neurons, *The Journal of Neuroscience* 10:4035–9. Reprinted by permission.)

estrous cycle. In golden hamsters, for example, food deprivation disrupts estrus, as does damage to the paraventricular nucleus of the hypothalamus (Schneider and Wade, 1989, 1990). Specifically, it is the lack of access to fatty acids and the state of compromised glycolysis that interfere with proper expression of the cycle (Schneider and Wade, 1989). All of this indicates that when animals are confronted with life-threatening situations, they tend to shift physiologic resources away from reproduction toward basic survival (Schneider and Wade, 1989).

High estrogen concentrations decrease the female's appetite for water and sodium, perhaps by influencing oxytocin concentration in the brain. Thirst and sodium appetite and their diminution during phases of the estrous cycle in the rat are linked to extracellular fluids, and to angiotensin in particular (see Chapter 2). For example, angiotensin-induced intake of water and sodium is reduced at estrus (Michell, 1976; Danielsen and Buggy, 1980; Findlay, Fitzsimons, and Kucharczyk, 1979), and angiotensin receptors in the brain are reduced at that time.

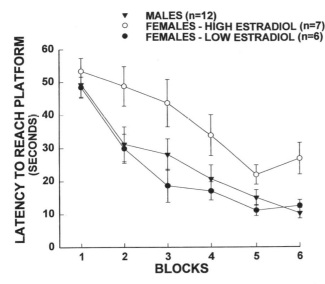

Figure 6.14. Group mean performance in a water-maze task over six blocks by male meadow voles and female meadow voles with high and low plasma concentrations of estradiol. (From L. A. Galea, M. Kavaliers, K. P. Ossenkopp, and E. Hampson, 1995, Gonadal hormone levels and spatial learning performance in the Morris water maze in male and female meadow voles, *Microtus pennsylvanicus, Hormones and Behavior* 29:106–25. Reprinted by permission.)

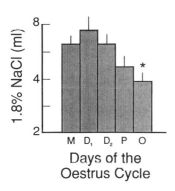

Days of the
Oestrus Cycle

Figure 6.15. Variations in ingestion of 1.8% saline across days of the estrous cycle. M, metestrus; D_1, diestrus 1; D_2, diestrus 2; P, proestrus; O, estrus. (Reprinted from *Brain Research Bulletin* 5, J. Danielson and J. Buggy, Depression ad lib and angiotensin-induced sodium intake at oestrus, pp. 501–4, copyright 1980, with kind permission from Elsevier Science.)

Estrogen-treated rats demonstrate decreases in angiotensin-induced intake of water and sodium (Fregly, 1980). Other signals not linked to the maintenance of extracellular fluid balance (see Chapter 2) are not different during this period (carbachol-induced and intracellularly induced water ingestion) (Kucharczyk, 1984). These effects in the rat appear to be mediated by the medial preoptic region. Direct application of estrogen to this region decreases angiotensin-induced ingestion of water and sodium, while not affecting carbachol-induced and intracellularly induced thirst (Jonklass and Buggy, 1984, 1985).

One way in which to envision this estrogen-related effect on ingestive behavior is by the activation of both oxytocinergic and cholecystokinin (CCK) signals that result in decreases in ingestive behavior (see Chapter 3). Interestingly, estrogen increases CCK activity in three sexually dimorphic brain regions in rats (medial amygdala, medial bed nucleus of the stria terminalis, and medial preoptic region) (Holland, Abelson, and Micevich, 1996). Moreover, central implants of estrogen will facilitate the satiety effects of CCK (Butera et al., 1996).

Monthly Clocks: The Menstrual Cycle in Humans

The menstrual cycle in women is a clear example of how changes in humoral status affect physiology and behavior. Paralleling hormonal changes in the hamster's 4-day clock, the human cycle consists of four stages: the menstrual (bleeding), preovulatory, midluteal, and premenstrual phases. Estrogen and progesterone vary throughout this monthly cycle, along with the general activation of the hypothalamic-pituitary-adrenal axis.

The menstrual cycle can also be characterized in terms of follicular, ovulatory, and luteal phases. Just before the onset of ovulation (roughly 36 hours before) there is a surge in luteinizing hormone (LH). During the next (follicular) phase, follicle-stimulating hormone (FSH) increases, resulting in increased amounts of estrogen being secreted into the systemic circulation. Progesterone concentrations are low at this time. Toward the end of the follicular period, estrogen rises, resulting in the activation of LH secretion, which in turn triggers ovulation – when the mature egg is released from the ovarian follicle. At the end of ovulation, both estrogen and progesterone concentrations are high, stimulating menstruation.

There is also evidence that the gonadal steroids may help regulate circadian rhythmicity in humans (Leibenluft, 1993). One hypothesis is that estrogen acts to shorten the circadian period and thereby extend the sleep phase of the cycle in females. For example, individuals infused with estrogen report sounder sleep, and menopausal women report greater disruption of sleep than premenopausal women (Leibenluft, 1993).

Human behaviors and various measures of performance also vary with the phases of the cycle. When estrogen concentrations are high, for example, researchers have found that women perform better on manual-dexterity tasks, but show decreased performance on spatial tasks (see Chapter 1). When both estrogen and progesterone are high, women perform better at

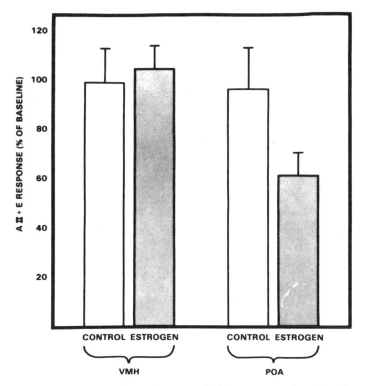

Figure 6.16. Effects of crystalline estradiol benzoate implants into the medial preoptic area (POA) or ventromedial hypothalamus (VMH) on angiotensin-II-stimulated fluid intake (mean ± SE) for 4 consecutive test days. (Reprinted from *Brain Research* 326, J. Jonklaas and J. Buggy, Angiotensin-estrogen central interaction: localization and mechanism, pp. 239–49, copyright 1985, with kind permission from Elsevier Science.)

certain motor tasks and show improved verbal abilities. The same investigators also found that postmenopausal women treated with estrogen expressed some of the same differences, that is, better performance at sensorimotor and verbal tasks and decreased performance at spatial tasks (Hampson, 1990; Kimura, 1992; see Chapter 1, this volume).

It is common for women to report feeling bad – physically and emotionally – during certain phases of the menstrual cycle (Rubinow and Schmidt, 1987). During the late luteal phase, for example, women vulnerable to affective disorders report greater occurrences of aversive life events (Rubinow et al., 1988). For clinicians, premenstrual syndromes are importantly linked with affective disorders in a subset of women they treat (Rubinow et al., 1988).

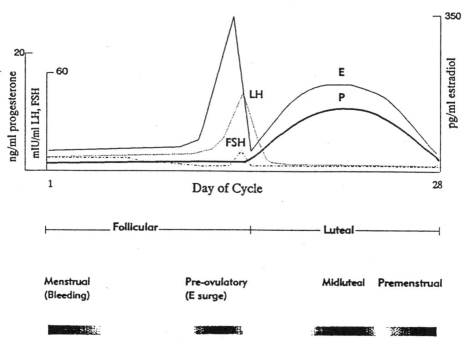

Figure 6.17. Monthly hormonal changes in women. (From E. Hampson and D. Kimura, 1992, Sex differences and hormonal influences on cognitive functions in humans, in *Behavioral Endocrinology,* ed. J. B. Becker, S. M. Breedlove, and D. Crews. MIT Press. Copyright © 1992 Massachusetts Institute of Technology. Reprinted by permission.

Seasonal Clocks: Photoperiods, Hormones, and Behavior

In many animal species, some of the most amazing evolutionary adaptations can be seen in behaviors that are altered with the changing seasons. Light influences a number of physical characteristics. In the meadow vole, for example, it can affect brain weight and DNA content – the longer the day length, the heavier the brain weight and the higher the DNA content (Dark, Dark, and Zucker, 1987). Long day length also increases myelination in developing meadow voles (Spears et al., 1990).

Seasons and Sexual Behavior

Seasonal elevations in gonadal steroid hormones during the spring and summer determine courtship displays and responses in a great number of species (Wingfield, 1994). Through the influences of hormones, courtship

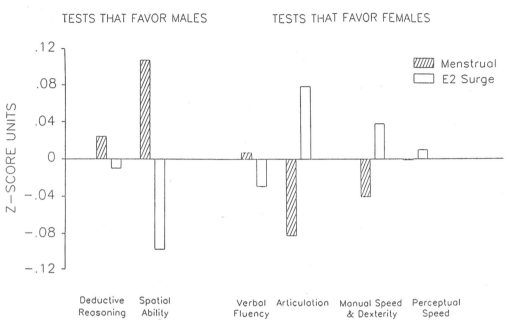

Figure 6.18. Mean scores on the six ability composites (deductive reasoning, spatial ability, verbal fluency, articulation, manual speed/coordination, perceptual speed) during the menstrual phase (shaded) and midluteal (E2) phase (high estrogen). (Reprinted from *Psychoneuroendocrinology* 15, E. Hampson, Estrogen-related variations in human spatial and articulatory-motor skills, pp. 97–111, copyright 1990, with kind permission from Elsevier Science Ltd, The Boulevard, Langford Lane, Kidlington Ox5 1GB, UK.)

and territorial behaviors are ultimately generated by changes in central motive states in the brain. In the ground squirrel – in which there are spring and summer elevations in testosterone and LH concentrations (Licht et al., 1982; Zucker and Licht, 1983) – circadian and seasonal changes in LH secretion precede the onset of locomotor activity (Turek and Van Cauter, 1994).

Photoperiod-induced changes are not restricted to mammals, but are also expressed in reptiles and birds (Wingfield, 1994). These hormonal events, in turn, affect sexual behavior. Responding to light, the circadian clock influences the secretion of luteinizing-hormone-releasing hormone (LHRH) (Turek and Van Cauter, 1994; Tubbiola and Bittman, 1995), which then influences estrogen and progesterone secretions, which in turn set the conditions for reproductive-related events (both behavior and physiology working toward the same end point).

Anatomic structures that mediate behaviors are also altered seasonally (e.g., Crews Robker, and Mendonca, 1993; Crews, Coomber, and Gonzalez-

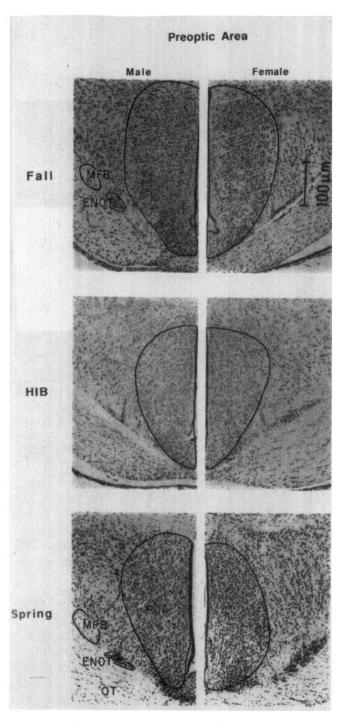

Figure 6.19. Brain areas measured in the red-sided garter snake (*Thamnophis sirtalis parietalis*): representative sections with the regions of interest outlined. Brains were from animals killed before entering hibernation (Fall), during hibernation (HIB), and on emergence from hibernation (Spring). In each panel, the male is on the left, and the female is on the right. Areas measured are outlined: POA, preoptic area; ENOT, external nucleus of the optic

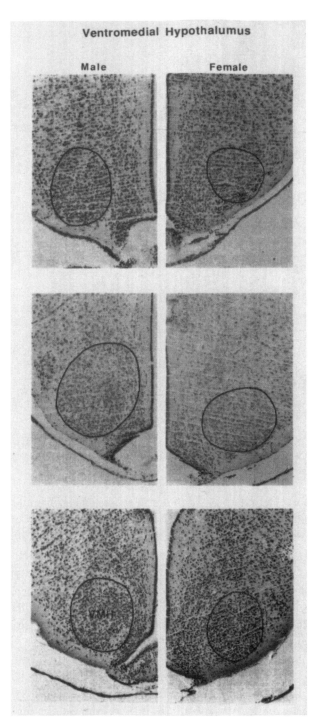

Ventromedial Hypothalumus

Male   Female

tract; VMH, ventromedial hypothalamus; MFB, medial forebrain bundle; OT, optic tract. (From D. Crews, R. Robker, and M. Mondonca, 1993, Seasonal fluctuations in brain nuclei in the red-sided garter snake and their hormonal control, *The Journal of Neuroscience* 13: 5356–64. Reprinted by permission.)

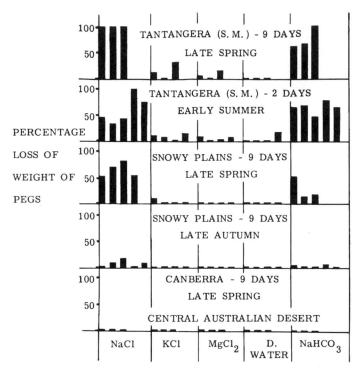

Figure 6.20. Salt pegs being ingested by rabbits during seasonal periods. (From D. A. Denton, 1982, *The Hunger for Salt,* New York: Springer-Verlag. Reprinted by permission.)

Lima, 1997). Interestingly, in the male red-sided garter snake, the hormones of courtship and sexual behaviors are not linked to elevated androgens at the same time as seasonal reproductive behaviors; instead, they are elevated months before that period (Crews, 1991), and thus they can induce changes in the brain regions that orchestrate the central states associated with sexual behaviors (Crews et al., 1993). Some hypothalamic neuropeptides change seasonally and affect behavior even without external environmental changes (Bissette et al., 1995). By inducing central states, these hormonal changes prepare the animal to behave in characteristic ways.

The research on amphibians (discussed in the Introduction and in Chapter 1) may have relevance in this context. Changes in vasotocin in specific regions of the brain underlie male reproductive behaviors (Deviche, Propper, and Moore, 1990). These changes in vasotocin in the brain are dependent upon testosterone and are under seasonal control (Moore et al., 1994). That is, the secretion of testosterone during the mating season influences central vasotocin production and the expression of sexual behavior.

Ingestion

Migration, a behavior demonstrated by animals ranging from salmon and birds to seals, turtles, and elk, is an impressive example of a seasonal change in behavior that is influenced by humoral signals. For instance, when the hormones of sodium homeostasis (e.g., aldosterone and angiotensin) are elevated, which often occurs during the reproductive season, many animals migrate toward salt licks (Denton, 1982), with females being more commonly spotted at these sodium sources. Researchers have observed seasonal, salt-related migrations in a variety of birds, reptiles, and mammals. In addition to directly inducing the hunger for sodium, the hormones of reproduction mediate seasonal changes in bone density (Edgren, 1960). The seasonal ingestion of salt pegs by rabbits is depicted in Figure 6.20 (Denton, 1982).

Syrian hamsters, when kept in the laboratory, develop seasonal obesity, a phenomenon that depends on both melatonin secretion (which is in turn related to light) and the fat content of the diet (Wade and Bartness, 1984; Wade and Schneider, 1992). In hamsters, the photoperiod has an impact on fat deposition and body weight (Bartness and Wade, 1984) and on the secretion of gonadal steroids, melatonin, and prolactin (Smale, Dark, and Zucker, 1988).

Seasonal Changes in Hormones and Dominance Behavior

Seasonal elevations in testosterone during spring and summer determine male courtship and territorial displays in a great number of species (Wingfield, 1994). Through the influence of gonadal steroid hormones on neuropeptide gene expression, courtship behavior is ultimately generated by changes in central states in the brain.

Elevated in many animals during the mating season, testosterone is related to an animal's size (e.g., iguana) (Alberts et al., 1992). In a number of species, including the red-winged blackbird (*Agelaius phoeniceus*), seasonal testosterone and LH concentrations vary with the dominance and density of individuals; testosterone is high in high-density areas of interactions (Beletsky, Orians, and Wingfield, 1992; Wingfield, Vleck, and Moore, 1992). Perhaps high concentrations of testosterone increase vasopressin concentrations in brain regions such as the bed nucleus of the stria terminalis and the medial region of the amygdala, influencing the anterior hypothalamus and dominant/subordinate behavioral responses (Albers et al., 1988; Ferris et al., 1997).

In subordinate sparrows, the seasonal change in corticosterone that takes place in mammals does not occur (Wingfield et al., 1992), although seasonal corticosterone concentrations can reflect metabolic changes in this species (Buttemer, Astheimer, and Wingfield, 1991). In a number of species, lower corticosterone concentrations are linked to a better chance of success at reproduction, as are high concentrations of testosterone and prolactin in males and females, respectively (Wingfield et al., 1992).

Hormones and Hibernation

The important seasonal events of entering into and emerging from hibernation are initiated by humoral signals activated by changes in temperature. Variations in temperature can change the onset of hibernation in many species (Crews et al., 1993). Testosterone influences the control of seasonal hibernation and daily torpor. Although the SCN is not essential for the expression of daily torpor in Siberian hamsters (Ruby and Zucker, 1992), testosterone and prolactin inhibit its expression. Male hamsters injected systemically with testosterone showed rapid reductions in torpor; in females, estrogen produced the same effect (Ruby et al., 1993). Elevated testosterone concentrations can also reduce the length of hibernation (Lee et al., 1990a), but are not essential for terminating the behavior (Darrow, Yogev, and Goldman, 1987). In the case of hibernation, therefore, hormones may be influencing biological clocks.

On the anatomic side, whereas SCN lesions can alter circannual rhythms, in several species (ground squirrel) (Bittman et al., 1991) such lesions can disrupt circadian rhythms while leaving annual rhythms relatively intact (Zucker, Boshes, and Dark, 1983). Hibernation rhythms may be somewhat disrupted by SCN lesions (Dark et al., 1990). One interesting idea is that the SCN monitors the length of hibernation bouts (Dark et al., 1990). Lesions of the SCN affect a variety of phenomena, including circadian rhythmicity, hibernation, body mass, and testosterone concentrations. For example, whereas testosterone secretion remains normal, hibernation can be affected by either partial or total SCN lesions. Lesions of the ventral medial hypothalamus, the lateral hypothalamus, and the paraventricular nucleus do not affect annual rhythms, but pinealectomy has a modest effect (Zucker, 1985).

Seasonal Changes in Neural Plasticity Reflect
Hormonal Effects on Brain and Behavior

Among the most interesting animal behaviors described in this book is the phenomenon of singing among male birds (see Chapter 1). A seasonal, tes-

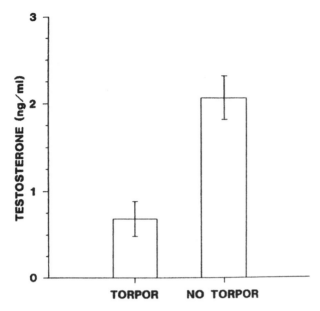

Figure 6.21. Testosterone concentrations ($\pm$ SE) in castrated squirrels displaying torpor were significanty lower than those in squirrels that did not display torpor. Testosterone-filled and blank capsules were implanted several weeks before the cold challenge. (From Lee et al., 1990a, with permission.)

tosterone-mediated behavior, singing has functional relationships to reproductive behavior (Marler et al., 1988). Males sing to attract females and to communicate readiness for aggression to other males (Wingfield et al., 1994).

Among different bird species there is a diversity of plasticity in singing behavior (Smith, 1977). Adult canaries, for example, slightly alter their songs each year, a change that depends on a brain nucleus called the HVC (see Chapter 1). Interestingly, each year the new song is produced by new neurons specifically recruited for this function. That is, neurons are replaced seasonally for the purpose of song (e.g., Kirn and Nottebohm, 1993; cf. Smith et al., 1997); these new neural connections last for several months and then decay.

It is now a well-known fact that neurons can regenerate. Seasonal changes in song-dependent brain events linked to testosterone certainly support this. Whereas the migration of neurons in the young is well understood, only in limited circumstances are we beginning to understand that something similar can occur in adult animals (e.g., the adult male canary) (Kirn and Nottebohm, 1993; see Chapter 1, this volume). The fact that there is more neural material than is needed for song reflects an important natural truth: Nature overproduces neurons; some of them survive, and others do not.

Seasonal neuronal replacement is not a phenomenon restricted to birds. Researchers have found that during hibernation (an adaptive behavior used by a diversity of animals), hippocampal pyramidal neurons are reduced in the ground squirrel. Specifically, dendrites are shortened during hibernation, but with the onset of awakening they assume their usual length (Popov, Bocharova, and Brain, 1992). A similar effect has been observed in the synaptic contacts between mossy fibers and pyramidal neurons in the hippocampus (Popov and Bocharova, 1992). In this case, we do not know if humoral signals play a role, although it is conceivable that low, "wake-up" concentrations of corticosterone may be involved.

Central States, Hormones, and Seasonal Sadness

Throughout recorded history, since the time of classical Greece, writers have made references to the "winter blues." More recently, clinicians have described a common pattern of features among individuals who become depressed in the winter; this is now called seasonal affective disorder (SAD) (Rosenthal et al., 1984). These features include disruptions of normal appetite and sleep patterns, as well as a feeling of sadness. Seasonal depression is quite different from the melancholic depression described in Chapter 5. In melancholic depression, patients are fearful and anxious. In contrast, seasonally depressed patients are lethargic and are hyperphagic for sweets (Rosenthal et al., 1984).

It has long been known that people in Sweden tend to commit suicide at higher rates than do people in Italy (Durkheim, 1951). Since at least the nineteenth century it has been believed that this phenomenon might be related to the seasons (Rosenthal et al., 1984). The great sociologist Durkheim once suggested that Scandinavian countries had higher suicide rates because of the religious rituals embodied in Protestantism. He claimed that Protestant religion places less emphasis on social cohesiveness and more on autonomy, on being separate individuals, than does the Catholic religion. He thought that the greater isolation lent itself to a greater suicide rate. But we are now coming to believe that the phenomenon may, at least in part, be related to exposure to light.

Researchers now know that light can help people who suffer from seasonal depression (Rosenthal et al., 1984). The ameliorative effects are more powerful when light is received through the eye, as opposed to the skin. Perhaps the seasonally depressed patient's tendency to eat more sweets is

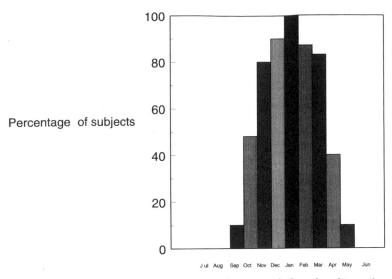

Figure 6.22. Percentage of subjects depressed per month (based on history) compared with mean photoperiod in Rockville, MD (Smithsonian Radiation Biology Laboratory), and average daily temperature at Dulles Airport, Virginia (National Climatic Center, Asheville, NC). (Adopted from Rosenthal et al., 1984.)

one way of trying to get a quick fix of energy. Moreover, carbohydrate intake is linked to serotonin concentrations (Wurtman, 1993), and manipulations of serotonin concentrations are also known to influence SAD (Rosenthal et al., 1989).

As in many other forms of depression, SAD sufferers tend to be female much more often than male. This may reflect the fact that brain regions such as the SCN are sexually dimorphic. One wonders if this hypothalamic site and perhaps others that underlie hormonally induced behaviors may be involved in the circuit underlying SAD and other mood disorders in which there are different rates of expression between men and women.

Richter (1965) proposed the existence of hormonal and calcium mechanisms that regulate mood, suggesting that mood swings are associated with imbalances in these mechanisms.

Another part of the circuit underlying SAD may lie in regions of the amygdala that concentrate vitamin D (Stumpf and O'Brien, 1987). Production of 1,25-dihydroxyvitamin D_3 is decreased during the winter, a phenomenon that is correlated with bone density and calcium metabolism (Hollick, 1994; DeLuca, 1987). Figure 6.23 depicts the seasonal differences in humans between vitamin D_3 concentrations in July and October (Hollick, 1994). Perhaps these seasonal differences matter in ameliorating SAD, but that is not certain.

Whereas those who study SAD have concentrated on retinal light, presumably because of the melatonin connection, perhaps changes in vitamin D production induced by ultraviolet (UV) light may also play a role. UV-β light induces vitamin D synthesis in the skin by a conversion process that results in vitamin D_3, the active metabolite (Hollick, 1994). Known to be involved in calcium metabolism, the hormone also has receptors in two key brain regions: the bed nucleus of the stria terminalis and the central amygdala. Both regions have been implicated in other hormone-regulated behaviors. Moreover, vitamin D is known to have effects on neurotransmitters (e.g., catecholamine) and calcium-binding proteins (Strauss et al., 1995). Perhaps it also influences other neurotransmitters or neuropeptides that are involved in seasonal depression (e.g., serotonin, corticotropin-releasing hormone).

Reasoning that because light makes people feel better, and because UV light activates vitamin D synthesis, we tested the hypothesis that seasonally depressed patients may have low concentrations of circulating vitamin D_3 (Stumpf and Privette, 1989; Oren et al., 1994). Moreover, perhaps light affects vitamin D_3 concentrations, which in turn act on brain regions like the amygdala that influence mood. As shown in Figure 6.24, we did find a tendency for seasonally depressed patients to exhibit less vitamin D activation than normal patients. It still remains possible that vitamin D_3 receptors are different in seasonally depressed and nondepressed subjects. Future studies using magnetic-resonance imaging may tease out which parts of the brain are different in these two populations, as well as identify hormonal markers that

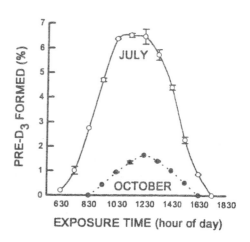

Figure 6.23. Photosynthesis of pre-cholecalciferol (previtamin D) at various times on cloudless days in Boston in October and July. (From M. F. Hollick, 1994, Vitamin D – new horizons for the 21st century, *American Journal of Clinical Nutrition* 60:619–30. Copyright © Am. J. Clin. Nutr., American Society for Clinical Nutrition. Reprinted by permission.)

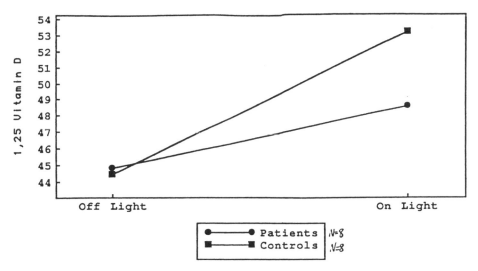

Figure 6.24. Plasma concentrations of vitamin D in seasonally depressed patients and controls with and without UV light. (Adapted from Oren et al., 1994.)

can discern regional brain differences. But with regard to Vitamin 1,25D₃, a recent report has demonstrated that systemic injections of this hormone can enhance mood states in normal subjects during the winter months (Lansdowne and Provost, 1998). For now, one is left with the possibility that calcium homeostasis and changes in the hormones that regulate calcium homeostasis may be linked to seasonal depression. But this is highly speculative.

Conclusion

Living organisms contain a number of biological clocks that regulate their behaviors. Many of these clocks, as well as the behaviors they control, are also influenced by hormonal events. Within the context of biological clocks, this chapter has emphasized reactive versus predictive homeostasis, the latter allowing for anticipatory control over behavior and the environment in which one is trying to adapt. Synchronized with one another, multiple biological clocks play a profound role in a wide variety of behaviors. Like space, time is an important determinant of what animals, including humans, do during their lives. Hormones are known to influence brain and behavioral events that are regulated by circadian, monthly, and seasonal clocks. Moreover, hormones such as estrogen influence the circadian clock. Behavior is affected by the expressions of the endogenous clocks, by the circulating concentrations of systemic hormones, and by neuropeptide gene expression.

Conclusion

A set of core concepts has organized the contents of this book, providing a perspective on the science of hormones, brain, and behavior. One of these is the notion that underlying many behaviors carried out by animals, including ourselves, are central motive states that are influenced by hormones acting on the brain (Lashley, 1938; Beach, 1942; Stellar, 1954). That is, hormones induce and sustain central states that prepare an animal to perceive the world in characteristic ways and then act accordingly.

Recall the Introduction: Hormonally facilitated central motive states can be divided into two phases. The first is the appetitive phase – searching for what is desired (e.g., sodium) and avoiding what is aversive (e.g., a predator). The second is the consummatory phase, in which the animal satisfies its desire (e.g., by ingesting sodium or by feeling safe). The organization of the behaviors is reflected in the functional roles that hormones play in the brain to generate behaviors that will maintain the internal milieu and respond to problems in the external world. Steroid and peptide hormones activate and sustain central motive states.

Herbert (1993) has produced a beautiful monograph depicting different contexts in which steroids and peptides interact to regulate behavior (see also Hoebel, 1988). Most of these have been discussed in this book. They range from ingesting food, water, and sodium to maternal behavior, fear, and aggression. Being economical, nature uses the same hormones to generate a variety of different central states. These, in turn, generate behaviors that, through both anticipatory and reactive mechanisms, help the organism maintain homeostasis.

In each chapter, I have highlighted several model systems for understanding relationships among hormones, brain, and behavior. In concluding this book, I want to review several examples from different chapters describing the interactions of hormones, brain, and behavior. From these examples, two of the book's central themes emerge: The first is that steroids and peptides or neuropeptides interact to influence behavior by their actions in the brain. In many chapters I provided examples in which hormones that regulate homeostasis physiologically also affect behaviors that serve the same goal. The second theme is that common and separate neural circuits underlie a variety of different central motive states.

Estrogen Effects on Oxytocin-Regulated Sexual Behavior

Perhaps one of the best-known examples of steroids and peptides acting together to generate behavior is that of estrogen-primed rats given progesterone and oxytocin to induce sexual receptivity (Pfaff et al., 1994). A variety of animals treated with systemic estrogen and then with progesterone demonstrate similar sexual receptivity (see Chapter 1).

One neuropeptide influenced by estrogen concentrations is oxytocin, which plays an essential role in female sexual behavior. Estrogen increases oxytocin expression in cells in the ventral medial hypothalamus. Oxytocinergic pathways that emanate from this region are malleable: Without sufficient estrogen, they decline, being restored only when estrogen is again elevated (e.g., Schumacher et al., 1990). Oxytocin infused within this region of the brain elicits sexual receptivity, and in estrogen- and progesterone-primed rats the dose of oxytocin needed to elicit the behavior is decreased. Thus, by increasing oxytocin expression in the brain's ventral medial hypothalamus, estrogen facilitates the likelihood of sexual receptivity.

Corticosteroid Effects on Angiotensin-II-induced Water and Sodium Appetite

Two kinds of adrenal steroid hormones (glucocorticoids and mineralocorticoids) affect both systemic physiology and behaviors linked to the body's fluid balance. During loss of the body's fluids, which can occur from sweating, bleeding, or dehydration due to water deprivation, the hormones of body-fluid homeostasis rise. In addition to angiotensin and aldosterone, these hormones include vasopressin and, to a lesser extent, corticosterone (Stricker et al., 1979 Denton, 1982). Physiologically, the hormones act to regulate the body's fluid volume by reducing water and sodium losses and redistributing the body's sodium to maintain extracellular volume.

A peptide produced in both the brain and periphery, angiotensin also has a major impact on behavior. Injected centrally, angiotensin II increases both water intake and sodium intake, independent of sodium loss. Mineralocorticoid hormones increase sodium ingestion. Mineralocorticoids increase

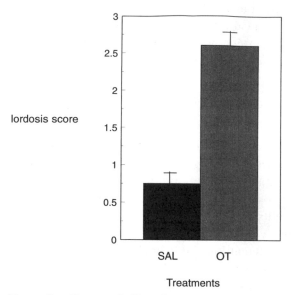

Figure C.1. Estrogen-facilitated oxytocin-induced lordosis behavior. (Adapted from Schumacher et al., 1990.)

angiotensin-II receptors in the brain and in cell-line cultures. Mineralocorticoids also potentiate angiotensin-II-induced sodium intake (Fluharty and Epstein, 1983; Sakai, 1986; Massi and Epstein, 1990).

Glucocorticoids are known to increase both vasopressin and angiotensin receptors in several brain regions, while having no effect in others (see Chapter 2). The same treatment is known to increase angiotensin mRNA in cell-line cultures, in addition to mobilizing intracellular calcium and second-messenger systems (S. J. Fluharty, unpublished observations). Elevated glucocorticoids potentiate angiotensin-II-induced drinking. The background of glucocorticoids that are normally elevated following depletion of the body's fluids induces angiotensin-II cells, thereby potentiating the hormone's dipsogenic (Ganesan and Sumners, 1989; Sumners et al., 1991a) and natriorexegenic actions. This interaction generates the central states of thirst and sodium hunger, in which both appetitive and consummatory behaviors are expressed. The behavior serves the same end point as the renin-angiotensin-aldosterone system, namely, the body's fluid homeostasis.

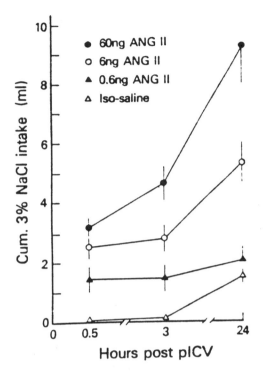

Figure C.2. Sodium ingestion by rats treated with systemic aldosterone (40 μg per day) and then later treated with intraventricular infusions of angiotensin. (From R. R. Sakai, 1986, The hormones of renal sodium conservation act synergistically to arouse a sodium appetite in the rat, in *The Physiology of Thirst and Sodium Appetite,* ed. G. de Caro, A. N. Epstein, and M. Massi, New York: Plenum. Reprinted by permission.)

Glucocortoid and Neuropeptide-Y-induced Food Intake

The hormone of energy metabolism, corticosterone plays roles in the activation of a great many different behaviors, including thirst and sodium appetite, amphetamine self-administration, and fear. In each case, energy homeostasis underlies the behaviors in which glucocorticoids play a fundamental role. Fear, for example, places great demands on an animal's energy reserves.

At low doses, corticosterone also stimulates food intake. It appears to do so through activation of neuropeptide Y in the brain. Produced in gastrointestinal sites, as well as the central nervous system (Gray et al., 1986), neuropeptide Y is activated functionally in the arcuate nucleus and the paraventricular nucleus by food deprivation and corticosterone (Brady et al., 1990; Dallman et al., 1993). This increase in food intake depends on an intact adrenal gland, as does the hyperphagia that results from damage to the ventral medial hypothalamus and from activation of type-2 corticosteroid receptors (Tempel and Leibowitz, 1994).

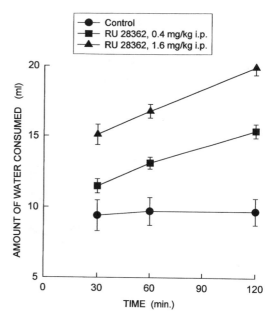

Figure C.3. Effects of a glucocorticoid agonist (RU 28362) on drinking induced by intraventricular infusions of angiotensin II (ANG II). Cumulative water intake after administration ANG II (10 ng/2 μl) to rats from different groups: control, and two different doses (0.4 and 1.6 mg/kg) of the glucocorticoid agonist 3 hours prior to the ANG II injections. (Reprinted from *Brain Research* 552, C. Sumners, T. R. Gault, and M. J. Fregly, Potentiation of angiotensin II-induced drinking by glucocorticoids is a specific glucocorticoid type II receptor (GR)-mediated event, pp. 283–90, copyright 1991, with kind permission from Elsevier Science.)

Importantly, low doses of glucocorticoids can potentiate neuropeptide-Y-induced food ingestion (Heinrichs et al., 1992). Specifically, corticosterone and neuropeptide Y generate carbohydrate ingestion, or fast energy pickups (Leibowitz, 1995). This steroid and neuropeptide hormone facilitate the central motive state of craving carbohydrates. Adrenalectomy abolishes these effects, and corticosterone will restore them. By contrast, another peptide that generates food intake, galanin, does not depend on circulating corticosterone for its effects if it is infused in either the paraventricular nucleus of the hypothalamus or the central nucleus of the amygdala (Corwin et al., 1993).

Neuropeptide Y is distributed along the central visceral axis, which includes the central nucleus of the amygdala, the bed nucleus of the stria terminalis, the paraventricular nucleus of the hypothalamus, and brainstem sites such as the parabrachial and solitary nuclei (Gray et al., 1986). This is the same pathway that organizes motivated behaviors in general (Pfaffmann et al., 1977; Stellar and Stellar, 1985). It is part of the neural system – described by Herrick (1905, 1948) and expanded upon by Nauta (1961) (Norgren, 1995) – that underlies central excitatory states.

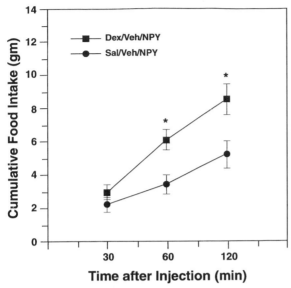

Figure C.4. Cumulative food intake at three time points following pretreatment with dexamethasone (100 μg/kg i.p.) or saline, followed 6 hours later by bilateral injection of CRF antagonist vehicle and 15 minutes later by a 500-ng dose of neuropeptide Y (NPY) into the PVN of the hypothalamus. (Reprinted from *Brain Research* 581, S. C. Heinrichs, E. M. Pich, K. A. Miczek, K. T. Britton, and G. F. Koob, Corticotropin-releasing factor antagonist reduces emotionality in socially defeated rats via direct neurotropic action, pp. 190–7, copyright 1992, with kind permission from Elsevier Science.)

Estrogen Effects on Oxytocin, Prolactin, and Maternal Behaviors

Recall that oxytocin is a peptide in both the pituitary gland and the central nervous system. Its physiologic role in milk letdown is well known, as is its role in sexual receptivity. Like other hormones, oxytocin works through both physiology and behavior. Both strategies serve the same end point: successful reproduction. By increasing oxytocin-producing cells in the bed nucleus of the stria terminalis (Insel, 1992), at least in voles, estrogen also increases the bonding between individuals of different sexes and between both genders and their offspring (see Chapter 4) (Insel, 1992). By increasing the neuropeptide, a background of estrogen increases the likelihood that all these adaptive behaviors will be expressed (e.g., Carter, 1992).

Like oxytocin and a number of other peptides, prolactin is a pituitary hormone and a neuropeptide with a diversity of functions. Central infusions of prolactin facilitate maternal behavior (Bridges and Mann, 1994) – another key adaptation for successful reproduction – but only if there is a sufficient level of background estrogen. As is the case with oxytocin, estrogen facilitates the likelihood of maternal behavior by increasing prolactin expression in the brain. Prolactin generates maternal behavior through both physiologic and behavioral mechanisms.

Testosterone's Effects on Vasopressin and Territorial Expression, Aggression, and Parental Behavior

As noted in earlier chapters, testosterone is not the hormone of aggression and can (in some circumstances) be linked to well-being and mood elevation. Nonetheless, in a number of species, aggression is linked to testosterone concentrations. Among song sparrows, for example, the highest testosterone concentrations are reached when the birds are establishing territory (Wingfield, 1994). The greater the aggressiveness, the higher the testosterone concentration and the larger the territory the bird must defend.

In most species, testosterone concentrations vary with the seasons (e.g., Wingfield, 1993, 1994). One testosterone-mediated behavior linked to territorial behavior, scent marking, occurs via the gonadal steroid's activation of vasopressin in the brain, particularly in the bed nucleus of the stria terminalis. Infused into this region, vasopressin facilitates the expression of scent marking. A background treatment of testosterone enhances this response (Albers et al., 1988). The same holds for male parental behavior among prairie voles. Presumably, testosterone facilitates this behavior by sustaining and increasing central vasopressin synthesis (DeVries, 1995).

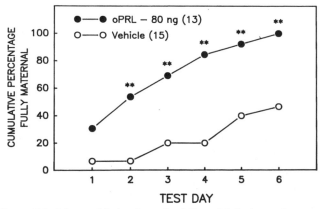

Figure C.5. Maternal behaviors after bilateral infusions of 40 ng of prolactin (PRL) into the medial preoptic nucleus in estrogen-treated and nontreated rats. (From Bridges et al. 1990, with permission.)

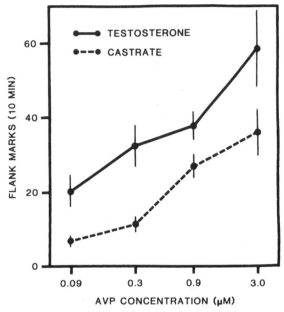

Figure C.6. Effects of testosterone and central injections of AVP on scent-marking behavior in hamsters. (Reprinted from *Brain Research* 456, H. E. Albers, S. Y. Liou, and C. F. Ferris, Testosterone alters the behavioral response of the medial preoptic – anterior hypothalamus to microinjection of arginine vasopressin in the hamster, pp. 382–6, copyright 1988, with kind permission from Elsevier Science.)

A circuit for these hormone-facilitated behaviors includes the medial amygdala as well as the medial preoptic region and perhaps the bed nucleus of the stria terminalis. This is the same forebrain circuit that underlies many of the hormone-induced behaviors discussed throughout this book.

Effects of Gonadal Steroid Hormones on Vasotocin Expression in Newts, Frogs, and Songbirds

Male and female sexual behaviors are sustained by gonadal steroid hormones through the induction of vasotocin expression in rough-skinned newts (Moore et al., 1992, 1994). The gonadal steroid hormones facilitate male sexual behavior and female egg-laying behavior. Vasotocin receptors are reduced in the amygdala by castration and can be reinstated with gonadal steroid, as can the behaviors (Boyd and Moore, 1991). The female clasping behavior that is linked to egg laying is depicted in Figure C.7.

The gonadal steroid hormones also facilitate sound production by frogs

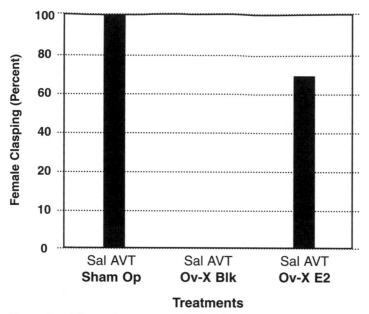

Figure C.7. Effects of ovariectomy (Ov) and steroid implants in vasotocin-treated (Sal AVT) newts on clasping behavior. (Adapted from F. L. Moore, R. E. Wood, and S. K. Boyd, 1992, Sex steroids and vasotocin interact in a female amphibian to elicit female-like egg laying behavior of male-like courtship, *Hormones and Behavior* 26:156–66. Reprinted by permission.)

and song by birds, which also may be mediated by central vasotocin (Marler et al., 1995; Maney et al., 1997).

Glucocorticoid Effects on Corticotropin-releasing Hormone: Anxiety and Fear

Corticotropin-releasing hormone (CRH) plays a fundamental role in the regulation of the pituitary-adrenal axis by increasing secretion of pituitary adrenocorticotropic hormone (ACTH), which in turn stimulates glucocorticoids such as corticosterone from the adrenal glands. Corticosterone then regulates its own production by decreasing CRH output from the paraventricular nucleus of the hypothalamus and ACTH output from the pituitary gland (see Chapters 4 and 5).

Corticosterone can increase CRH mRNA concentrations in the central nucleus of the amygdala and the lateral bed nucleus of the stria terminalis. States of fear are associated with these brain sites, as well as with CRH activation itself. Corticosterone is in fact elevated during conditions of adversity that generate fear (e.g., Kalin and Shelton, 1989). Given systemically or infused directly into the central nucleus of the amygdala, the hormone can potentiate fear-induced freezing behavior. Elevated corticosterone in rats, monkeys, and humans who experience fear in adverse conditions may cause sustained activation of CRH in the central nucleus and the bed nucleus. Moreover, corticosterone can facilitate CRH-induced startle responses (Lee et al., 1994). Together, the steroid and the peptide sustain the central motive state of fear; the appetitive phase is avoiding the fearful situation, and the consummatory phase is the relief that comes from being out of danger.

Two functions stand out in corticosterone's impact on CRH systems. The first restrains the hypothalamic-pituitary-adrenal axis (e.g., Munck et al., 1984). The second is corticosterone's induction of CRH gene expression outside of the paraventricular nucleus (Swanson and Simmons, 1989; Makino et al., 1994a, b). The first system is involved in regulating systemic physiologic processes from pituitary activation (Dallman et al., 1993), and the second affects behavior. One system is linked to the hypothalamic-pituitary-adrenal axis (in addition to hippocampal control of the hypothalamic-pituitary-adrenal axis) (Sapolsky, 1992), and the other to brainstem sites that orchestrate the behavioral responses. This latter circuit underlies the central motive states that generate a number of hormone-induced behaviors (e.g., Schulkin et al., 1994a).

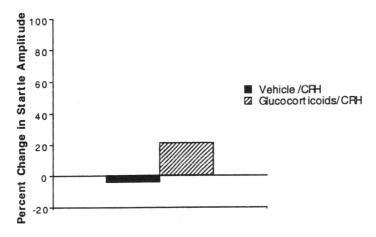

Figure C.8. Effects of chronic corticosterone injection on CRH-induced changes in startle amplitudes: mean percentage change in startle amplitude computed over the mean of the last 10 trials of the pre-drug-infusion period and the 48 5-trial blocks of the post-drug-infusion period: [(post − pre)/pre] × 100; 0.25 μg of CRH given intracerebroventricularly following 5 days of corticosterone (7.5 mg or 15 mg) or vehicle. (Adapted from Lee et al., 1994.)

Culture and Biology

When it comes to cravings for food, water, and sodium, nobody gets excited about the roles of hormones in determining these events. When it comes to sex, aggression, or intelligence, the debate heats up, and rightfully so. But how does one determine reasonable grounds on which to think about the impacts of hormones on brain and behavior? The issues are so loaded, and the ground so replete with cultural baggage.

The fact that we use hormones in so many of our products (e.g., growth hormone in milk production) is an example of the problems of infusion of culture into biology. What do I mean? Nature and culture, once so disparate, now converge in much of our lives, as we transform our natural surroundings by our artifacts and inventions. From gene therapy to brain implants, the culture of science infuses the practices of life. The Frankenstein image looms harrowingly. But recall Mary Shelley's book. Frankenstein's creature was not bad at first, but was ill-treated by humans and became changed into a monster. Maybe that is the point: Our inventions, though not neutral, are always subject to good use or bad. So, too, for the discoveries in the science of hormones, brain, and behavior.

We need a vision that can account for the possibilities that nature affords, not simply the constraints it imposes. Too many look on nature as a realm of constraints, rather than possibilities.

Artificial Intelligence and Simulation of Mind

Perhaps one of the reasons that the artificial heart did not work is that the heart is not just a pump but also is an endocrine gland. It secretes atrial natriuretic factor, which regulates the amount of blood that perfuses the heart, which then regulates the concentrations of angiotensin and aldosterone that maintain the extracellular fluid volume, in addition to decreasing both water ingestion and sodium ingestion when they act in the brain. The simulation that was attempted with the artificial heart perhaps would have to be linked to the hormones that also emerge from normal heart functions. Similarly, we have come a long way toward models that can simulate fear, which is in part mediated by hormonal signals acting in the brain. But models must be coupled with means to process information about dangerous events if we are to simulate the "wet stuff."

Wet Mind (Kooslyn and Koenig, 1992) concerns the idea that simulation of the human mind needs to take into account the function of the brain. Perhaps one should extend the simulation of the mind to capturing humoral signals that pervade normal brain function. When it is argued that the brain secretes intentions somewhat as the pancreas secretes insulin (Searle, 1986), to me this means that simulations of mental functions need to take into account humoral signals in the brain. This is analogous to arguments in the field of artificial intelligence that one needs to take into account the visual environment in which we evolved (Marr, 1982) in simulating visual functions in artificial systems.

Perhaps not until we can design a machine to simulate avoid and approach reactions and to secrete humoral messages that can induce fear will we understand how to get machines to be smart like us. We have made progress toward simulating thought, but not emotional thought patterns. Underlying the central state of fear, for instance, is the activation of peptides in the brain, in many circumstances against a background of steroid hormones. The result is functional behaviors that evolved for problem-solving, which is what the brain does.

Future Directions

Ours is the age of biology. The latter part of the twentieth century has been to biology what the first part was to physics. It is truly amazing. Biological machinations traverse our cultural space and imagination as at no time before. I have emphasized behavioral, physiologic, and anatomic levels of analysis. But this is the age of molecular biology. The advances in molecular biology will need to be integrated into the behavioral neuroscience of hormones, brain, and behavior (e.g., Keverne, 1994; Pfaff, Dellovade, and Zhu, 1997; Nelson, 1997; Young et al., 1997).

Molecular biology is a great technological tool, whether it is coupled with the human-genome project or used in everyday neuroscience. The fact that we can now control the design of genes that produce hormones and their receptor sites is an exciting development in this science that, as most of us realize, we have just begun to develop, one whose scientific trajectory is still quite uncertain.

But molecular biology will need to be integrated into a systems approach to the neuroscience of hormones, function, and brain. What we need is a positive view of biological events, an approach that can harness the miracles of biology in line with an enlightened view of the human condition, and in this endeavor we see considerable reasons to be optimistic.

References

Abercrombie, H. C., Larso, C. L., Ward, R. T., Schaefer, S. M., Hjolden, J. E., Perlman, S. B., Turski, P. A., Krahn, D. D., and Davidson, R. J. (in press). Metabolic rate in the amygdala predicts negative affect and depression severity in depressed patients: An FDG-PET study.

Adamec, R. E., and McKay, D. (1993). Amygdala kindling, anxiety, and corticotrophin releasing factor. *Physiol. Behav.* 54:423–31.

Adels, L. E., Leon, M., Wiener, S. G., and Smith, M. S. (1986). Endocrine response to acute cold exposure by lactating and non-lactating Norway rats. *Physiol. Behav.* 36: 179–81.

Adkins, E. K., and Pniewski, E. E. (1978). Control of reproductive behavior by sex steroids in male quail. *J. Comp. Physiol. Psychol.* 92:1169–78.

Adkins-Regan, E. (1989). Sex hormones and sexual orientation in animals. *Psychobiology* 16:335–47.

Adkins-Regan, E., and Ascenci, M. (1987). Social and sexual behavior of male and female zebra finches treated with oestradiol during the nestling period. *Anim. Behav.* 35:1100–12.

Adler, N. T. (1981). *Neuroendocrinology of Reproduction.* New York: Plenum.

Adler, N. T., Tesko, J. A., and Goy, R. W. (1970). The effect of copulatory behavior on hormonal change in the female rat prior to implantation. *Physiol. Behav.* 5:1003–7.

Adolphs, R., Tranel, D., Damasio, H., and Damasio, A. (1994). Impaired recognition of emotion in facial expressions following bilateral damage to the human amygdala. *Nature* 372:669–72.

Adolphs, R., Tranel, D., Damasio, H., and Damasio, A. R. (1995). Fear and the human amygdala. *J. Neurosci.* 15:5879–91.

Aggleton, J. P., and Mishkin, M. (1986). The amygdala: sensory gateway to the emotions. In: *Emotion: Theory, Research and Experience,* vol. 3, ed. R. Plutchik and H. Kellerman, pp. 981–99. Orlando: Academic Press.

Ahima, R. S., Garcia, M. M., and Harlan, R. E. (1992). Glucocorticoid regulation of preproenkephalin gene expression in the rat forebrain. *Mol. Brain Res.* 16:119–27.

Ahima, R. S., and Harlan, R. E. (1992). Glucocorticoid receptors in LHRH neurons. *Neuroendocrinology* 56:845–50.

Ainsworth, M. D. S. (1969). Object relations, dependency, and attachment: a theoretical review of the infant – mother relationship. *Child Dev.* 40:969–1025.

Akabayashi, A., Wahlestedt, C., Alexander, J. T., and Leibowitz, S. F. (1994). Specific inhibition of endogenous neuropeptide Y synthesis in arcuato nucleus by antisense oligonucleotides suppresses feeding behavior and insulin secretion. *Mol. Brain Res.* 21:55–61.

Akana, S. F., Strack, A. M., Hanson, E. S., and Dallman, M. F. (1994). Regulation of activity in the hypothalamic-pituitary-adrenal axis is integral to a larger hypothalamic system that determines caloric flow. *Endocrinology* 135:1125–34.

Akil, H., Haskett, R. F., Young, E. A., Grunhaus, L., Kotun, J., Weinberg, V., Greden, J., and Watson, S. J. (1993). Multiple HPA profiles in endogenous depression: effect of age and sex on cortisol and beta-endorphin. *Biol. Psychiatry* 33:73–85.

Akil, H., Shiomi, H., and Matthews, J. (1985). Induction of the intermediate pituitary by stress: synthesis and release of a nonopiod form of β-endorphin. *Science* 227: 424–6.

Albeck, D. S., McKittrick, C. R., Blanchard, D. C., Blanchard, R. J., Nikulina, J., McEwen, B. S., and Sakai, R. R. (1997). Chronic social stress alters levels of corticotropin-releasing factor and arginine vasopressin mRNA in rat brain. *J. Neurosci.* 17: 4895–903.

Albers, H. E., (1981). Gonadal hormones organize and modulate the circadian system of the rat. *Am. J. Physiol.* 241:R63–6.

Albers, H. E., and Cooper, T. T. (1995). Effects of testosterone on the behavioral response to arginine vasopressin microinjected into the central gray and septum. *Peptides* 16:269–73.

Albers, H. E., Gerall, A. A., and Axelson, J. F., (1981). Effect of reproductive state on circadian periodicity in the rat. *Physiol. Behav.* 26:21–5.

Albers, H. E., Karom, M., and Whitman, D. C. (1996). Ovarian hormones alter the behavioral response of the medial preoptic anterior hypothalamus to argininevasopressin. *Peptides* 17:1359–63.

Albers, H. E., Liou, S. Y., and Ferris, C. F. (1988). Testosterone alters the behavioral response of the medial preoptic–anterior hypothalamus to microinjection of arginine vasopressin in the hamster. *Brain Res.* 456:382–6.

Albers, H. E., and Rawls, S. (1989). Coordination of hamster lordosis and flank marking behavior: role of arginine vasopressin within the medial preoptic–anterior hypothalamus. *Brain Res. Bull.* 23:105–9.

Albert, D. J., Jonik, R. H., and Walsh, M. L. (1992). Hormone-dependent aggression in male and female rats: experiential, hormonal, and neural foundations. *Neurosci. Biobehav. Rev.* 16:177–92.

Albert, D. J., Walsh, M. L., and Jonik, R. H. (1993). Aggression in humans: What is its biological foundation? *Neurosci. Biobehav. Rev.* 17:405–25.

Alberts, A. C., Jackinetell, L. A., and Phillips, J. A. (1994). Effects of chemical and visual exposure to adults on growth, hormones, and behavior of juvenile green iguanas. *Physiol. Behav.* 55:987–92.

Alberts, A. C., Pratt, N. C., and Phillips, J. A. (1992). Seasonal productivity of lizard femoral glands: relationship to social dominance and androgen levels. *Physiol. Behav.* 51:729–33.

Alexander, G. M., Packard, M. G., and Hines, M. (1994). Testosterone has rewarding affective properties in male rats: implications for the biological basis of sexual motivation. *Behav. Neurosci.* 108:424–8.

Alheid, G. F., de Olmos, J., and Beltramino, C. A. (1996). Amygdala and extended amygdala. In: *The Rat Nervous System,* 2nd ed., ed. G. Paxinos. San Diego: Academic Press.

Alheid, G. F., and Heimer, L. (1988). New perspectives in basal forebrain organization of special relevance for neuropsychiatric disorders: the striatopallidal, amygdaloid and corticopetal components of substantia innominata. *Neuroscience* 27:1–39.

Alheid, G. F., Schulkin, J., and Epstein, A. N. (1993). Amygdala pathways involved in sodium appetite in the rat, anatomical considerations. *Society for Neuroscience Abstracts.*

Allen, L. A., Hines, M., Shryne, J. E., and Gorski, R. A. (1989). Two sexually dimorphic cell groups in the human brain. *J. Neurosci.* 9:497–506.

Allison, C., Alberts, N., Pratt, N. C., and Phillips, J. A. (1992). Seasonal productivity of lizard femoral glands: relationship to social dominance and androgen levels. *Physiol. Behav.* 51:729–33.

Altemus, M., Deuster, P. A., Galliven, E., Carter, C. S., and Gold, P. W. (1995). Suppression of hormonal stress responses in lactating women. *J. Clin. Endocrinol. Metab.* 10:2954–9.

Altemus, M., Pigott, T., Kalogeras, K. T., Demitrack, M., Dubbert, B., Murphy, D. L., and Gold, P. W. (1992). Abnormalities in the regulation of vasopressin and corticotropin-releasing-factor secretion in obsessive compulsive disorder. *Arch. Gen. Psychiatry* 49:9–20.

Altschuler, S. M., Rinaman, L., and Miselis, R. R. (1992). Viscerotopic representation of the alimentary tract in the dorsal and ventral vagal complexes in the rat. In: *Neuroanatomy and Physiology of Abdominal Vagal Afferents,* ed. S. Ritter, R. C. Ritter, and C. D. Barnes. Boca Raton: CRC Press.

Alvarez-Buvila, A., Ling, C.-Y., and Nottebohm, F. (1992). High vocal center growth and its relation to neurogenesis, neuronal replacement and song acquisition in juvenile canaries. *J. Neurobiol.* 23:396–406.

Anderson, I. M., Ware, C. J., DaRoza, J. M., and Cowen, P. J. (1992). Decreased 5-HT-mediated prolactin release in major depression. *Br. J. Psychiatry* 160:372–8.

Andrease, L. C., and Herbert, J. (1993). Expression of c-fos in restricted areas of the basal forebrain and brainstem following single or combined intraventricular infusions of vasopressin and corticotropin-releasing factor. *Neuroscience* 53:735–48.

Angulo, J. A., Schulkin, J., and McEwen, B. S. (1988). Effect of sodium depletion and aldosterone treatment on angiotensinogen mRNA in the brain of the rat. *Society for Neuroscience Abstracts.*

Antin, J., Gibbs, J., Holt, J., Young, R. C., and Smith, G. P. (1975). Cholecystokinin elicits the complete behavioral sequence of satiety in rats. *J. Comp. Physiol. Psychol.* 89:784–90.

Antoch, M. P., Song, E.-J., Chang, A.-M., Vitaterna, M. H., Zhao, Y., Wilsbacher, L. D., Sangoram, A. M., King, D. P., Pinto, L. H., and Takahashi, J. S. (1997). Functional identification of the mouse circadian clock gene by transgenic BAC rescue. *Cell* 69:16–28.

Antunes-Rodrigues, J., McCann, S. M., and Samson, W. K. (1986). Central administration of atrial natriuretic factor inhibits saline preference in rats. *Endocrinology* 118:1726–8.

Arnold, A. P., and Breedlove, S. M. (1985). Organizational and activational effects of sex steroids on brain and behavior: a reanalysis. *Horm. Behav.* 19:469–98.

Arnold, A. P., and Schlinger, B. A. (1993). Sexual differentiation of brain and behavior: the zebra finch is not just a flying rat. *Brain Behav. Evol.* 42:231–41.

Aronson, L. R. (1959). Hormones and reproductive behavior: some phylogenetic considerations. In: *Comparative Endocrinology,* ed. A. Gorbman. New York: Wiley.

Arriaza, J. L., Simerly, R., Swanson, L. W., and Evans, R. M. (1988). The neuronal mineralocorticoid receptor as a mediator of glucocorticoid response. *Neuron* 1:887–900.

Arriaza, J. L., Weinberger, C., Cerelli, G., Glaser, T. M., Handelin, B. L., Housman, D. E., and Evans, R. M. (1987). Cloning of human mineralocorticoid receptor complementary DNA: structural and functional kinship with the glucocorticoid receptor. *Science* 237:268–75.

Aschoff, J. (1994). The timing of defecation within the sleep–wake cycle of humans during temporal isolation. *J. Biol. Rhythms* 9:43–50.

Asmus, S. E., and Newman, S. W. (1993). Tyrosine hydroxylase neurons in the male hamster chemosensory pathway contain androgen receptors and are influenced by gonadal hormones. *J. Comp. Neurol.* 33:445–7.

Badura, L. L., and Goldman, B. D. (1992a). Central sites mediating reproductive responses to melatonin in juvenile male Siberian hamsters. *Brain Res.* 598:98–106.

Badura, L. L., and Goldman, B. D. (1992b). Prolactin-dependent seasonal changes in pelage: role of the pineal gland and dopamine. *J. Exp. Zool.* 261:27–33.

Badura, L. L., and Goldman, B. D. (1994). Prolactin secretion in female Siberian hamsters following hypothalamic deafferentation: role of photoperiod and dopamine. *Neuroendocrinology* 59:49–56.

Bakowska, J. C., and Morrell, J. I. (1997). Atlas of the neurons that express mRNA for the long form of the prolactin receptor in the forebrain of the female rat. *J. Comp. Neurol.* 386:161–77.

Ball, G. F., Faris, P. L., Hartman, B. K., and Wingfield, J. C. (1988). Immunohistochemical localization of neuropeptides in the vocal control regions of two songbird species. *J. Comp. Neurol.* 265:171–80.

Balleine, B., Davies, A., and Dickinson, A. (1995). Cholecystokinin attenuates incentive learning in rats. *Behav. Neurosci.* 109:312–19.

Balthazart, J., Castagna, C., and Ball, G. F. (1997). Aromatase inhibition blocks the activation and sexual differentiation of appetitive male sexual behavior in Japanese quail. *Behav. Neurosci.* 111:281–97.

Balthazart, J., Reid, J., Absil, P., and Foidart, R. (1995). Appetitive as well as consummatory aspects of male sexual behavior in quail are activated by androgens and estrogens. *Behav. Neurosci.* 109:485–501.

Bamshad, M., and Albers, H. E. (1996). Neural circuitry controlling vasopressin-stimulated scent marking in Syrian hamsters. *J. Comp. Neurol.* 369:262–3.

Bamshad, M., Novak, M. A., and DeVries, G. J. (1994). Cohabitation alters vasopressin innervation and paternal behavior in prairie voles (*Microtus ochrogaster*). *Physiol. Behav.* 56:751–8.

Baram, T. Z., Hirsch, E., Snead, O. C., and Schultz, L. (1992). Corticotropin-releasing hormone-induced seizures in infant rats originate in the amygdala. *Ann. Neurol.* 31: 489–94.

Bard, P. (1939). Central nervous mechanisms for emotional behavior patterns in animals. *Res. Nerv. Ment. Dis.* 29:190–216.

Bard, P. (1940). The hypothalamus and sexual behavior. *Res. Nerv. Ment. Dis.* 20:551–79.

Barnet, S. A. (1963). *The Rat: A Study in Behaviour.* Chicago: Aldine.

Barraclough, C. A. (1961). Production of anovulatory, sterile rats by single injections of testosterone propionate. *Endocrinology* 68:62–7.

Barraclough, C. A. (1962). Studies on mating behaviour in the androgen-sterilized female rat in relation to the hypothalamic regulation of sexual behaviour. *J. Endocrinol.* 25:175–82.

Bartness, T. J., Powers, J. B., Hastings, M. H., Bittman, E. L., and Goldman, B. D. (1993).

The timed infusion paradigm for melatonin delivery: What has it taught us about the melatonin signal, its reception, and the photoperiodic control of seasonal responses? *J. Pineal Res.* 15:161–90.

Bartness, T. J., and Wade, G. N. (1984). Photoperiodic control of body weight and energy metabolism in Syrian hamsters (*Mesocricetus auratus*): role of pineal gland, melatonin, gonads, and diet. *Endocrinology* 114:492–8.

Bartoshuk, L. M. (1988). Clinical psychophysics of taste. *Gerodontics* 4:249–55.

Bassett, J. R., Cairncross, K. D., and King, M. G. (1973). Parameters of novelty, shock predictability and response contingency in corticosterone release in the rat. *Physiol. Behav.* 10:901–7.

Bauer, M. S., Droba, M., and Whybrow, P.(1987). Disorders of the thyroid and parathyroid. In: *Handbook of Psychoneuroendocrinology,* ed. C. B. Nemeroff and P. T. Loosen. New York: Guilford Press.

Baulieu, E.-E., and Robel, P. (1990). Neurosteroids: a new brain function? *J. Steroid Biochem. Mol. Biol.* 37:395–403.

Baum, M. J. (1992). Neuroendocrinology of sexual behavior in the male. In: *Behavioral Endocrinology,* ed. J. B. Becker, S. M. Breedlove, and D. Crews. MIT Press.

Beach, F. A. (1942). Central nervous mechanisms involved in the reproductive behaviors of vertebrates. *Psychol. Bull.* 26:200–25.

Beach, F. A. (1948). *Hormones and Behavior.* New York: Hoeber.

Beach, F. A. (1967). Cerebral and hormonal control of reflexive mechanisms involved in copulatory behavior. *Physiol. Rev.* 47:289–316.

Beach, F. A. (1984). Hormonal modulation of genital reflexes in male and masculinized female dogs. *Behav. Neurosci.* 98:325–32.

Beauchamp, G. K., Bertino, M., Burke, D., and Engelman, K. (1990). Experimental sodium depletion and salt taste in normal human volunteers. *Am. J. Clin. Nutr.* 51: 881–9.

Beaulieu, S., DiPaolo, T., Cot, J., and Barden, N. (1987). Participation of the central amygdaloid nucleus in the response of adrenocorticotropin secretion to immobilization stress: opposing roles of the noradrenergic and dopaminergic systems. *Neuroendocrinology* 45:37–46.

Beaulieu, S., Rousse, I., Gratton, A., Barden, N., and Rochford, J. (1994). Behavioral and endocrine impact of impaired type II glucocorticoid receptor function in a transgenic mouse model. *Ann. N.Y. Acad. Sci.* 746:388–91.

Beck, A. T. (1978). *Depression.* University of Pennsylvania Press. (Originally published 1967.)

Beck, U. (1981). Hormonal secretion during sleep in man. *Int. J. Neurol.* 15:17–29.

Becker, J. B. (1992). Hormonal influences on extrapyramidal sensorimotor function and hippocampal plasticity. In: *Behavioral Endocrinology,* ed. J. B. Becker, S. M. Breedlove, and D. Crews. MIT Press.

Beletsky, L. D., Orians, G. H., and Wingfield, J. C. (1992). Year-to-year patterns of circulating levels of testosterone and corticosterone in relation to breeding density, experience, and reproductive success of the polygynous red-winged blackbird. *Horm. Behav.* 26:420–32.

Bell, I. R., Martino, G. M., Meredith, K. E., Schwartz, G. E., Siani, M. M., and Morrow, F. D. (1993). Vascular disease risk factors, urinary free cortisol, and health histories in older adults: shyness and gender interactions. *Biol. Psychol.* 35:37–49.

Belovsky, G. E. (1978). Diet optimization in a generalist herbivore: the moose. *Theor. Popul. Biol.* 14:105–34.

Benler, S. L., and Johnson, A. K. (1979). Sodium consumption following lesions surrounding the anteroventral third ventricle. *Brain Res. Bull.* 4:287–90.

Bentley, P. J. (1982). *Comparative Vertebrate Endocrinology.* Cambridge University Press.

Berenbaum, S. A., and Snyder, E. (1995). Early hormonal influences on childhood sex-typed activity and playmate preferences: implications for the development of sexual orientation. *Dev. Psychol.* 31:31–42.

Bern, H. A. (1967). Hormones and endocrine glands of fishes. *Science* 158:455–69.

Bernard, C. A. (1957). *An Introduction to the Study of Experimental Medicine.* New York: Dover. (Originally published 1865.)

Berridge, K. C. (1996). Food reward: brain substrates of wanting and liking. *Neurosci. Biobehav. Rev.* 20:1–25.

Berridge, K. C., Flynn, F. W., Schulkin, J., and Grill, H. J. (1984). Sodium depletion enhances salt palatability in rats. *Behav. Neurosci.* 98:652–60.

Berridge, K. C., and Grill, H. J. (1984). Isohedonic tastes support a two-dimensional hypothesis of palatability. *Appetite* 5:221–31.

Berridge, K. C., Grill, H. J., and Norgren, R. (1981). Relation of consummatory responses and preabsorptive insulin release to palatability and taste aversions. *J. Comp. Physiol. Psychol.* 95:363–82.

Berridge, K. C., and Schulkin, J. (1989). Palatability shift of a salt-associated incentive drive during sodium depletion. *Q. J. Exp. Psychol.* 41B:121–38.

Berridge, K. C., and Valenstein, E. S. (1991). What psychological process mediates feeding evoked by electrical stimulation of the lateral hypothalamus? *Behav. Neurosci.* 105:3–14.

Bidmon, H. J., and Stumpf, W. E. (1994). Distribution of target cells for 1,25-dihydroxyvitamin D_3 in the brain of the yellow bellied turtle *Trachemys scripta*. *Brain Res.* 640:277–85.

Billington, C. J., Briggs, J. E., Harker, S., Grace, M., and Levine, A. S. (1994). Neuropeptide Y in hypothalamic paraventricular nucleus: a center coordinating energy metabolism. *Am. J. Physiol.* 266:R1765–70.

Birmingham, M. K., Sar, M., and Stumpf, W. E. (1984). Localization of aldosterone and corticosterone in the central nervous system, assessed by quantitative autoradiography. *Neurochem. Res.* 9:333–50.

Bissette, G., Griff, D., Carnes, M., Goodman, B., Lavine, M., and Levant, B. (1995). Apparent seasonal rhythms in hypothalamic neuropeptides in rats without photoperiod changes. *Endocrinology* 136:622–7.

Bittman, E. L., Bartness, T. J., Goldman, B. D., and DeVries, G. J. (1991). Suprachiasmatic and paraventricular control of photoperiodism in Siberian hamsters. *Am. J. Physiol.* 260:R90–101.

Bittman, E. L., Thomas, E. M., and Zucker, I. (1994). Melatonin binding sites in sciurid and hystricomorph rodents: studies on ground squirrels and guinea pigs. *Brain Res.* 648:73–9.

Black, R. M., Weingarten, H. P., Epstein, A. N., Maki, R., and Schulkin, J. (1992). Transection of the stria terminalis without damage to the medial amygdala does not alter sodium regulation in rats. *Acta Neurobiol. Exp.* (*Warsz.*) 52:9–15.

Blackburn, R. E., Demko, A. D., Hoffman, G. E., Stricker, E. M., and Verbalis, J. G. (1992). Central oxytocin inhibition of angiotensin-induced salt appetite in rats. *Am. J. Physiol.* 2633:R1347–53.

Blackburn, R. E., Samson, W. K., Fulton, R. J., Stricker, E. M., and Verbalis, J. G. (1993).

Central oxytocin inhibition of salt appetite in rats: evidence for differential sensing of plasma sodium and osmolality. *Proc. Natl. Acad. Sci. U.S.A.* 90:10380–4.

Blaine, E. H., Covelli, M. D., Denton, D. A., Nelson, J. F., and Schulkes, A. A. (1975). The role of ACTH and adrenal glucocorticoids in the salt appetite of wild rabbits. *Coryctolagus cuniculus. Endocrinology* 97:793–801.

Blair-West, J. R., Coghlan, J. P., Denton, D. A., Goding, J. R., and Wright, R. D. (1963). The effect of aldosterone, cortisol, and corticosterone upon the sodium and potassium content of sheep's parotid saliva. *J. Clin. Invest.* 4:484–92.

Blair-West, J. R., Denton, D. A., McBurnie, M., Tarjan, E., and Weisinger, R. S. (1995). Influence of adrenal steroid hormones on sodium appetite of Balb/c mice. *Appetite* 24:11–24.

Blair-West, J. R., Denton, D. A., McBurnie, M. F., and Weisinger, R. S. (1996). The effect of ACTH on water intake in mice. *Physiol. Behav.* 60:1053–6.

Blanchard, D. C., and Blanchard, R. J. (1972). Innate and conditioned reactions to threat in rats with amygdaloid lesions. *J. Comp. Physiol. Psychol.* 81:281–90.

Blass, E. M. (1994). Behavioral and physiological consequences of suckling in rat and human newborns. *Acta Paediatr. [Suppl.]* 397:71–6.

Blass, E. M., and Shide, D. J. (1993). Endogenous cholecystokinin reduces vocalization in isolated 10-day-old rats. *Behav. Neurosci.* 107:484–92.

Blaustein, J. D., King, J. C., Toft, D. O., and Turcotte, J. (1988). Immunocytochemical localization of estrogen-induced progestin receptors in guinea pig brain. *Brain Res.* 474:1–15.

Blaustein, J. D., Tetel, M. J., and Meredith, J. M. (1995). Neurobiological regulation of hormonal response by progestin and estrogen receptors. In: *Neurobiological Effects of Sex Steroid Hormones*, ed. P. E. Micevych and R. P. Hammer, Jr. Cambridge University Press.

Blaustein, J. D., Tetel, M. J., Ricciardi, K. H., Delville, Y., and Turcotte, J. C. (1994). Hypothalamic ovarian steroid hormone-sensitive neurons involved in female sexual behavior. *Psychoneuroendocrinology* 19:505–16.

Bligh, M. E., Douglass, L. W., and Castonguay, T. W. (1993). Corticosterone modulation of dietary selection patterns. *Physiol. Behav.* 53:975–82.

Bohner, J., Harding, C. F., and Marler, P. (1992). Androstenedione therapy reinstates normal, not supernormal, song structure in castrated adult male zebra finches. *Horm. Behav.* 26:136–42.

Boling, J. L., and Blandau, R. J. (1939). The estrogen-progesterone induction of mating responses in the spayed female rat. *Endocrinology* 25:359–64,

Borer, K. T., Bestervelt, L. L., Mannheim, M., Brosamer, M. B., Thomson, M., Swamy, U., and Piper, W. N. (1992). Stimulation by voluntary exercise of adrenal glucocorticoid secretion in mature female hamsters. *Physiol. Behav.* 51:713–18.

Borer, K. T., Ketch, R. P., and Corley, K. (1982). Hamster prolactin: physiological changes in blood and pituitary concentrations as measured by a homologous radioimmunoassay. *Neuroendocrinology* 35:13–21.

Borromeo, V., Cremonesi, F., Perucchetti, E., Berrini, A., and Secchi, C. (1994). Circadian and circannual plasma secretory patterns of growth hormone and prolactin Frisian heifers: hormonal profiles and signal analysis. *Comp. Biochem. Physiol.* 107A:313–21.

Botchin, M. B., Kaplan, J. R., Manuck, S. B., and Mann, J. J. (1993). Low versus high prolactin responders to fenfluramine challenge: marker of behavioral differences in adult male cynomolgus macaques. *Neuropsychopharmacology* 9:93–9.

Bowlby, J. (1973). *Attachment and Loss. Vol. 11: Separation*. New York: Basic Books.

Boyd, S. K., and Moore, F. L. (1991). Gonadectomy reduces the concentrations of putative receptors for arginine vasotocin in the brain of an amphibian. *Brain Res.* 541: 193–7.

Brady, L. S., Lynn, A. B., Herkenham, M., and Gottesfeld, Z. (1994). Systemic interleukin-1 induces early and late patterns of c-fos mRNA expression in brain. *J. Neurosci.* 14:4951–64.

Brady, L. S., Smith, M. A., Gold, P. W., and Herkenham, M. (1990). Altered expression of hypothalamic neuropeptide mRNAs in food-restricted and food-deprived rats. *Neuroendocrinology* 52:441–7.

Brainard, G. C., Asch, R. H., and Reiter, R. J. (1981). Circadian rhythms of serum melatonin and prolactin in the rhesus monkey (*Macaca mulatta*). *Biomed. Res.* 2:291–7.

Brands, M. W., and Freeman, R. H. (1988). Aldosterone and renin inhibition by physiological levels of atrial natriuretic factor. *Am. J. Physiol.* 23:R1011–16.

Brantley, R. K., Wingfield, J. C., and Bass, A. H. (1993). Sex steroid levels in *Porichthys novatus,* a fish with alternative reproductive tactics, and a review of the hormonal bases for male dimorphism among teleost fishes. *Horm. Behav.* 27:332–47.

Braun, C. B., Wicht, H., and Northcutt, R. G. (1995). Distribution of gonadotropin releasing hormone immunoreactivity in the brain of the Pacific hagfish, *Eptatretus stouti* (Craniata: Myxinoidea). *J. Comp. Neurol.* 253:464–76.

Braun-Menendez, E. (1953). Modificadores del apetito especifico para la sal en ratas blancas. *Rev. Soc. Argent. Biol.* 11:92–102.

Braun-Menendenz, E., and Brandt, P. (1952). Aumento del apetito especifico para lasal provocada por la desoxicorticosterona. 1 caraacteristicas. *Rev. Soc. Argent. Biol.* 28: 15–23.

Bray, G. A. (1992). Peptides affect the intake of specific nutrients and the sympathetic nervous system. *Am. J. Clin. Nutr.* 55:265S–71S.

Breedlove, S. M. (1992). Sexual dimorphism in the vertebrate nervous system. *J. Neurosci.* 12:4133–42.

Brenner, L., and Ritter, R. C. (1995). Peptide cholecystokinin receptor agonist increases food intake in rats. *Appetite* 24:1–9.

Brenowitz, E. A. (1991). Altered perception of species-specific song by female birds after lesions of a forebrain nucleus. *Science* 251:303–5.

Brenowitz, E. A., Nails, B., Wingfield, J. C., and Kroodsma, D. E. (1991). Seasonal changes in avian song nuclei without seasonal changes in song repertoire. *J. Neurosci.* 11:1367–74.

Bridges, R. S., and Freemark, M. S. (1995). Human placental lactogen infusions into the medial preoptic area stimulate maternal behavior in steroid-primed, nulliparous female rats. *Horm. Behav.* 29:216–26.

Bridges, R. S., and Mann, P. E. (1994). Prolactin–brain interactions in the induction of maternal behavior in rats. *Psychoneuroendocrinology* 19:611–22.

Bridges, R. S., Numan, M., Ronsheim, P. M., Mann, P. E., and Lupini, C. E. (1990). Central prolactin infusions stimulate maternal behavior in steroid-treated, nulliparous female rats. *Proc. Natl. Acad. Sci. U.S.A.* 87:8003–7.

Bridges, R. S., Robertson, M. C., Shiu, R. P. C., Friesen, H. G., Stuer, A. M., and Mann, P. E. (1996). Endocrine communication between conceptus and mother: placental lactogen stimulation of maternal behavior. *Neuroendocrinology* 64:57–64.

Bridges, R. S., and Ronsheim, P. M. (1990). Prolactin (PRL) regulation of maternal behavior in rats: bromocriptine treatment delays and PRL promotes the rapid onset of behavior. *Endocrinology* 126:837–48.

Brier, A. (1989). Experimental approaches to human stress research: assessment of neurobiological mechanisms of stress in volunteers and psychiatric patients. *Biol. Psychiatry* 26:438–62.

Brier, A., Albus, M., Pickar, D., Zahn, T. P., and Wolkowitz, O. M. (1987). Controllable and uncontrollable stress in humans: alterations in mood and neuroendocrine and psychophysiological function. *Am. J. Psychiatry* 144:1419–25.

Brinton, R. E., and McEwen, B. S. (1988). Regional distinctions in the regulation of type I and type II adrenal steroid receptors in the central nervous system. *Neurosci. Res. Commun.* 2:37–45.

Brisson, G. R., Boisvert, P., Peronnet, F., Quirion, A., and Senecal, L. (1989). Face cooling-induced reduction of plasma prolactin response to exercise as part of an integrated response to thermal stress. *Eur. J. Appl. Physiol.* 58:816–20.

Britton, K. T., Wylie Vale, G. L., Rivier, J., and Koob, G. F. (1986). Corticotropin releasing factor (CRF) receptor antagonist blocks activating actions of CRF in the rat. *Brain Res.* 369:303–6.

Brommage, R., and DeLuca, H. F. (1984). Self-selection of a high calcium diet by vitamin D-deficient lactating rats increases food consumption and milk production. *Am. Inst. Nutr.* 114:1377–85.

Brookhart, J. M., and Dey, F. L. (1941). Reduction of sexual behavior in male guinea pigs by hypothalamic lesions. *Am. J. Physiol.* 133:551–4.

Brown, J. E., and Toma, R. B. (1986). Taste changes during pregnancy. *Am. J. Clin. Nutr.* 43:414.

Brown, R. P., Stoll, P. M.., Strokes, P. E., Frances, A., Sweeny, J., Kocals, J. H., and Mann, J. J. (1988). Adrenocortical hyperactivity in depression: effects of agitation, delusions, melancholia and other illness variables. *Psychiatry Res.* 23:176–8.

Bruce, B. K., King, B. M., Phelps, G. R., and Veitia, M. C. (1982). Effects of adrenalectomy and corticosterone administration on hypothalamic obesity in rats. *Am. J. Physiol.* 243:E152–7.

Bruhn, T. O., Plotsky, P. M., and Vale, W. W. (1984). Effects of paraventricular lesions on corticotropin-releasing-factor-like immunoreactivity in the stalk–median eminence: studies on the adrenocorticotrophic hormone response to either stress and exogenous CRF. *Endocrinology* 114:57–65.

Bryant, R. W., Epstein, A. N., Fitzsimons, J. T., and Fluharty, S. J. (1980). Arousal of a specific and persistent sodium appetite in the rat with continuous intracerebroventricular infusion of angiotensin II. *J. Physiol. (Lond.)* 301:365–82.

Buggy, J., and Fisher, A. E. (1974). Evidence for a dual central role for angiotensin in water and sodium intake. *Nature* 250:733–5.

Buggy. J., Huot, S., Pamnani, M., and Haddy, F. (1984). Periventricular forebrain mechanisms for blood pressure regulation. *Fed. Proc.* 43:25–31.

Buggy, J., and Johnson, A. K. (1977). Preoptic-hypothalamic periventricular lesions: thirst deficits and hypernatremia. *Am. J. Physiol.* 5:R44–52.

Bullock, T. H. (1984). Comparative neuroethology of startle, rapid escape, and giant fiber-mediated responses. In: *Neural Mechanisms of Startle Behavior,* ed. R. C. Eaten. New York: Plenum.

Bünger, L., and Hill, W. G. (1997). Effects of leptin administration on long-term selected fat mice. *Genet. Res. (Camb.)* 69:215–25.

Buntin, J. D. (1992) Neural substrates for prolactin-induced changes in behavior and neuroendocrine function. *Poultry Science Reg.* 4:275–87.

Buntin, J. D., Ruzycki, E., and Witebsky, J. (1993). Prolactin receptors in dove brain: autoradiographic analysis of binding characteristics in discrete brain regions and accessibility to blood-borne prolactin. *Neuroendocrinology* 57:738–50.

Burmester, J. K., Wiese, R. J., Maeda, N., and DeLuca, H. F. (1988). Structure and regulation of the rat 1,25-dihydroxyvitamin. *Proc. Natl. Acad. Sci. U.S.A.* 85:9499–502.

Burris, A. S., Banks, S. M., Carter, C. S., Davidson, J. M., and Sherins, R. J. (1992). A long-term, prospective study of the physiologic and behavioral effects of hormone replacement in untreated hypogonadal men. *J. Andrology* 13:297–304.

Butera, P. C., Xiong, M., Davis, R. J., and Platania, S. P. (1996). Central implants of dilute estradiol enhance the satiety effect of CCK-8. *Behav. Neurosci.* 110:823–30.

Buttemer, W. A., Astheimer, I. E., and Wingfield, J. D. (1991). The effect of corticosterone on standard metabolic rates of small passerine birds. *J. Physiol. (Lond.)* 161:427–31.

Cabanac, M. (1971). Physiological role of pleasure. *Science* 173:1403–7.

Cagnacci, A., Soldani, R., Romagnolo, C., and Yen, S. S. C. (1994). Melatonin-induced decrease of body temperature in women: a threshold event. *Neuroendocrinology* 60:549–52.

Cahill, L., Prins, B., Weber, M., and McGaugh, J. L. (1994). β-adrenergic activation and memory for emotional events. *Nature* 371:702–4.

Caldwell, J. D., Jirikowski, G. F., Greer, E. R., Stumpf, W. E., and Pedersen, C. A. (1988). Ovarian steroids and sexual interaction alter oxytocinergic content and distribution in the basal forebrain. *Brain Res.* 446:236–44.

Caldwell, J. D., Johns, J. M., Faggin, B. M., Senger, M. A., and Pedersen, C. A. (1994). Infusion of an oxytocin antagonist into the medial preoptic area prior to progesterone inhibits sexual receptivity and increases rejection in female rats. *Horm. Behav.* 28:288–302.

Caldwell, J. D., Walker, C. H., Pedersen, C. A., Barakat, A. S., and Mason, G. A. (1994). Estrogen increases affinity of oxytocin receptors in the medial preoptic area–anterior hypothalamus. *Peptides* 15:1079–84.

Caldwell, J. D., Walker, C. H., Pedersen, C. A., and Mason, G. A. (1992). Effects of steroids and mating on central oxytocin receptors. *Ann. N.Y. Acad. Sci.* 652:433–6.

Camacho, A., and Phillips, M. I. (1981). Horseradish peroxidase study in rat of the neural connections of the organum vasculosum of the lamina terminalis. *Neurosci. Lett.* 25:201–4.

Camfield, L. A., Smith, F. J., Guisez, Y., Devos, R., and Burn, P. (1995). Recombinant mouse OB protein: evidence for a peripheral signal linking adiposity and central neural networks. *Science* 269:546–9.

Cannon, W. B. (1932). *The Wisdom of the Body.* New York: Norton.

Canteras, N. S., Simerly, R. B., and Swanson, L. W. (1992). Connections of the posterior nucleus of the amygdala. *J. Comp. Neurol.* 324:143–79.

Cantin, M., and Genest, J. (1985). The heart and the atrial natriuretic factor. *Endocrinol. Rev.* 6:107–27.

Carlberg, C., Bendik, I., Wyss, A., Meier, E., Sturzenbecker, L. J., Grippo, J. F., and Hunziker, W. (1993). Two nuclear signalling pathways for vitamin D. *Nature* 361:657–60.

Carrick, S., and Balment, R. J. (1983). The renin-angiotensin system and drinking in the Euryhaline flounder, *Platichthys flesus. Gen. Comp. Endocrinol.* 51:423–33.

Carroll, B. J., Curtis, G. C., Davies, B. M., Mendels, J., and Sugarman, A. A. (1976). Urinary free cortisol excretion in depression. *Psychol. Med.* 6:43–50.

Carter, C. S. (1992). Oxytocin and sexual behavior. *Neurosci. Biobehav. Rev.* 16:131–44.

Carter, C. S. (1994). Integrative functions of lactational hormones in social behavior and stress management: animal and human models. Presented at the International and Interdisciplinary Seminar on Women, Stress and Heart Disease, Stockholm, Sweden.

Carter, C. S., and Altemus, M. (1997). Integrative functions of lactational hormones in social behavior and stress management. *Ann. N.Y. Acad. Sci.* 807:164–74.

Carter, C. S., DeVries, A. C., and Getz, L. L. (1995). Physiological substrates of mammalian monogamy: the prairie vole model. *Neurosci. Biobehav. Rev.* 19:303–14.

Cashdan, E. (1995). Hormones, sex, and status in women. *Horm. Behav.* 29:354–66.

Cassone, V. M. (1990). Effects of melatonin on vertebrate circadian systems. *Trends Neurosci.* 13:457–64.

Cassone, V. M. (1991). Melatonin and suprachiasmatic nucleus function. In: *Suprachiasmatic Nucleus: The Mind's Clock,* ed. P. Klein. Oxford University Press.

Cassone, V. M., Brooks, D. S., and Kelm, T. A. (1995). Comparative distribution of melatonin binding in the brains of diurnal birds: outgroup analysis with turtles. *Brain Behav. Evol.* 45:241–56.

Castonguay, T. W., Dallman, M. F., and Stern, J. S. (1986). Some metabolic and behavioral effects of adrenalectomy on obese Zucker rats. *Am. J. Physiol.* 20:R923–9.

Catz, D. S., Fischer, L. M., Moschella, M. C., Tobias, M. L., and Kelley, D. B. (1992). Sexually dimorphic expression of a laryngeal-specific, androgen-related myosin heavy chain gene during *Xenopus laevis* development. *Dev. Biol.* 154:366–76.

Chalmers, D. T., Kwak, S. P., Mansour, A., Akil, H., and Watson, S. J. (1993). Corticosteroids regulate brain hippocampal D-HT receptor mRNA. *J. Neurosci.* 13:914–23.

Champoux, M. B., Coe, C. L., Schanberg, S. M., Kihn, C. M., and Suomi, S. J. (1989). Hormonal effects of early rearing conditions in the infant rhesus monkey. *Am. J. Primatol.* 19:111–18.

Chan, R. K. W., Brown, E. R., Ericsson, A., Kovacs, K. J., and Sawchenko, P. E. (1993). A comparison of two immediate-early genes, c-fos and NGFI-B, as markers for functional activation in stress-related neuroendocrine circuitry. *J. Neurosci.* 13:5126–38.

Chao, H. M., Blanchard, D. C., Blanchard, R. J., McEwen, B. S., and Sakai, R. R. (1993). The effect of social stress on hippocampal gene expression. *Mol. Cell. Neurosci.* 4:543–8.

Chao, H. M., and McEwen, B. S. (1990). Glucocorticoid regulation of preproenkephalin messenger ribonucleic acid in the rat striatum. *Endocrinology* 126:3124–30.

Chappell, P. B., Smith, M. A., Kilts, C. D., Bissett, G., Ritchie, J., Anderson, C., and Nemeroff, C. B. (1986). Alterations in corticotropin-releasing factor like immunoreactivity in discrete rat brain regions after acute and chronic stress. *J. Neurosci.* 10:2908–14.

Chavez, M., Kaiyala, K., Madden, L. J., Schwartz, M. W., and Woods, S. C. (1995). Intraventricular insulin and the level of maintained body weight in rats. *Behav. Neurosci.* 109:528–31.

Chen, R., Lewis, K. A., Perrin, M. H., and Vale, W. W. (1993). Expression cloning of a human corticotropin-releasing-factor receptor. *Biochemistry* 90:8967–71.

Cheney, D. L., and Seyfarth, R. M. (1990). *How Monkeys See the World.* University of Chicago Press.

Cheng, M.-F., and Silver, R. (1975). Estrogen-progesterone regulation of nest-building and incubation behavior in ovariectomized ring doves (*Streptopelia risoria*). *J. Comp. Physiol. Psychol.* 88:256–63.

Chiaraviglio, E. (1971). Amygdaloid modulation of sodium chloride and water intake in the rat. *J. Comp. Physiol. Psychol.* 76:401–7.

Chiaraviglio, E. (1976). Effects of renin-angiotensin system on sodium intake. *J. Physiol. (Lond.)* 25D:57–66.

Chiaraviglio, E., and Perez Guaita, M. F. (1984). Anterior third ventricle (A3v) lesions and homeostasis-regulation. *J. Physiol. (Lond.)* 79:446–52.

Chomsky, N. (1972). *Language and Mind.* New York: Harcourt Brace Jovanovich.

Chow, S. Y., Sakai, R. R., Witcher, J. A., Adler, N. T., and Epstein, A. N. (1992). Sex and sodium intake in the rat. *Behav. Neurosci.* 106:172–80.

Chrousos, G. P. (1995). The hypothalamic-pituitary-adrenal axis and immune-mediated inflammation. *N. Engl. J. Med.* 332:1351–62.

Cintra, A., Fuxe, K., Solfrini, V., Agnati, L. F., Tinner, B., Wilkstrom, A. C., Staines, W., Okret, S., and Gustafsson, J. A. (1991). Central peptidergic neurons as targets for glucocorticoid action: evidence for the presence of glucocorticoid receptor immunoreactivity in various types of classes of peptidergic neurons. *J. Steroid Biochem. Mol. Biol.* 40:93–103.

Cizza, G., Calogero, A. E., Brady, L. S., Bagdv, G., Bergamini, E., Blackman, M. R., Chrousos, G. P., and Gold, P. W. (1994). Male Fisher 344/N rats show a progressive central impairment of the hypothalamic-pituitary-adrenal axis with advancing age. *Endocrinology* 134:1611–20.

Clark, A. S., and Goldman-Rakic, P. S. (1989). Gonadal hormones influence the emergence of cortical function in nonhuman primates. *Behav. Neurosci.* 103:1287–95.

Clark, J. T., Kalra, P. S., Crowley, W. R., and Kalra, S. P. (1984). Neuropeptide Y and human pancreatic polypeptide stimulate feeding behavior in rats. *Endocrinology* 115:427–32.

Clark, J. T., Kalra, P. S., and Kalra, S. P. (1985). Neuropeptide Y stimulates feeding but inhibits sexual behavior in rats. *Endocrinology* 117:2435–42.

Clark, M. M., Bishop, A. M., Vom Saal, F. S., and Galef, B. G., Jr. (1993). Responsiveness to testosterone of male gerbils from known intrauterine positions. *Physiol. Behav.* 53:1183–7.

Clark, M. M., and Galef, B. G., Jr. (1995). Prenatal influences on reproductive life history strategies. *Trends Ecol. Evol.* 10:151–3.

Clark, M. M., Vom Saal, F. S., and Galef, B. G., Jr. (1992). Intrauterine positions and testosterone levels of adult male gerbils are correlated. *Physiol. Behav.* 51:957–60.

Cloues, R. F., Ramos, C., and Silver, R. (1990). Vasoactive intestinal polypeptide-like immunoreactivity during reproduction in doves: influence of experience and number of offspring. *Horm. Behav.* 24:215–31.

Coe, C. L., Weiner, C. G., Rosenberg, L. T., and Levine, S. L. (1985). Physiological consequences of maternal separation and loss in the squirel monkey. In: *Handbook of the Squirrel Monkey,* ed. L. A. Rosenbloom and C. L. Coe. New York: Plenum Press.

Coghill, G. E. (1929). *Anatomy and the Problem of Behavior.* Cambridge University Press.

Coirini, H., Marusie, E. T., DeNicola, A. F., Rainbow, T. C., and McEwen, B. S. (1983). Identification of mineralocorticoid binding sites in rat brain by competition studies and density gradient centrifugation. *Neuroendocrinology* 37:354–60.

Coirini, H., Schulkin, J., and McEwen, B. S. (1988). Behavioral and neuroendocrine regulation of mineralocorticoid and glucocorticoid action. *Society for Neuroscience Abstracts.*

Cole, B. J., Cador, M., Stinus, L., Rivier, J., Vale, W., and Koob, G. F. (1990). Central administration of a CRF antagonist blocks the development of stress-induced behavioral sensitization. *Brain Res.* 512:343–6.

Colson, P., Ibarondo, J., Devilliers, G., and Balestre, M. N. (1992). Upregulation of vasopressin receptors by glucocorticoids. *Am. J. Physiol.* 26:E1054–62.

Considine, R. V., Sinha, M. K., Heiman, M. L., Kriauciunas, A., Stephens, T. W., Nyce, M. R., Ohannesian, J. P., Marco, C. C., McKee, L. J., Bauer, T. L., and Caro, J. F. (1996). Serum immunoreactive-leptin concentrations in normal-weight and obese humans. *N. Engl. J. Med.* 334:292–5.

Conti, L. H., and Foote, S. L. (1995). Effects of pretreatment with corticotropin-releasing factor on the electrophysiological responsivity of the locus coeruleus to subsequent corticotropin-releasing factor challenge. *Neuroscience* 69:209–19.

Contreras, R. J. (1977). Changes in gustatory nerve discharges with sodium deficiency: a single unit analysis. *Brain Res.* 121:373–8.

Contreras, R. J. (1993). High NaCl intake of rat dams alters maternal behavior and elevates blood pressure of adult offspring. *Am. J. Physiol.* 264:R296–304.

Cooney, A. S., and Fitzsimons, J. T. (1993). The effect of the putative AT2 agonist on angiotensin II, on thirst and sodium appetite in rats. *Exp. Physiol.* 78:767–74.

Coplan, J. D., Andrews, M. W., Rosenblum, L. A., Owens, M. J., Friedman, S., Gorman, J. M., and Nemeroff, C. B. (1996). Persistent elevations of cerebrospinal fluid concentrations of corticotrophin-releasing factor in adult nonhuman primates exposed to early-life stressors: implications for the pathophysiology of mood and anxiety disorders. *Proc. Natl. Acad. Sci. U.S.A.* 93:1619–23.

Corodimas, K. P., LeDoux, J. E., Gold, P., and Schulkin, J. (1994). Corticosterone potentiation of conditioned fear. In: *Brain Corticosteroid Receptor,* ed. R. DeKloet, E. C. Azmita, and P. W. Landfield. New York Academy of Sciences Press.

Corodimas, K. P., Rosenblatt, J. S., Canfield, M. E., and Morrell, J. I. (1993). Neurons in the lateral subdivision of the habenular complex mediate the hormonal onset of maternal behavior in rats. *Behav. Neurosci.* 107:827–43.

Corp, E. S., Woods, S. C., Porte, D., Jr., Dorsa, D. M., Figlewiez, D. P., and Baskin, D. G. (1986). Localization of insulin binding sites in the rat hypothalamus by quantitative autoradiography. *Neurosci. Lett.* 70:17–22.

Corwin, R. L., Robinson, J. K., and Crawley, J. N. (1993). Galanin antagonists block galanin induced feeding in the hypothalamus and amygdala of the rat. *Eur. J. Neurosci.* 5:1528–33.

Crabbe, J. (1961). Stimulation of active sodium transport by the isolated toad bladder with aldosterone in vitro. *J. Clin. Invest.* 40:2103–10.

Craig, W. C. (1918). Appetites and aversions as constituents of instincts. *Biol. Bull.* 34:91–107.

Cratty, M. S., Ward, H. E., Johnson, E. A., Azzaro, A. J., and Birkle, D. L. (1995). Prenatal stress increases corticotropin-releasing factor (CRF) content and release in rat amygdala minces. *Brain Res.* 675:297–302.

Crawley, J. N., Austin, M. C., Fiske, S. M., Martin, B., Consolo, S., Berthold, M., Langel, U., Fisone, G., and Bartfai, T. (1990). Activity of centrally administered galanin fragments on stimulation of feeding behavior and on galanin receptor binding in the rat hypothalamus. *J. Neurosci.* 10:3695–700.

Crawley, J. N., Fiske, S. M., Durieux, C., Derrien, M., and Roques, B. P. (1991). Centrally administered cholecystokinin suppresses feeding through a peripheral-type receptor mechanism. *J. Pharmacol. Exp. Ther.* 257:1076–80.

Crawley, J. N., and Kiss, J. Z. (1985). Paraventricular nucleus lesions abolish the inhibition of feeding induced by systemic cholecystokinin. *Peptides* 6:927–35.

Crenshaw, B. J., DeVries, G. J., and Yahr, P. (1992). Vasopressin innervation of sexually dimorphic structures of the gerbil forebrain under various hormonal conditions. *J. Comp. Neurol.* 322:589–98.

Crews, D. (1991). Trans-seasonal action of androgen in the control of spring courtship behavior in male red-sided garter snakes. *Proc. Natl. Acad. Sci. U.S.A.* 88:3545–8.

Crews, D. (1993). The organizational concept and vertebrates without sex chromosomes. *Brain Behav. Evol.* 42:202–14.

Crews, D. (1997). Species diversity and the evolution of behavioral controlling mechanisms. In: *The Integrative Neurobiology of Affiliation. Ann. N.Y. Acad. Sci.* 807.

Crews, D., Bull, J. J., and Wibbels, T. (1991). Estrogen and sex reversal in turtles: a dose-dependent phenomenon. *Gen. Comp. Endocrinol.* 81:357–64.

Crews, D., Coomber, P., Baldwin, R., Azad, N., and Gonzalez-Lima, F. (1996). Brain organization in a reptile lacking sex chromosomes: effects of gonadectomy and exogenous testosterone. *Horm. Behav.* 30:474–86.

Crews, D., Coomber, P., and Gonzalez-Lima, F. (1997). Effects of age and sociosexual experience on the morphology and metabolic capacity of brain nuclei in the leopard gecko (*Eublepharis macularius*), a lizard with temperature-dependent sex determination. *Brain Res.* 758:169–79.

Crews, D., and Moore, M. D. (1986). Evolution of mechanisms controlling mating behavior. *Science* 231:121–5.

Crews, D., Robker, R., and Mendonca, M. (1993). Seasonal fluctuations in brain nuclei in the red-sided garter snake and their hormonal control. *J. Neurosci.* 13:5356–64.

Cullinan, W. E., Herman, J. P., and Watson, S. J. (1993). Ventral subicular interaction with the paraventricular nucleus of the hypothalamus: evidence for a relay in the bed nucleus of the stria terminalis. *J. Comp. Neurol.* 332:1–20.

Cynx, J., and Nottebohm, F. (1992a). Role of gender, season, and familiarity in discrimination of conspecific song by zebra finches (*Taeniopygia guttata*). *Neurobiology* 89:1368–71.

Cynx, J., and Nottebohm, F. (1992b). Testosterone facilitates some conspecific song discriminations in castrated zebra finches (*Taeniopygia guttata*). *Neurobiology* 89:1376–8.

Cynx, J., Williams, H., and Nottebohm, F. (1992). Hemispheric differences in avian song discrimination. *Proc. Natl. Acad. Sci. U.S.A.* 89:1372–5.

Dachir, S., Kadar, T., Robinson, B., and Levy, A. (1993). Cognitive deficits induced in young rats by long-term corticosterone administration. *Behav. Neural Biol.* 60:103–9.

Dallman, M. F., Akana, S. F., Scribner, K. A., Bradbury, M. J., Walker, C. D., Strack, A. M., and Cascio, C. S. (1992). Stress, feedback and facilitation in the hypothalamic-pituitary-adrenal axis. *J. Neuroendocrinol.* 4:517–26.

Dallman, M. F., Akana, S. F., Strack, A. M., Hanson, E. S., and Sebastian, R. J. (1995). The neural network that regulates energy balance is responsive to glucocorticoids and insulin and also regulates HPA axis responsivity at a site proximal to CRF neurons. *Ann. N.Y. Acad. Sci.* 771:730–42.

Dallman, M. F., Strack, A. M., Akana, S. F., Bradbury, M. J., Hanson, E. S., Scribner, K. A., and Smith, M. (1993). Feast and famine: critical role of glucocorticoids with insulin in daily energy flow. *Front. Neuroendocrinol.* 14:303–47.

Danielson, J., and Buggy, J. (1980). Depression ad lib and angiotensin-induced sodium intake at oestrus. *Brain Res. Bull.* 5:501–4.

Dark, J., Dark, K. A., and Zucker, I. (1987). Long day lengths increase brain weight and DNA content in the meadow vole, *Microtus pennsylvanicus. Brain Res.* 409:302–7.

Dark, J., Kilduff, T. S., Heller, H. C., Licht, P., and Zucker, I. (1990). Suprachiasmatic nuclei influence hibernation rhythms of golden-mantled ground squirrels. *Brain Res.* 509:111–18.

Dark, J., and Zucker, I. (1985). Circannual rhythms of ground squirrels: role of the hypothalamic paraventricular nucleus. *J. Biol. Rhythms* 1:17–23.

Darrow, J. M., Yogev, L., and Goldman, B. D. (1987). Patterns of reproductive hormone secretion in hibernating Turkish hamsters. *Am. J. Physiol.* 253:R329–36.

Darwin, C. (1901). *Descent of Man.* London: Rand McNally. (Originally published 1871.)

Darwin, C. (1958). *The Origin of Species.* New York: New American Library (Mentor Books). (Originally published 1859.)

Darwin, C. (1965). *The Expression of Emotions in Man and Animals.* University of Chicago Press. (Originally published 1872.)

Daugherty, D. R., and Callard, I. P. (1972). Plasma corticosterone levels in the male iguanid lizard *Sceloporus cyanogenys* under various physiological conditions. *Gen. Comp. Endocrinol.* 19:69–79.

Davidson, J. M. (1966). Activation of the male rat's sexual behavior by intracerebral implantation of androgen. *Endocrinology* 84:1365–72.

Davidson, J. M. (1976). The physiology of meditation and mystical states of consciousness. *Perspect. Biol. Med.* 19:345–79.

Davidson, R. J. (1994). Asymmetric brain function, affective style, and psychopathology: the role of early experience and plasticity. *Dev. Psychopathol.* 6:741–58.

Davidson, R. J., Ekman, P., Saron, C. D., Senulis, J. A., and Friesen, W. V. (1990). Approach – withdrawal and cerebral asymmetry: emotional expression and brain physiology. I. *J. Person. Social Psychol.* 58:330–41.

Davis, C. (1939). Results of the self-selection of diets by young children. *Can. Med. Assoc.* pp. 257-61.

Davis, G. A., and Moore, F. L. (1996). Neuroanatomical distribution of androgen and estrogen receptor-immunoreactive cells in the brain of the male roughskin newt. *J. Comp. Neurol.* 372:294–308.

Davis, M., Falls, W. A., Campeau, S., and Kim, M. (1993). Fear-potentiated startle: a neural and pharmacological analysis. *Behav. Brain Res.* 58:175–98.

Davis, M., Walker, D. L., and Lee, Y. (1997). Roles of the amygdala and bed nucleus of the stria terminalis in fear and anxiety measured with the acoustic startle reflex. *Ann. N.Y. Acad. Sci.* 821.

DeBold, A. J., Brownstein, H. B., Veress, A. T., and Sonnenberg, H. (1981). A rapid potent natriuetic response to intravenous injection of atrial myocardial extracts in rats. *Life Sci.* 28:89–94.

De Kloet, E. R. (1991). Brain corticosteroid receptor balance and homeostatic control. *Front. Neuroendocrinol.* 19:95–164.

De Kloet, E. R., Joels, J., Oitzl, M., and Sutanto, W. (1991). Implication of brain corticosteroid receptor diversity for the adaptation syndrome concept. In: *Stress Revisited. Vol. 1: Neuroendocrinology of Stress,* ed. G. Jasmin and M. Cantin, pp. 104–55. Basel: Karger.

De Lacoste-Utamsing, C., and Holloway, R. L. (1982). Sexual dimorphism in the human corpus callosum. *Science* 216:1431–2.

de la Iglesia, H. O., Blaustein, J. D., and Bittman, E. L. (1995). The suprachiasmatic area in the female hamster projects to neurons containing estrogen receptors and GnRH. *Neuroreport* 6:1715–22.

Delmonte, M. M. (1985). Biochemical indices associated with meditation practice: a literature review. *Neurosci. Biobehav. Rev.* 9:557–61.

DeLuca, H. F. (1987). The vitamin D story: a collaborative effort of basic science and clinical medicine. *FASEB J.* 2:224–36.

DeLuca, L. A., Galaverna, O., Schulkin, J., Yao, S. Z., and Epstein, A. N. (1992). The anteroventral wall of the third ventricle and the angiotensinergic component of need-induced sodium intake in the rat. *Brain Res. Bull.* 8:73–87.

Delville, Y., Koh, E. T., and Ferris, C. F. (1994). Sexual differences in the magnocellular vasopressinergic system in golden hamsters. *Brain Res. Bull.* 33:535–40.

Delville, Y., Mansour, K. M., and Ferris, C. F. (1996). Testosterone facilitates aggression by modulating vasopressin receptors in the hypothalamus. *Physiol. Behav.* 60:25–9.

Denef, C., Magnus, C., and McEwen, B. S. (1974). Sex-dependent changes in pituitary 5α-dihydrotestosterone and 3α-androstanediol formation during postnatal development and puberty in the rat. *Endocrinology* 94:1265–74.

Dennett, D. C. (1991). *Consciousness Explained.* Boston: Little, Brown.

Denton, D. A. (1965). Evolutionary aspects of the emergence of aldosterone secretion and salt appetite. *Physiol. Rev.* 45:245–95.

Denton, D. A. (1982). *The Hunger for Salt.* Berlin: Springer-Verlag.

Denton, D. A., Blair-West, J. R., McBurnie, M., Osborne, P. G., Tarjan, E., Williams, R., and Weisinger, R. S. (1990). Angiotensin and salt appetite of BALB/c mice. *Am. J. Physiol.* 28:R729–35.

Denton, D. A., Eichberg, J. W., Shade, R., and Weisinger, R. S. (1993). Sodium appetite in response to sodium deficiency in baboons. *Am. J. Physiol.* 264:R539–43.

Denton, D. A. McKinley, M. J., and Weisinger, R. S. (1996). Hypothalamic integration of body fluid regulation. *Proc. Natl. Acad. Sci. U.S.A.* 93:7397–404.

Denton, D. A., and Nelson, J. F. (1971). Effects of pregnancy and lactation on the mineral appetites of wild rabbits (*Oryctolagus cuniculus L.*). *Endocrinology* 88:31–40.

Denton, D. A., Nelson, J. F., and Tarjan, E. (1985). Water and salt intake of wild rabbits (*Oryctolagus cuniculus L.*) following dipsogenic stimuli. *J. Physiol.* 362:285–301.

DeRijk, R., and Berkenbosch, R. (1994). Suppressive and permissive actions of glucocorticoids; a way to control innate immunity and to facilitate specificity of adaptive immunity? In: *Bilateral Communication Between the Endocrine and Immune Systems,* ed. C. J. Grossman, pp. 73–95. Berlin: Springer-Verlag.

DeRijk, R., and Sternberg, E. M. (1994). Corticosteroid action and neuroendocrine–immune interactions. *Ann. N.Y. Acad. Sci.* 746:33–41.

Deroche, V., Marinelli, M., Maccari, S., LeMoal, M., Simon, H., and Piazza, P. V. (1995). Stress-induced sensitization and glucocorticoids. 1. Sensitization of dopamine-dependent locomotor effects of amphetamine and morphine depends on stress-induced corticosterone secretion. *J. Neurosci.* 15:7181–8.

DeSouza, E. B., Insel, T. R., Perrin, M. H., Rivier, J., Ude, W. W., and Kuhar, W. J. (1985a). Corticotropin-releasing factor receptors are widely distributed within the rat central nervous system: an autoradiographic study. *J. Neurosci.* 5:3189–203.

DeSouza, E. B., Insel, T. R., Perrin, M. H., Rivier, J., Vale, W. W., and Kuhar, M. J. (1985b). Differential regulation of corticotropin-releasing factor receptors in anterior and intermediate lobes of pituitary and in brain following adrenalectomy in rats. *Neurosci. Lett.* 56:121–8.

Dethier, V. G. (1968). Chemosensory input and taste discrimination in the blowfly. *Science* 161:389–91.

Devenport, L., and Stith, R. (1992). Mimicking corticosterone's daily rhythm with specific receptor agonists: effects on food, water, and sodium intake. *Physiol. Behav.* 51: 1247–55.

Devenport, L., Torres, A., and Murray, C. G. (1983). Effects of aldosterone and deoxycorticosterone on food intake and body weight. *Behav. Neurosci.* 97:667–9.

Deviche, P., and Moore, F. L. (1988). Steroid control of sexual behavior in the roughskinned newt: effects of testosterone, estradiol, dihydrotestosterone. *Horm. Behav.* 22:26–34.

Deviche, P., Propper, C. R., and Moore, F. L. (1990). Neuroendocrine, behavioral and morphological changes associated with the termination of the reproductive period in a natural population of male rough skinned newts. *Horm. Behav.* 24:284–300.

DeVries, G. J. (1995). Studying neurotransmitter systems to understand the development and function of sex differences in the brain; the case of vasopressin. In: *Neurobiological Effects of Sex Steroid Hormones,* ed. P. E. Micevych and R. P. Hamer, Jr. Cambridge University Press.

DeVries, G. J., and Buijs, R. M. (1983). The origin of the vasopressinergic and oxytocinergic innervation of the rat brain with special reference to the lateral septum. *Brain Res.* 273:307–17.

DeVries, G. J., Buijs, R. M., and Sluiter, A. A. (1984). Gonadal hormone actions on the morphology of the vasopressinergic innervation of the adult rat brain. *Brain Res.* 298:141–5.

DeVries, G. J., Buijs, R. M., Van Leeuwen, F. W., Caffe, A. R., and Swaab, D. F. (1985). The vasopressinergic innervation of the brain in normal and castrated rats. *J. Comp. Neurol.* 233:236–54.

DeVries, G. J., Wang, Z., Bullock, N. A., and Numan, S. (1994). Sex differences in the effects of testosterone and its metabolites on vasopressin messenger RNA levels in the bed nucleus of the stria terminalis of rats. *J. Neurosci.* 14:1789–94.

Dewey, J. (1894). The theory of emotion. I. Emotional attitudes. *Psychol. Rev.* 1:553–69.

Dewey, J. (1895). The theory of emotion. II. The significance of emotions. *Psychol. Rev.* 2:13–32.

Dewey, J. (1989). *Experience and Nature.* LaSalle, IL: Open Court. (Originally published 1925.)

Diamond, M. C. (1991). Hormonal effects on the development of cerebral lateralization. *Psychoneuroendocrinology* 16:121–9.

Dickinson, A. (1980). *Contemporary Animal Learning.* Cambridge University Press.

Dijkstra, H., Tilders, F. J. H., Hiehle, M. A., and Smelik, P. G. (1992). Hormonal reactions to fighting in rat colonies: prolactin rises during defence, not during offence. *Physiol. Behav.* 51:961–8.

Ditkoff, E. C., Crary, W. G., Cristo, M., and Lobo, R. A. (1991). Estrogen improves psychological function in asymptomatic postmenopausal women. *Obstet. Gynecol.* 78: 992–5.

Dixson, A. F., and George, L. (1982). Prolactin and parental behaviour in a male New World primate. *Nature* 299:551–2.

Dobie, D. J., Miller, M. A., Raskind, M. A., and Dorsa, D. M. (1992). Testosterone reverses a senescent decline in extrahypothalamic vasopressin mRNA. *Brain Res.* 583: 247–52.

Dollins, A. B., Lynch, H. J., Wurtman, R. J., Deng, M. H., Kischka, K. U., Gleason, R. E., and Woerman, H. R. (1993). Effect of pharmacological daytime doses of melatonin on human mood and performance. *Psychopharmacology* 112:490–6.

Dollins, A. B., Zhdanova, W., Wurtman, R. J., Lynch, H. J., and Deng, M. H. (1994). Effect of inducing nocturnal serum melatonin concentrations in daytime on sleep, mood, body temperature and performance: *Proc. Natl. Acad. Sci. U.S.A.* 91:1824–8.

Dores, R. M., Khachaturian, H. J., Watson, S. J., and Akil, H. (1984). Localization of neurons containing pro-opiomelanocortin-related peptides in the hypothalamus and midbrain of the lizard, *Anolis carolinensis:* evidence for region-specific processing of β-endorphin. *Brain Res.* 324:384–9.

Drago, F. (1989). Prolactin as a protective factor in stress-induced biological changes. *J. Clin. Lab. Anal.* 3:340–4.

Drago, F. (1990). Behavioral effects of prolactin. *Curr. Top. Neuroendocrinol.* 10:263–90.

Drago, F., Pulvirenti, L., Spadaro, F., and Pennisi, G. (1991). Effects of TRH and prolactin in the behavioral despair (swim) model of depression in rats. *Psychoneuroendocrinology* 15:349–56.

Drevets, W. Y., Videen, T. O., Price, J. L., Preskorn, H., Carmichael, S. T., and Raichelle, M. E. (1992). A functional anatomical study of unipolar depression. *J. Neurosci.* 12:3628–41.

DuLac, S., and Knudsen, E. I. (1991). Early visual deprivation results in a degraded motor map in the optic tectum of barn owls. *Neurobiology* 88:3426–30.

Duncan, M. J., and Goldman, B. D. (1985). Physiological doses of prolactin stimulate pelage pigmentation in Djungarian hamster. *Am. J. Physiol.* 248:R664–7.

Duncan, M. J., Goldman, B. D., DiPinto, M. N., and Stetson, M. H. (1985). Testicular function and pelage color have different critical day lengths in the Djungarian hamster, *Phodopus sungorus sungorus. Endocrinology* 116:424–30.

Dunn, A. J. (1989). Psychoneuroimmunology for the psychoneuroendocrinologist: a review of animal studies of nervous system–immune system interactions. *Psychoneuroendocrinology* 14:251–74.

Dunn, A. J. (1993). Infection as a stressor: a cytokine-mediated activation of the hypothalamopituitary-adrenal axis? In: *Corticotropin-releasing Factor,* ed. D. J. Chadwick, J. Marsh, and K. Ackrill, pp. 226–42. New York: Wiley.

Dunn, A. J., and Wang, J. (1995). Cytokine effects on CNS biogenic amines. *Neuroimmunomodulation* 2:319–28.

Durkheim, E. (1951). *Suicide.* New York: Free Press. (Originally published 1987.)

Dutt, A., Kaplitt, M. G., Kow, L.-M., and Pfaff, D. W. (1994). Prolactin, central nervous system and behavior: a critical review. *Neuroendocrinology* 59:413–19.

Edelman, I. S. (1978). Candidate mediators in the action of aldosterone on Na⁺ transport. In: *Membrane Transport Processes,* vol. 1, ed. J. F. Hoffman. New York: Raven Press.

Edgar, D. M., Dement, W. C., and Fuller, C. A. (1992). Effect of SCN lesions on sleep in squirrel monkeys: evidence for opponent processes in sleep–wake regulation. *J. Neurosci.* 13:1065–79.

Edgren, R. A. (1960). A seasonal change in bone density in female musk turtles, *Sternothaerus odoratus (Latreille). Comp. Biochem. Physiol.* 1:213–16.

Ehlers, C. L., Henriksen, S. J., Wang, M., Rivier, J., Vale, W., and Bloom, F. E. (1983). Corticotropin releasing factor produces increases in brain excitability and convulsive seizures in rats. *Brain Res.* 278:332–6.

Ehrlich, K. J., and Fitts, D. A. (1990). Atrial natriuretic peptide in the subfornical organ reduces drinking induced by angiotensin or in response to water deprivation. *Behav. Neurosci.* 2:366–79.

Ekman, P. (1972). *Emotion in the Human Face.* Cambridge University Press.

Epstein, A. N. (1982a). Instinct and motivation. In: *The Physiological Basis of Motivation,* ed. D. Pfaff. Berlin: Springer-Verlag.

Epstein, A. N. (1982b). Mineralocorticoids and cerebral angiotensin may act to produce sodium appetite. *Peptides* 3:493–4.

Epstein, A. N., Fitzsimons, J. T., and Rolls, B. J. (1970). Drinking induced by injection of angiotensin into the brain of the rat. *J. Physiol. (Lond.)* 210:457–75.

Epstein, A. N., and Massi, M. (1987). Salt appetite in the pigeon to pharmacological treatments. *J. Physiol. (Lond.)* 393:555–68.

Epstein, A. N., and Stellar, E. (1955). The control of salt preference in the adrenalectomized rat. *J. Comp. Physiol. Psychol.* 48:167–72.

Ermisch, A., and Ruhle, H.-J. (1978). Autoradiographic demonstration of aldosterone-concentrating neuron populations in rat brain. *Brain Res.* 147:154–8.

Erskine, M. S. (1992). Pelvic and pudendal nerves influence the display of paced mating behavior in response to estrogen and progesterone in the female rat. *Behav. Neurosci.* 106:690–7.

Evaard, J. M. (1915). Is the appetite of swine a reliable indication of physiological need? *Proc. Iowa Acad. Sci.* 22:375–402.

Evans, R. M. (1988). The steroid and thyroid hormone receptor superfamily. *Science* 240:889–95.

Everitt, B. J., and Stacey, P. (1987). Studies of instrumental behavior with sexual reinforcement in male rats (*Rattus norvegicus*): II. Effects of preoptic area lesions, castration and testosterone. *J. Comp. Psychol.* 101:407–19.

Fahrbach, S. E., Morrell, J. I., and Pfaff, D. W. (1984). Oxytocin induction of short-latency maternal behavior in nulliparous, estrogen-primed female rats. *Horm. Behav.* 18:267–86.

Fahrbach, S. E., Morrell, J. I., and Pfaff, D. W. (1985). Possible role of endogenous oxytocin in estrogen-facilitated maternal behavior in rats. *Neuroendocrinology* 40:526–32.

Falasco, J. D., Smith, G. P., and Gibbs, J. (1979). Cholecystokinin suppresses sham feeding in the rhesus monkey. *Physiol. Behav.* 23:887–90.

Faulkes, C. G., Abbott, D. H., and Jarvis, J. U. (1991). Social suppression of reproduction in male naked mole-rats, *Heterocephalus glaber. J. Reprod. Fertil.* 91: 593–604.

Fawcett, J. (1992). Suicide risk factors in depressive disorders and in panic disorder. *J. Clin. Psychiatry* 53:9–13.

Feder, H. H., Blaustein, J. D., and Nock, B. L. (1979). Oestrogen-progestin regulation of female sexual behavior in guinea pigs. *J. Steroid Biochem.* 11:873–7.

Feder, H. H., and Whalen, R. E. (1965). Feminine behavior in neonatally castrated and estrogen-treated male rats. *Science* 147:306–7.

Felicio, L. F., and Nasello, A. G. (1989). Effect of acute bromopride treatment on rat prolactin levels and sexual behavior. *Braz. J. Med. Biol. Res.* 22:1011–14.

Fernandez, X., Meunier-Salaun, M.-C., and Mormede, P. (1994). Agonistic behavior, plasma stress hormones, and metabolites in response to dyadic encounters in domestic pigs: interrelationships and effect of dominance status. *Physiol. Behav.* 56: 841–7.

Ferris, C. (1992). Role of vasopressin in aggressive and dominant/subordinate behaviors. *Ann. N. Y. Acad. Sci.* 652:212–26.

Ferris, C., Axelson, J. F., Martin, A. M., and Roberge, L. F. (1989). Vasopressin immunoreactivity in the anterior hypothalamus is altered during the establishment of dominant/subordinate relationships between hamsters. *Neuroscience* 29:675–83.

Ferris, C., Gold, L., DeVries, G. J., and Potegal, M. (1990). Evidence for a functional and anatomical relationship between the lateral septum and the hypothalamus in the control of flank marking behavior in golden hamsters. *J. Comp. Neurol.* 293:476–85.

Ferris, C., Singer, E. A., Meenan, D. M., and Albers, H. E. (1988). Inhibition of vasopressin stimulated flank marking behavior by V1-receptor antagonists. *Eur. J. Pharmacol.* 154:153–9.

Ferris, C. F., Delville, Y., Irvin, R. W., and Poteaal, M. (1994). Septo-hypothalamic organization of a stereotyped behavior controlled by vasopressin in golden hamsters. *Physiol. Behav.* 55:755–9.

Ferris, C. F., Melloni, R. H., Koppel, G., Perry, K. W., Fuller, R. W., and Delville, Y. (1997). Vasopressin/serotonin interactions in the anterior hypothalamus control aggressive behavior in golden hamsters. *J. Neurosci.* 17:4331–40.

Figlewicz, D. P., Sipols, A. J., Seeley, R. J., Chavez, M., Woods, S. C., and Porte, D., Jr. (1995). IVT insulin enhances the meal-suppressive efficacy of IVT CCK-8 in the baboon. *Behav. Neurosci.* 109:567–9.

Findlay, A. L. R., Fitzsimons, J. T., and Kucharczyk, J. (1979). Dependence of spontaneous and angiotensin-induced drinking in the rat upon the oestrous cycle and ovarian hormones. *J. Endocrinol.* 82:215–25.

Finger, T. E., and Kanwal, J. S. (1992). Ascending general visceral pathways within the brainstems of two teleost fishes: *Ictalurus punctatus* and *Carassius auratus. J. Comp. Neurol.* 320:509–20.

Finn, P. D., and Yahr, P. (1994). Projection of the sexually dimorphic area of the gerbil hypothalamus to the retrorubal field is essential for male sexual behavior: role of A8 and other cells. *Behav. Neurosci.* 108:362–78.

Fischer, D., Patchev, V. K., Hellbach, S., Hassan, A. H. S., and Alimeda, O. F. X. (1995a). Lactation as a model of naturally reversible hypercorticalism plasticity in the mechanisms governing hypothalamic-pituitary-adrenocortical activity in rats. *J. Clin. Invest.* 96:1208–15.

Fischer, L. M., Catz, D., and Kelley, D. B. (1993). An androgen receptor mRNA isoform associated with hormone-induced cell proliferation. *Dev. Biol.* 90:8254–8.

Fischer, L. M., Catz, D., and Kelley, D. B. (1995b). Androgen-directed development of the *Xenopus laevis* larynx: control of androgen receptor expression and tissue differentiation. *Dev. Biol.* 170:115–26.

Fitts, D. A. (1993). *Effects of ANP Infusions into the OVLT on Salt Appetite in Rats.* Washington, DC: Society for Neuroscience.

Fitts, D. A., and Mason, D. B. (1989a). Subfornical organ connectivity and drinking to captopril or carbachol in rats. *Behav. Neurosci.* 103:873–80.

Fitts, D. A., and Mason, D. B. (1989b). Forebrain sites of action for drinking and salt appetite to angiotensin or captopril. *Behav. Neurosci.* 103:865–72.

Fitts, D. A., and Mason, D. B. (1990). Preoptic angiotensin and salt appetite. *Behav. Neurosci.* 104:643–50.

Fitts, D. A., Thunhorst, R. L., and Simpson, J. B. (1985). Diuresis and reduction of salt appetite by lateral ventricular infusions of atriopeptin II. *Brain Res.* 348:118–24.

Fitts, D. A., Tjepkes, D. S., and Bright, R. O. (1990). Salt appetite and lesions of the ventral part of the ventral medial preoptic nucleus. *Behav. Neurosci.* 103:818–97.

Fitts, D. A., Yang, O. O., Corp, E. S., and Simpson, J. B. (1983). Sodium retention and salt appetite following deoxycorticosterone in hamsters. *Am. J. Physiol.* 244:R78–83.

Fitzsimons, J. T. (1961). Drinking by rats depleted of body fluid without increase in osmotic pressure. *J. Physiol. (Lond.)* 159:297–309.

Fitzsimons, J. T. (1969). The role of a renal thirst factor in drinking induced by extracellular stimuli. *J. Physiol. (Lond.)* 201:349–68.

Fitzsimons, J. T. (1979). *The Physiology of Thirst and Sodium Appetite.* Cambridge University Press.

Fitzsimons, J. T., and Fuller, L. M. (1985). Effects of angiotensin or carbachol on sodium intake and excretion in adrenalectomized or deoxycorticosterone-treated rats. *J. Physiol. (Lond.)* 359:447–58.

Fitzsimons, J. T., and Kaufman, S. (1977). Cellular and extracellular dehydration and angiotensin as stimuli to drinking in the common iguana *Iguana iguana. J. Physiol. (Lond.)* 265:4443–63.

Flanagan, L. M., Dohanics, J., Verbalis, J. G., and Stricker, E. M. (1992a). Gastric motility and food intake in rats after lesions of hypothalamic paraventricular nucleus. *Am. J. Physiol.* 263:R39–44.

Flanagan, L. M., Olson, B. R., Sved, A. F., Verbalis, J. G., and Stricker, E. M. (1992b). Gastric motility in conscious rats given oxytocin and an oxytocin antagonist centrally. *Brain Res.* 578:256–60.

Fleming, A. S., Corter, C., and Steiner, M. (1995). Sensory and hormonal control of maternal behavior in rat and human mothers. In: *Motherhood in Human and Non-Human Primates,* ed. R. D. Martin and D. Skuse. Basel: S. Karger.

Fleming, A. S., Suh, E. J., Korsmit, M., and Rusak, B. (1994). Activation of fos-like immunoreactivity in the medial preoptic area and limbic structures by maternal and social interactions in rats. *Behav. Neurosci.* 108:724–34.

Fluharty, S. J., and Epstein, A. N. (1983). Sodium appetite elicited by intracerebroventricular infusion of angiotensin II in the rat: II. Synergistic interaction with systemic mineralocorticoids. *Behav. Neurosci.* 97:746–58.

Fluharty, S. J., and Manaker, S. (1983). Sodium appetite elicited by intracerebroventricular infusion of angiotensin II in the rat: I. Relation to urinary sodium excretion. *Behav. Neurosci.* 97:738–45.

Fluharty, S. J., and Sakai, R. R. (1995). Behavioral and cellular studies of corticosterone and angiotensin interaction in brain. In: *Progress in Psychobiology and Physiological Psychology,* ed. S. J. Fluharty, A. R. Morrison, J. M. Sprague, and E. Stellar. San Diego: Academic Press.

Flynn, F. W. (1992). Caudal brain stem systems mediate effects of bombesin-like peptides on intake in rats. *Am. J. Physiol.* 262:39–45.

Flynn, F. W., Berridge, K. C., and Grill, H. J. (1986). Pre- and postabsorptive insulin secretion in chronic decerebrate rats. *Am. J. Physiol.* 250:R539–48.

Flynn, F. W., Grill, H. J., Schulkin, J., and Norgren, R. (1992). Central gustatory lesions. II: Effects on sodium appetite, taste aversion learning, and feeding behaviors. *Behav. Neurosci.* 105:944–54.

Flynn, F. W., Schulkin, J., and Havens, M. (1993). Sex differences in salt preference and taste reactivity in rats. *Brain Res. Bull.* 32:91–5.

Flynn, J. P. (1969). Neural aspects of attack behavior in cats. *Ann. N.Y. Acad. Sci.* 159:1008–12.

Fonberg, E. (1974). Amygdala functions within the alimentary system. *Acta Neurobiol. Experimentalis* 22:51–7.

Ford, C. S., and Beach, F. A. (1951). *Patterns of Sexual Behavior.* New York: Harper & Row.

Forger, N. G., Roberts, S. L., Wong, V., and Breedlove, S. M. (1993). Ciliary neurotrophic factor maintains motoneurons and their target muscles in developing rats. *J. Neurosci.* 13:4720–6.

Francis, R. C., Jacobson, B., Wingfield, J. C., and Fernald, R. D. (1992). Hypertrophy of gonadotropin releasing hormone-containing neurons after castration in the teleost, *Haplochromis burtoni. J. Neurobiol.* 23:1084–93.

Frank, L. G., and Glickman, S. E. (1994). Giving birth through a penile clitoris: parturition and dystocia in the spotted hyaena (*Crocuta crocuta*). Communications from the Mammal Society. *J. Zool. (Lond.)* 234:659–90.

Frank, L. G., Glickman, S. E., and Licht, P. (1991). Fatal sibling aggression, precocial development, and androgens in neonatal spotted hyenas. *Science* 252:702–4.

Frank, L. G., Glickman, S. E., and Powch, I. (1990). Sexual dimorphism in the spotted hyaena (*Crocuta crocuta*). *Nature* 221:308–13.

Frankfurt, M., Gould, E., Woolley, C. S., and McEwen, B. S. (1990). Gonadal steroids modify dendritic spine density in ventromedial hypothalamic neurons: a Golgi study in the adult rat. *Neuroendocrinology* 51:530–5.

Frankmann, S. P., Ulrich, P., and Epstein, A. N. (1991). Transient and lasting effects of reproductive episodes on NaCl intake of the female rat. *Appetite* 16:193–204.

Frederick, R. C., Hamann, A., Anderson, S., Lollmann, B., Lowell, B. B., and Flier, J. S. (1995). Leptin levels reflect body lipid content in mice: evidence for diet-induced resistance to leptin action. *Nature Medicine* 12:1311–14.

Fregly, M. J. (1980). Effect of chronic treatment with estrogen on the dipsogenic response of rats to angiotensin. *Biochem. Behav.* 12:131–6.

Fregly, M. J., and Rowland, N. E. (1985). Role of renin-angiotensin-aldosterone system in NaCl appetite of rats. *Am. J. Physiol.* 248:R1–11.

Fregly, M. J., and Waters, W. (1966). Effect of mineralocorticoids on spontaneous sodium chloride appetite of adrenalectomized rats. *Physiol. Behav.* 1:65–74.

Freud, S. (1960). *A General Introduction to Psychoanalysis.* New York: Washington Square Press. (Originally published 1920.)

Friedman, M. I., and Ramirez, I. (1987). Insulin counteracts the satiating effect of a fat meal in rats. *Physiol. Behav.* 40:655–9.

Friedman, M. I., and Stricker, E. M. (1976). The physiological psychology of hunger: a physiological perspective. *Psychol. Rev.* 83:409–31.

Frijda, N. H. (1986). *The Emotions.* Cambridge University Press.

Frye, C. A., Mermelstein, P. G., and DeBold, J. F. (1992). Evidence for a non-genomic action of progestins on sexual receptivity in hamster ventral tegmental area but not hypothalamus. *Brain Res.* 578:87–93.

Fuchs, E., and Flugge, G. (1995). Modulation of binding sites for corticotropin-releasing hormone by chronic psychosocial stress. *Psychoneuroendocrinology* 20:33–51.

Funder, J. W. (1993). Mineralocorticoids, glucocorticoids, receptors and response elements. *Science* 259:1132–3.

Funder, J. W., Pearce, P. T., Smith, R., and Smith, A. I. (1988). Mineralocorticoid action: target tissue specificity is enzyme, not receptor mediated. *Science* 242:583–5.

Gabr, R. W., Birkle, D. L., and Azzaro, A. J. (1995). Stimulation of the amygdala by glutamate facilitates corticotropin-releasing factor release from the median eminence and activation of the hypothalamic-pituitary-adrenal axis in stressed rats. *Neuroendocrinology* 62:333–9.

Galaverna, O., DeLuca, L. A., Schulkin, J., Yal, S. Z., and Epstein, A. N. (1992). Deficits in NaCl ingestion after damage to the central nucleus of the amygdala in the rat. *Brain Res. Bull.* 28:89–98.

Galaverna O., Seeley, R. J., Berridge, K. C., Grill, H. J., Epstein, A. N., and Schulkin, J. (1993). Lesions of the central nucleus of the amygdala. 1: Effects on taste reactivity, taste aversion learning and sodium appetite. *Behav. Brain Res.* 59:11–17.

Galea, L. A., Kavaliers, M., and Ossenkopp, K. P. (1996). Sexually dimophric spatial learning in meadow voles and deer mice. *J. Exp. Biol.* 199:195–200.

Galea, L. A., Kavaliers, M., Ossenkopp, K. P., and Hampson, E. (1995). Gonadal hormone levels and spatial learning performance in the Morris water maze in male and female meadow voles, *Microtus pennsylvanicus. Horm. Behav.* 29:106–25.

Galef, B. G., Jr. (1986). Social interaction modifies learned aversions, sodium appetite and both palatability and handling-time induced dietary preference in rats. *J. Comp. Psychol.* 4:432–9.

Galef, B. G., Jr. (1991). A contrarian view of the wisdom of the body as it relates to dietary self-selection. *Psychol. Rev.* 98:218–23.

Galef, B. G., Jr. (1996). Food selection: problems in understanding how we choose foods to eat. *Neurosci. Biobehav. Rev.* 20:67–73.

Galeno, T. M., van Hoesen, G. W., Maixner, W., Johnson, A. K., and Brody, M. J. (1982). Contribution of the amygdala to the development of spontaneous hypertension. *Brain Res.* 246:1–6.

Gallagher, M., Graham, P. W., and Holland, P. C. (1990). The amygdala central nucleus and appetitive Pavlovian conditioning: lesions impair one class of conditioned behavior. *J. Neurosci.* 10:1906–11.

Gallagher, M., and Holland, P. C. (1994). The amygdala complex: multiple roles in associative learning and attention. *Proc. Natl. Acad. Sci. U.S.A.* 91:11771–6.

Gallistel, C. R. (1980). *The Organization of Action: A New Synthesis.* Hillsdale, NJ: Lawrence Erlbaum.

Gallistel, C. R. (1990). *The Organization of Learning.* MIT Press.

Gammie, S. G., and Truman, J. W. (1997). Neuropeptide hierarchies and the activation of sequential motor behaviors in the hawkmoth, *Manduca sexta. J. Neurosci.* 11: 4389–97.

Ganesan, R., and Sumners, C. (1989). Glucocorticoids potentiate the dipsogenic action of angiotensin II. *Brain Res.* 499:121–30.

Ganong, W. F. (1984). The brain renin-angiotensin system. *Annu. Rev. Physiol.* 46:17–31.

Ganten, D., Hutchinson, S., and Schelling, P. (1975). The intrinsic brain iso-renin-angiotensin system in the rat: its possible role in central mechanisms of blood pressure regulation. *Clin. Sci. Mol. Med.* 2:2655–85.

Garcia, J., Hankins, W. G., and Rusiniak, K. W. (1974). Behavioral regulation of the internal milieu in man and rat. *Science* 185:824–31.

Garcia, V., Jouventin, P., and Mauget, R. (1996). Parental care and the prolactin pattern in the king penguin: an endogenously timed mechanism? *Horm. Behav.* 30:259–65.

Gardiner, T. W., Verbalis, J. G., and Stricker, E. M. (1985). Impaired secretion of vasopressin and oxytocin in rats after lesions of nucleus medianus. *Am. J. Physiol.* 249: R681–8.

Gardner, D. G., Gertz, B. J., Deschepper, C. F., and Kim, D. Y. (1988). Gene for the rat atrial natriuretic peptide is regulated by glucocorticoids in vitro. *J. Clin. Invest.* 82: 1275–81.

Garel, J.-M. (1987). Hormonal control of calcium metabolism during the reproductive cycle in mammals. *Physiol. Rev.* 67:1–66.

Garrick, N. A., Hill, J. L., Szele, F. G., Tomai, T. P., Gold, P. W., and Murphy, D. L. (1987). Corticotropin-releasing factor: a marked circadian rhythm in primate cerebrospinal fluid peaks in the evening and is inversely related to the cortisol circadian rhythm. *Endocrinology* 121:1329–34.

Gerardo-Gettens, T., Moore, B. J., Stern, J. S., and Horwitz, B. A. (1989). Prolactin stimulates food intake in the absence of ovarian progesterone. *Am. J. Physiol.* 256:R701–6.

Geshwind, N. (1974). *Selected Papers on Language and the Brain.* Dordrecht: Reidel.

Giardino, L., Ceccatelli, S., Zanni, M., Hökfelt, T., and Calza, L. (1994). Regulation of VIP mRNA expression by thyroid hormone in different brain areas of adult rat. *Mol. Brain Res.* 27:87–94.

Gibbs, J., Kulkosky, P. J., and Smith, G. P. (1981). Effects of peripheral and central bombesin on feeding behavior of rats. *Peptides* 2:179–83.

Gibbs, J., Young, R. C., and Smith, G. P. (1973). Cholecystokinin decreases food intake in rats. *J. Comp. Physiol. Psychol.* 84:488–95.

Gibson, T. R., Wildey, G. M., Manaker, S., and Glembotski, C. C. (1986). Autoradiograph localization and characterization of atrial natriuretic peptide binding sites in the rat central nervous system and adrenal gland. *J. Neurosci.* 7:2004–11.

Giguere, V., Yang, N., Segui, P., and Evens, R. M. (1988). Identification of a new class of steroid hormone receptors. *Nature* 331:91–5.

Giordano, A. L., Siegel, H. I., and Rosenblatt, J. S. (1991). Nuclear estrogen receptor binding in microdissected brain regions of female rats during pregnancy: implications for maternal and sexual behavior. *Physiol. Behav.* 50:1262–7.

Giza, B. K., and Scott, T. R. (1987). Intravenous insulin infusions in rats decrease gustatory-evoked responses to sugars. *Am. J. Physiol.* 252:R994–1002.

Gladue, B. A. (1991). Aggressive behavioral characteristics, hormones, and sexual orientation in men and women. *Aggressive Behavior* 17:313–26.

Gladue, B. A. (1994). The biopsychology of sexual orientation. *Psychological Sciences* 3:150–4.

Gladue, B. A., and Beatty, W. W. (1990). Sexual orientation and spatial ability in men and women. *Psychobiology* 18:101–8.

Glickman, S. E., Frank, L. G., Licht, P., Yalcinkaya, T., Shiteri, P. K., and Davidson, J. (1992a). Sexual differentiation of the female spotted hyena: one of nature's experiments. *Ann. N.Y. Acad. Sci.* 662:135–59.

Glickman, S. E., Frank, L. G., Pavgi, S., and Licht, P. (1992b). Hormonal correlates of 'masculinization' in female spotted hyaenas (*Crocuta crocuta*). 1. Infancy to sexual maturity. *J. Reprod.* 95:451–62.

Glietman, L., and Wanner, E. (1982). *Language Acquisition.* Cambridge University Press.

Gloor, P. (1978). Inputs and outputs of the amygdala: what the amygdala is trying to tell the rest of the brain. In: *Limbic Mechanisms,* ed. K. E. Livingston and O. Hornykiewicz, pp. 189–206. New York: Plenum Press.

Glowa, J. R., Barrett, J. E., Russell, J., and Gold, P. (1992). Effects of corticotropin releasing hormone on appetitive behaviors. *Peptides* 13:609–21.

Gold, P. W., and Chrousos, G. P. (1985). Clinical studies with corticotropin releasing factor: implications for the diagnosis and pathophysiology of depression, Cushing's disease, and adrenal insufficiency. *Psychoneuroendocrinology* 10:401–19.

Gold, P. W., Chrousos, G., Kellner, C., Post, R., Avgerinos, P., Schulte, H., and Oldfield, D. (1984). Psychiatric implications of basic and clinic studies with CRF. *Am. J. Psychiatry* 141:619–29.

Gold, P. W., Goodwin, F. K., and Chrousos, G. P. (1988). Clinical and biochemical manifestations of depression. *N. Engl. J. Med.* 319:348–53, 413–20.

Goldman, B. D. (1991). Parameters of the circadian rhythm of pineal melatonin secretion affecting reproductive responses in Siberian hamsters. *Steroids* 56:218–25.

Goldsmith, A. R. (1983). Prolactin in avian reproductive cycles. In: *Hormones and Behaviour in Higher Vertebrates,* ed. B. J. Balthazart, E. Prove, and R. Gilles. Berlin: Springer-Verlag.

Goldsmith, A. R., Edwards, C., Koprucu, M., and Silver, R. (1981). Concentrations of prolactin and luteinizing hormone in plasma of doves in relation to incubation and development of the crop gland. *J. Endocrinol.* 90:437–43.

Goodall, J. (1986). *The Chimpanzee of Gombe.* Cambridge University Press.

Goodyer, I., Herbert, J., Moor, S., and Altham, P. (1991). Cortisol hypersecretion in depressed school-aged children and adolescents. *Psychiatry Res.* 37:237–44.

Gorewit, R. C., Svennersten, K., Butler, W. R., and Uvnas-Moberg, K. (1992). Endocrine responses in cows milked by hand and machine. *J. Dairy Sci.* 75:443–8.

Gorski, R. A., Gordon, J. H., Shryne, J. E., and Southam, A. M. (1978). Evidence for a morphological sex difference within the medial preoptic area of the rat brain. *Brain Res.* 148:333–46.

Gould, E., Westlind-Danielsson, A., Frankfurt, M., and McEwen, B. S. (1990). Sex differences and thyroid hormone sensitivity of hippocampal pyramidal cells. *J. Neurosci.* 10:996–1003.

Gould, S. J. (1977). *Ontogeny and Phylogeny.* Harvard University Press.

Goy, R. W., and McEwen, B. S. (1980). *Sexual Differentiation of the Brain.* MIT Press.

Grahn, R. E., Kalman, B. A., Sutton, L., Silbert, L. H., Wiertelak, E. P., Watkins, L. R., and Maier, S. F. (1992). The effect of inescapable shock on shuttle-box escape performance, conditioned fear, and anxiety in rats with lesions of the dorsal raphe nucleus of central nucleus of the amygdala. *Society for Neuroscience Abstracts* 18: 534.

Gratto-Trevor, C. L., Oring, L. W., Fivizzani, A. J., El Halawani, M. E., and Cooke, F. (1990). The role of prolactin in parental care in a monogamous and a polyandrous shorebird. *The Auk* 107:718–29.

Gray, J. (1987). *The Psychology of Fear and Stress.* Cambridge University Press. (Originally published 1971.)

Gray, T. S. (1990). The organization and possible function of amygdaloid corticotropin-releasing factor pathways. In: *Corticotropin-releasing Factor: Basic and Clinic Studies of a Neuropeptide,* ed. E. B. DeSousa and C. B. Nemeroff. Boca Raton: CRC Press.

Gray, T. S., and Bingaman, E. W. (1996). The amygdala: corticotropin-releasing factor, steroids and stress. *Crit. Rev. Neurobiol.* 10:155–68.

Gray, T. S., Carney, M. E., and Magnuson, D. J. (1989). Direct projections from the central amygdaloid nucleus to the hypothalamic paraventricular nucleus: possible role in stress-induced adrenocorticotropin release. *Neuroendocrinology* 50:433–46.

Gray, T. S., and Magnuson, D. J. (1987). Neuropeptide neuronal efferents from the bed nucleus of the stria terminalis and central amygdaloid nucleus to the dorsal vagal complex in the rat. *J. Comp. Neurol.* 262:365–74.

Gray, T. S., O'Donohue, T. L., and Magnuson, D. J. (1986). Neuropeptide Y innervation of amygdaloid and hypothalamic neurons that project to the dorsal vagal complex in rat. *Peptides* 7:341–9.

Greenberg, N., Chen, T., and Crews, D. (1984). Social status, gonadal state, and the adrenal stress response in the lizard, *Anolis carolinensis*. *Horm. Behav.* 18:1–11.

Greer, N. L., Bartolome, J. V., and Schanberg, S. M. (1990). Further evidence for the hypothesis that beta-endorphin mediates maternal deprivation effects. *Life Sci.* 48: 643–8.

Griffond, B., Deray, A., Jacquemard, C., Fellmann, D., and Bugnon, C. (1994). Prolactin immunoreactive neurons of the rat lateral hypothalamus: immunocytochemical and ultrastructural studies. *Brain Res.* 635:179–86.

Grigson, P. S., Shimura, T., and Norgren, R. (1997). Brainstem lesions and gustatory function. II: The role of the nucleus of the solitary tract in Na^+ appetite, conditioned taste aversion, and conditioned odor aversion in rats. *Behav. Neurosci.* 111:169–79.

Grill, H. J. (1980). Production and regulation of consummatory behavior in the chronic decrebrate rat. *Brain Res. Bull.* 5:79–87.

Grill, H. J., Berridge, K. C., and Ganster, D. J. (1984). Oral glucose is the prime elicitor of preabsorptive insulin secretion. *Am. J. Physiol.* 246:R88–95.

Grill, H. J., and Norgren, R. (1978a). The taste reactivity test. I. Oral-facial responses to gustatory stimuli in neurologically normal rats. *Brain Res.* 143:263–79.

Grill, H. J., and Norgren, R. (1978b). The taste reactivity test. II. Mimetic responses to gustatory stimuli in chronic thalamic and chronic decerebrate rats. *Brain Res.* 143: 281–97.

Grill, H. J., Schulkin, J., and Flynn, F. W. (1986). Sodium homeostasis in chronic decerebrate rats. *Behav. Neurosci.* 100:536–43.

Grill, H. J., and Smith, G. P. (1988). Cholecystokinin decreases sucrose intake in chronic decerebrate rats. *Am. J. Physiol.* 254:R853–6.

Grillo, C., Coirini, H., McEwen, B. S., and De Nicola, A. F. (1989). Changes of salt intake salt intake and of (Na + K)-ATPase activity in brain after high dose treatment of dexoycorticosterone. *Brain Res.* 499:225–33.

Grillon, C., Ameli, R., Foot, M., and Davis, M. (1993). Fear potentiated startle: relationship to the level of state/trait anxiety in healthy subjects. *Biol. Psychiatry* 33:566–74.

Grisham, W., Mathews, G. A., and Arnold, A. P. (1994). Local intracerebral implants of estrogen masculinize some aspects of the zebra finch song system. *J. Neurobiol.* 25: 185–96.

Groosman, S. P. (1960). Eating or drinking elicited by direct adrenergic or cholinergic stimulation of hypothalamus. *Science* 225:218–25.

Groosman, S. P. (1990). *Thirst and Sodium Appetite*. San Diego: Academic Press.

Grunt, J. A., and Young, W. C. (1952). Differential reactivity of individuals and the response of the male guinea pig to testosterone propionate. *Endocrinology* 51:237–48.

Gu, G., and Simerly, R. B. (1994). Hormonal regulation of opioid peptide neurons in the anteroventral periventricular nucleus. *Horm. Behav.* 28:503–11.

Gubernick, D. J., and Nelson, R. J. (1989). Prolactin and paternal behavior in the biparental California mouse, *Peromyscus californicus*. *Horm. Behav.* 23:203–10.

Gubernick, D. J., Sengelaub, D. R., and Kurz, E. M. (1993). A neuroanatomical correlate of paternal and maternal behavior in the biparental California mouse *(Peromyscus californicus)*. *Behav. Neurosci.* 107:194–201.

Gubernick, D. J., Winslow, J. T., Jensen, P., Jeanotte, L., and Bowen, J. (1995). Oxytocin changes in males over the reproductive cycle in the monogamous, biparental California mouse, *Peromyscus californicus. Horm. Behav.* 29:59–73.

Gunnar, M. R., Isensee, J., and Fust, S. (1987). Adrenocortical activity and the Brazelton neonatal assessment scale: moderating effects of the newborn's biomedical status. *Child Dev.* 5:1448–58.

Gunnar, M. R., Mangelsdorf, S., Larson, M., and Hertsgaard, L. (1989). Attachment, temperament, and adrenocortical activity in infancy: a study of psychoendocrine regulation. *Dev. Psychology* 23:355–63.

Gurney, M. E. (1982). Behavioral correlates of sexual differentiation in the zebra finch song system. *Brain Res.* 231:153–72.

Gurney, M. E., and Konishi, M. (1980). Hormone-induced sexual differentiation of brain and behavior in zebra finches. *Science* 208:1380–3.

Gyurko, R., Wielbo, D., and Phillips, M. I. (1993). Antisense inhibition of AT1 receptor mRNA and angiotensinogen MRNA in the brain of spontaneously hypertensive rats reduces hypertension of neurogenic origin. *Regul. Pept.* 49:167–74.

Haimov, I., and Lavie, P. (1996). Melatonin-A soporific hormone. *Psychol. Sci.* 5:106–10.

Halaas, J. L., Gajiwala, K. S., Mattei, M., Cohen, S. L., Chait, B. T., Rabinowitz, D., Lallone, R. L., Burley, S. C., and Friedman, J. M. (1995). Weight reducing effects of the plasma protein encoded by the obese gene. *Science* 269:543–6.

Hall, W. G. (1990). The ontogeny of ingestive behavior. Changing control of components in the feeding sequence. In: *Handbook of Behavioral Neurobiology,* vol. 10, ed. E. M. Stricker. New York: Plenum.

Halpern, B. P. (1983). Tasting and smelling as active, exploratory sensory processes. *Am. J. Otolaryngol.* 4:246–9.

Hamelink, C. R., Currie, P. J., Chambers, J. W., Castonguay, T. W., and Coscina, D. V. (1994). Corticosterone-responsive and unresponsive metabolic characteristics of adrenalectomized rats. *Am. J. Physiol.* 36:R799–804.

Hamlin, M. N., Webb, R. C., Ling, W. D., and Bohr, D. F. (1988). Parallel effects of DOCA on salt appetite, thirst, and blood pressure in sheep (42705). *Exp. Biol. Med.* 188:46–51.

Hammer, R. P., Jr., and Cheung, S. (1995). Sex steroid regulation of hypothalamic opioid function. In: *Neurobiological Effects of Sex Steroid Hormones,* ed. P. E. Micevych and R. P. Hammer, Jr. Cambridge University Press.

Hampson, E. (1990). Estrogen-related variations in human spatial and articulatory-motor skills. *Psychoneuroendocrinology* 15:97–111.

Hampson, E. (1995). Spatial cognition in humans: possible modulation by androgens and estrogens. *J Psychiatr. Neurosci.* 20:397–404.

Hampson, E., and Kimura, D. (1992). Sex differences and hormonal influences on cognitive functions in humans. In: *Behavioral Endocrinology,* ed. J. B. Becker, S. M. Breedlove, and D. Crews. MIT Press.

Hanson, E. S., and Dallman, M. F. (1995). Neuropeptide Y (NPY) may integrate responses of hypothalamic feeding systems and the hypothalamo-pituitary-adrenal axis. *J. Neuroendocrinol.* 7:273–9.

Harlan, R. E. (1988). Regulation of neuropeptide gene expression by steroid hormones. *Mol. Neurobiol.* 2:193–9.

Harlow, H. F. (1949). The nature of love. *Am. Psychol.* 13:673–85.

Harris, L. J., Clay, J., Hargreaves, F. J., and Ward, A. L. (1933). Appetite and choice of diet. The ability of the vitamin B deficient rat to discriminate between diets containing and lacking the vitamin. *Proc. R. Soc. Lond. B Biol. Sci.* 113:161–90.

Hauser, M. D. (1996). *The Evolution of Communication.* MIT Press.

Hayden-Hixson, D. M., and Ferris, C. F. (1991). Steroid-specific regulation of agonistic responding in the anterior hypothalamus of male hamsters. *Physiol. Behav.* 50:793–9.

Hebb, D. O. (1946). On the nature of fear. *Psychol. Rev.* 53:259–76.

Hebb, D. O. (1949). *The Organization of Behavior: A Neuropsychological Theory.* New York: Wiley.

Hebert, D., and Cowan, I. (1971). Natural salt licks as a part of the ecology of the mountain goat. *Can. J. Zool.* 49:605–10.

Heilig, M., Koob, G. F., Ekman, R., and Britton, K. T. (1994). Corticotropin-releasing factor and neuropeptide Y: role in emotional integration. *Trends Neurosci.* 17:80–5.

Heilig, M., McLeod, S., Brot, M., Heinrichs, S. C., Menzaghi, F., Koob, G. F., and Britton, K. T. (1993). Anxiolytic-like action of neuropeptide Y: mediation by Y1 receptors in amygdala, and dissociation from food intake effects. *Neuropsychopharmacology* 8: 357–63.

Heinrichs, S. C., Pich, E. M., Miczek, K. A., Britton, K. T., and Koob, G. F. (1992). Corticotropin-releasing factor antagonist reduces emotionality in socially defeated rats via direct neurotropic action. *Brain Res.* 581:190–7.

Heinrichs, S. C., Menzaghi, F., Pich, E. M., Hauger, R. L., and Koob, G. F. (1993). Corticotropin-releasing factor in the paraventricular nucleus modulates feeding induced neuropeptide Y. *Brain Res.* 611:18–24.

Helmstetter, F. (1992). The amygdala is essential for the expression of conditioned hypoalgesia. *Behav. Neurosci.* 106:518–28.

Henke, P. G. (1982). The telencephalic limbic system and experimental pathology: a review. *Neurosci. Biobehav. Rev.* 6:381–97.

Hennessey, A. C., Huhman, K. L., and Albers, H. E. (1994). Vasopressin and sex differences in hamster flank marking. *Physiol. Behav.* 55:905–11.

Hennessey, A. C., Whitman, D. C., and Albers, H. E. (1992). Microinjection of arginine-vasopressin into the periaqueductal gray stimulates flank marking in Syrian hamsters (*Mesocricetus auratus*). *Brain Res.* 569:136–40.

Hennessy, M. B., Heybach, J. P., Vernikos, J., and Levine, S. (1979). Plasma corticosterone concentrations sensitively reflect levels of stimulus intensity in the rat. *Physiol. Behav.* 22:821–5.

Henry, C., Kabbaj, M., Simon, H., LeMoal, M., and Maccari, S. (1994). Prenatal stress increases the hypothalamo-pituitary-adrenal axis response in young and adult rats. *J. Neuroendocrinol.* 6:341–5.

Herbert, J. (1993). Peptides in the limbic system: neurochemical codes for co-ordinated adaptive responses to behavioural and physiological demand. *Neurobiology* 41:723–91.

Herbert, J. (1997). Stress, the brain, and mental illness. *Br. Med. J.* 315:530–5.

Herbert, J., Forsling, M. L., Howes, S. R., Stacey, P. M., and Shiers, H. M. (1992). Regional expression of c-fos in the basal forebrain following intraventricular infusions of angiotensin and its modulation by drinking either water or saline. *Neuroscience* 51:867–82.

Herkenham, M. (1987). Mismatches between neurotransmitter and receptor localizations in brain: observations and implications. *Neuroscience* 23:1–38.

Herman, J. P., and Cullinan, W. E. (1997). Neurocircuitry of stress: central control of the hypothalamic-pituitary-adrenocortical axis. *Trends Neurosci.* 20:103–10.

Herman, J. P., Cullinan, W. E., Morano, M. I. K., Akil, H., and Watson, S. J. (1995). Contribution of the ventral subiculum to inhibitory regulation of the hypothalamo-pituitary adrenocortical axis. *J. Neuroendocrinol.* 7:475–82.

Herman, J. P., Cullinan, W. E., and Watson, S. J. (1994). Involvement of the bed nucleus of the stria terminalis in tonic regulation of paraventricular hypothalamic CRH and AVP mRNA expression. *J. Neuroendocrinol.* 6:433–42.

Herman, J. P., Cullinan, W. E., Young, E. A., Akil, H., and Watson, S. (1992). Selective forebrain fiber tract lesions implicate ventral hippocampal structures in tonic regulation of paraventricular nucleus corticotropin-releasing hormone (CRH) and arginine vasopressin (AVP) mRNA expression. *Brain Res.* 592:228–38.

Herman, J. P., Schafer, M. K.-H., Young, E. Z., Thompson, R., Douglass, J., Akil, H., and Watson, S. J. (1989). Evidence for hippocampal regulation of neuroendocrine neurons of the hypothalamic-pituitary-adrenocortical axis. *J. Neurosci.* 9:3079–82.

Hernstein, R. J., Loveland, D. H., and Cable, C. (1976). Natural concepts in pigeons. *J. Exp. Psychol.: Anim. Behav. Proc.* 2:285–302.

Herrick, C. J. (1905). The central gustatory pathway in the brain of bony fishes. *J. Comp. Neurol.* 15:375–456.

Herrick, C. J. (1948). *The Brain of the Tiger Salamander.* University of Chicago Press.

Hertsgaard, L., Gunnar, M., Erickson, M. F., and Nachmias, M. (1995). Adrenocortical responses to the strange situation in infants with disorganized disoriented attachment relationships. *Child Dev.* 66:1100–6.

Hertsgaard, L., Gunnar, M., Larson, M., Brodersen, L., and Lehman, H. (1992). First time experiences in infancy: When they appear to be pleasant, do they activate the adrenocortical stress response? *Dev. Psychobiol.* 25:319–33.

Hews, D. K., and Moore, M. C. (1996). A critical period for the organization of alternative male phenotypes of tree lizards by exogenous testosterone? *Physiol. Behav.* 60: 425–9.

Higley, J. D., Suomi, S. J., and Linnoila, M. (1992). A longitudinal assessment of CSF monoamine metabolite and plasma cortisol concentrations in young rhesus monkeys. *Biol. Psychiatry* 32:127–45.

Hilakivi-Clarke, L. (1994). Overexpression of transforming growth factor in transgenic mice alters nonreproductive, sex-related behavioral differences: interaction with gonadal hormones. *Behav. Neurosci.* 108:410–17.

Hill, D. L. (1987). Development of taste responses in the rat parabrachial nucleus. *J. Neurophysiol.* 57:481–95.

Himick, B. A., and Peter, R. E. (1994). CCK/gastrin-like immunoreactivity in brain and gut, and CCK suppression of feeding in goldfish. *Am. J. Physiol.* 267:R841–51.

Hinde, R. A. (1982). *Ethology.* Oxford University Press.

Hinde, R. A., and Stevenson, J. G. (1970). Goals and response control. In: *Development and Evolution of Behavior,* ed. L. R. Aronson, E. Tobach, D. S. Lehrman, and J. S. Rosenblatt. San Francisco: Freeman.

Hines, M. (1982). Prenatal gonadal hormones and sex differences in human behavior. *Psychol. Bull.* 92:56–80.

Hines, M., Allen, L. S., and Gorski, R. A. (1992). Sex differences in subregions of the medial nucleus of the amygdala and the bed nucleus of the stria terminalis of the rat. *Brain Res.* 579:321–6.

Hirshfeld, D. R. (1992). Stable behavioral inhibition and its association with anxiety disorder. *J. Am. Acad. Child Adolesc. Psychiatry* 31:103–11.

Hitchcock, J., and Davis, M. (1986). Lesions of the amygdala, but not of the cerebellum or red nucleus, block conditioned fear as measured with potential startle paradigm. *Behav. Neurosci.* 100:11–22.

Hnatczuk, O. C., Lisciotto, C. A., Don Carlos, L. L., Carter, C. S., and Morrell, J. I. (1994). Estrogen receptor immunoreactivity in specific brain areas of the prairie vole (*Microtus ochrogaster*) is altered by sexual receptivity and genetic sex. *J. Neuroendocrinol.* 6:89–100.

Hodges, L. L., Jordan, C. L., and Breedlove, S. M. (1993). Hormone-sensitive periods for the control of motoneuron number and soma size in the dorsolateral nucleus of the rat spinal cord. *Brain Res.* 602:187–90.

Hoebel, B. (1988). Neuroscience of motivation: peptides and pathways that define motivational systems. In: *Handbook of Experimental Psychology,* ed. S. S. Stevens. New York: Wiley.

Hofer, M. A. (1994). Early relationships as regulators of infant physiology and behavior. *Acta Paediatr. [Suppl.]* 397:1–8.

Hoffman, G. E., Smith, M. S., and Verbalis, J. G. (1993). C-fos and related immediate early gene products as markers of activity in neuroendocrine systems. *Front. Neuroendocrinol.* 14:173–213.

Hoffman, R. A., and Robinson, P. F. (1966). Changes in some endocrine glands of white-tailed deer as affected by season, sex and age. *J. Mammal.* 47:266–80.

Hofman, M. A., and Swaab, D. F. (1993). Diurnal and seasonal rhythms of neuronal activity in the suprachiasmatic nucleus of humans. *J. Biol. Rhythms* 8:283–95.

Hogan-Warburg, A. J., and Hogan, J. A. (1981). Feeding strategies in the development of food recognition in young chicks. *Anim. Behav.* 29:143–54.

Holland, K. L., Abelson, K. L., and Micevich, P. E. (1996). Estrogen stimulates CCK and preproenkephalin mRNA in the female rat limbic-hypothalamic circuit during the peripubertal period. *Neuroscience Abstracts.*

Hollick, M. F. (1994). Vitamin D – new horizons for the 21st century. *Am. J. Clin. Nutr.* 60:619–30.

Holsboer, F., Muller, O. A., Doerr, H. G., Sippell, W. G., Stalla, G. K., Gerken, A., Steiger, A., Boll, E., and Benkert, O. (1984). ACTH and multisteroid responses to CRF factor in depressive illness: relationship to multisteroid responses after ACTH stimulation and dexamethasone suppression. *Psychoneuroendocrinology* 9:147–60.

Honkaniemi, J., Huikko, M. P., Recjardt, L., Isola, J., Lammi, A., Fuxe, K., Gustafsson, J. A., Wikstrom, A. C., and Hökfelt, T. (1992). Colocalization of peptide and glucocorticoid receptor immunoreactivities in rat central amygdaloid nucleus. *Neuroendocrinology* 55:451–9.

Hubel, D. H., and Weisel, T. H. (1972). Laminar and columnar distribution of geniculocortical fibers in the macaque monkey. *J. Comp. Neurol.* 146:421–50.

Hughes, B. O., and Wood-Gush, D. G. M (1972). Hypothetical mechanisms underlying calcium appetite in fowls. *Rev. Comp. Animal.*, pp. 95–106.

Huhman, K. L., and Albers, H. E. (1993). Estradiol increases the behavioral response to arginine vasopressin (AVP) in the medial preoptic-anterior hypothalamus. *Peptides* 14:1049–54.

Huhman, K. L., Babagbemi, T. O., and Albers, H. E. (1995). Bicuculline blocks neuropeptide Y-induced phase advances when microinjected in the suprachiasmatic nucleus of Syrian hamsters. *Brain Res.* 675:333–6.

Hull, C. L. (1943). *Principles of Psychology*. New York: Appleton-Century-Crofts.

Iigo, M., Kobayashi, M., Ohtani-Kaneko, R., Hara, M., Hattori, A., Suzuki, T., and Aida, K. (1994). Characteristics, day–night changes, subcellular distribution and localization of melatonin binding sites in the goldfish brain. *Brain Res.* 64:213–20.

Ikonomov, O. C., Stoyneve, A. G., Vrabchev, N. C., Shisheva, A. C., and Tarkolev, N. T. (1983). Circadian rhythms of food and 1% NaCl intake, urine and electrolyte excretion, plasma renin activity and insulin concentration in adrenalectomized rats. *Acta Physiol. Hung.* 65:181–98.

Imaki, J., Nahon, J. L., Rivier, C., Sawchenko, P. E., and Vale, W. (1991). Differential regulation of corticotropin releasing factor mRNA in rat brain cell types by glucocorticoid and stress. *J. Neurosci.* 11:585–99.

Insel, T. R. (1992). Oxytocin – a neuropeptide for affiliation: evidence from behavioral, receptor autoradiographic, and comparative studies. *Psychoneuroendocrinology* 17: 3–35.

Insel, T. R., and Shapiro, L. E. (1992). Oxytocin receptor distribution reflects social organization in monogamous and polygamous voles. *Neurobiology* 89:5981–5.

Israel, A., Garrido, M. R., Barbella, Y., and Becemberg, I. (1988). Rat atrial natriuretic peptide (99–126) stimulates guanylate cyclase activity in rat subfornical organ and choroid plexus. *Brain Res. Bull.* 20:253–6.

Jackson, H. (1958). Evolution and dissolution of the nervous system. In: *Selected Writings of John Hughlings Jackson*, vol. 2, ed. J. Taylor, pp. London: Staples Press. (Originally published 1884.)

Jacobs, K., Mark, G. P., and Scott, T. R. (1988). Taste responses in the nucleus tractus solitarius of sodium-deprived rats. *J. Physiol.* (*Lond.*) 406:393–410.

Jacobson, L., and Sapolsky, R. (1991). The role of the hippocampus in feedback regulation of the hypothalamic-pituitary-adrenocortical axis. *Endocrine Rev.* 12:118–34.

Jalowiec, J. E., Stricker, E. M., and Wolf, G. (1970). Restoration of sodium balance in hypophysectomized rats after acute sodium deficiency. *Physiol. Behav.* 5:1145–9.

James, W. (1952). *The Principles of Psychology*, 2 vols. New York: Dover. (Originally published 1890.)

Janowsky, J. S., Oviatt, S. K., and Orwoll, E. S. (1994). Testosterone influences spatial cognition in older men. *Behav. Neurosci.* 108:325–32.

Janzen, D. (1977). Why fruits rot, seeds mold and meat spoils. *American Naturalist* 111: 691–713.

Jenab, C. Q., Jenab, S., Ogawa, S., Adan, R. A., Burbach, J. P., and Pfaff, D. W. (1997). Effects of estrogen on oxytocin receptor messenger ribonucleic acid expression in the uterus, pituitary and forebrain of the female rat. *Neuroendocrinology* 65:9–17.

Jevning, R., Wilson, A. F., and VanderLaan, E. F. (1978). Plasma prolactin and growth hormone during meditation. *Psychosom. Med.* 40:329 33.

Jewett, D. C., Clearly, J., Levine, A. S., Schaal, D. W., and Thompson, T. (1995). Effects of neuropeptide Y, insulin, 2-deoxyglucose and food deprivation on food-motivated behavior. *Psychopharmacology* 120:267–71.

Jirikowski, G. F., Caldwell, J. D., Haussler, H. U., and Pedersen, C. A. (1991). Mating alters topography and content of oxytocin immunoreactivity in male mouse brain. *Cell Tissue Res.* 266:399–403.

Joels, M., and De Kloet, R. (1994). Mineralocorticoid and glucocorticoid receptors in the brain. Implications for ion permeability and transmitter systems. *Prog. Neurobiol.* 43:1–36.

Johnson, A. E., Barberis, C., and Albers, H. E. (1995). Castration reduces vasopressin receptor binding in the hamster hypothalamus. *Brain Res.* 674:153–8.

Johnson, A. K. (1985). The periventricular anteroventral third ventrical (AV3V): its relationship with the subfornical organ and neural systems involved in maintaining body fluid homeostasis. *Brain Res. Bull.* 15:595–601.

Johnson, A. K., and Thunhorst, R. L. (1995). Sensory mechanisms in the behavioral control of body fluid balance: thirst and salt appetite. *Prog. Psychobiol. Physiol. Psychol.* 16:145–76.

Johnston, J. B. (1923). Further contributions to the study of the evolution of the forebrain. *J. Comp. Neurol.* 5:337–81.

Jones, D., Gonzalez-Lima, F., Crews, D., Galef, B. G., Jr., and Clark, M. M. (1997). Effects of intrauterine position on the metabolic capacity of the hypothalamus of female gerbils. *Physiol. Behav.* 61:513–19.

Jones, R. B., Beuving, G., and Blokhuis, H. J. (1988). Tonic immobility and heterophil/ lymphocyte responses of the domestic fowl to corticosterone infusion. *Physiol. Behav.* 42:249–53.

Jones, R. L., and Hanson, H. C. (1985). *Mineral Licks: Geography and Biogeochemistry of North American Ungulates.* Iowa State University Press.

Jonklaas, J., and Buggy, J. (1984). Angiotensin–estrogen interaction in female brain reduces drinking and pressor responses. *Am. J. Physiol.* 247:R167–72.

Jonklaas, J., and Buggy, J. (1985). Angiotensin-estrogen central interaction: localization and mechanism. *Brain Res.* 326:239–49.

Joyner, K., Smith, G. P., and Gibbs, J. (1993). Abdominal vagotomy decreases the satiating potency of CCK-8 in sham and real feeding. *Am. J. Physiol.* 264:R912–16.

Ju, G., and Swanson, L. W. (1989). Studies on the cellular architecture of the bed nuclei of the stria terminalis in the rat. I. Cytoarchitecture. *J. Comp. Neurol.* 180:587–602.

Ju, G., Swanson, L. W., and Simerly, R. B. (1989). Studies on the cellular architecture of the bed nuclei of the stria terminalis in the rat. II. Chemoarchitecture. *J. Comp. Neurol.* 280:603–21.

Kagan, J. (1984). *The Nature of the Child.* New York: Basic Books.

Kagan, J. (1989). *Unstable Ideas.* Harvard University Press.

Kagan, J., Reznick, J. S., and Snidman, N. (1988). Biological bases of childhood shyness. *Science* 250:167–71.

Kagan, J., and Schulkin, J. (1995). On the concepts of fear. *Harvard Rev. Psychiatry* 3: 231–4.

Kainu, T., Honkaiemi, J., Gustafsson, J. A., and Petro-Huikko, M. (1993). Co-localization of peptide-like immunoreactives with glucocorticoid receptor and Fos-like immunoreactivity in the rat parabrachial nucleus. *Brain Res.* 615:245–51.

Kaiyala, K. J., Woods, S. C., and Schwartz, M. W. (1995). New model for the regulation of energy balance and adiposity by the central nervous system. *Am. J. Clin. Nutr.* 62: 11235–45.

Kalin, N. H., and Shelton, S. E. (1989). Defensive behaviors in infant rhesus monkeys: environmental cues and neurochemical regulation. *Science* 243:1718–21.

Kalin, N. H., Shelton, S. E., Rickman, M., and Davidson, R. J. (1998). Individual differences in freezing and cortisol in infant and mother rhesus monkeys. *Behav. Neurosci.* 112:251–4.

Kalin, N. H., Takahashi, L. K., and Chen, F. L. (1994). Restraint stress increases corticotropin-releasing hormone mRNA content in the amygdala and paraventricular nucleus. *Brain Res.* 656:182–6.

Kalra, S. P. (1993). Mandatory neuropeptide-steroid signaling for the preovulatory LHRH discharge. *Endocr. Rev.* 14:507–38.

Kalra, S. P., Dube, M. G., Sahu, A., Phelps, C. P., and Kalra, P. S. (1991). Neuropeptide Y secretion increases in the paraventricular nucleus in association with increased appetite for food. *Proc. Natl. Acad. Sci. U.S.A.* 88:10931–5.

Kalra, S. P., and Kalra, P. S. (1996). Is NPY a naturally occurring appetite transducer? *Endocrinol. Metab.* 3:157–63.

Kamara, K. S., Kamara, A. K., and Castonguay, T. W. (1992). A reexamination of the effects of intracerebroventricular glucocorticoids in adrenalectomized rats. *Brain Res. Bull.* 29:355–8.

Kant, G. J., Bauman, R. B., Anderson, S. M., and Moughey, E. H. (1992). Effects of controllable vs uncontrollable chronic stress on stress-responsive plasma hormones. *Physiol. Behav.* 51:1285–8.

Kaplan, J. M., Seeley, R. J., and Grill, H. J. (1993). Daily caloric intake in intact and chronic decerebrate rats. *Behav. Neurosci.* 107:876–81.

Kapp, B. S., Frysinger, R. C., Gallagher, M., and Haselton, J. (1979). Amygdala central nucleus lesions: effects on heart rate conditioning in the rabbit. *Physiol. Behav.* 23:1109–17.

Kaufman, S. (1980). A comparison of the dipsogenic responses of male and female rats to a variety of stimuli. *Can. J. Physiol. Pharmacol.* 58:1180–3.

Kaufman, S. (1981). Control of fluid intake in pregnant and lactating rats. *J. Physiol. (Lond.)* 318:9–16.

Kaufman, S., MacKay, B. J., and Scott, J. Z. (1981). Daily water and electrolyte balance in chronically hyperprolactinaemic rats. *J. Physiol. (Lond.)* 321:19.

Kawauchi, H., Yasuda, A., and Rand-Weaver, M. (1990). Evolution of prolactin and growth hormone family. *Prog. Comp. Endocrinol.* 342:47–53.

Kehoe, P., and Boylan, C. B. (1994). Behavioral effects of kappa-opioid-receptor stimulation on neonatal rats. *Behav. Neurosci.* 108:418–23.

Kelley, D. B. (1986). The genesis of male and female brains. *Trends Neurosci.* 9:499–502.

Kelley, D. B. (1992). Opening and closing a hormone-regulated period for the development of courtship song. A cellular and molecular analysis of vocal neuroeffectors. *Dev. Psychobiol.* 662:178–88.

Kelley, D. B., and Denison, J. (1990). The vocal motor neurons of *Xenopus laevis*: development of sex differences in axon number. *J. Neurobiol.* 21:869–82.

Kelley, D. B., Fenstemaker, S., Hannigan, P., and Shih, S. (1988). Sex differences in the motor nucleus of cranial nerve IX–X in *Xenopus laevis*: a quantitative Golgi study. *J. Neurobiol.* 19:413–29.

Kelly, P. A., Djiane, J., Postel-Vinay, M.-C., and Edery, M. (1991). The prolactin/growth hormone receptor family. *Endocr. Rev.* 12:235–51.

Kendrick, K. M., Keverne, E. B., Hinton, M. R., and Goode, J. A. (1992). Oxytocin, amino acid and monoamine release in the region of the medial preoptic area and bed nucleus of the stria terminalis of the sheep during parturition and suckling. *Brain Res.* 569:199–209.

Kendrick, K. M., Rand, M. S., and Crews, D. (1995). Electrolytic lesions to the ventromedial hypothalamus abolish receptivity in female whiptail lizards, *Cnemidophorus uniparens*. *Brain Res.* 680:226–8.

Kent, S., Rodriguez, F., Kelley, K. W., and Dantzer, R. (1994). Reduction in food and water intake induced by microinjection of interleukin-1B in the ventromedial hypothalamus of the rat. *Physiol. Behav.* 56:1031–6.

Ketter, T. A., George, M. S., Ring, H. A., Pazzaglia, P., Marangell, L., Kimbrell, T. A., and Post, R. M. (1994). Primary mood disorders: structural and resting functional studies. *Psychiatric Annals* 24:637–42.

Keverne, E. B. (1994). Molecular genetic approaches to understanding brain development and behaviour. *Psychoneuroendocrinology* 19:407–14.

Kimura, D. (1992). Sex differences in the brain. *Sci. Am.* 267:118–25.

Kimura, D. (1995). Estrogen replacement therapy may protect against intellectual decline in postmenopausal women. *Horm. Behav.* 29:312–21.

King, B. M. (1988). Glucocorticoids and hypothalamic obesity. *Neurosci. Biobehav. Rev.* 12:29–37.

King, B. M. (1993). Level of corticosterone replacement determines body weight gain in adrenalectomized rats with VMH lesions. *Physiol. Behav.* 54:1187–90.

King, B. M., Banta, A. R., Tharel, G. N., Bruce, B. K., and Frohman, L. A. (1983). Hypothalamic hyperinsulinemia and obesity: role of adrenal glucocorticoids. *Am. J. Physiol.* 245:E194–9.

King, B. M., Zansler, C. A., Richard, S. M., Gutierrez, C., and Dallman, M. F. (1992). Paraventricular hypothalamic obesity in rats: role of corticosterone. *Physiol. Behav.* 51:1207–12.

King, D. P., Zhao, Y., Sangoram, A. M., Wilsbacher, L. D.., Tanaka, M., Antoch, M. P., Steeves, T. D. L., Vitaterna, M. H., Kornhauser, J. M., Lowrey, P. L., Turek, F. W., and Takahashi, J. S. (1997). Positional cloning of the mouse circadian clock gene. *Cell* 89:641–53.

King, S. J., Harding, J. W., and Moe, K. E. (1988). Elevated salt appetite and brain binding of angiotensin II in mineralocorticoid-treated rats. *Brain Res.* 448:140–9.

Kingston, P. A., and Crews, D. (1994). Effects of hypothalamic lesions on courtship and copulatory behavior in sexual and unisexual whiptail lizards. *Brain Res.* 643: 349–51.

Kirn, J. R., Alvarez-Buylla, A., and Nottebohm, F. (1991). Production and survival of projection neurons in a forebrain vocal center of adult male canaries. *J. Neurosci.* 11:1756–62.

Kirn, J. R., and Nottebohm, F. (1993). Direct evidence for loss and replacement of projection neurons in adult canary brain. *J. Neurosci.* 13:1654–63.

Kirschbaum, C., Pirke, K. M., and Hellhammer, D. H. (1993). The 'Trier social stress test' – a tool for investigating psychobiological stress responses in a laboratory setting. *Neuropsychobiology* 28:76–81.

Kirschbaum, C., Pirke, K. M., and Hellhammer, D. H. (1995). Preliminary evidence for reduced cortisol responsivity to psychological stress in women using oral contraceptive medication. *Psychoneuroendocrinology* 20:509–14.

Klein, D. C. (1983). Lesions of the PVN disrupt suprachiasmatic–spinal cord circuit in the melatonin rhythm generating system. *Brain Res. Bull.* 10:647–52.

Kling, M. A., Roy, A., Doran, A. R., Calabrese, J. R., Rubinow, D. R., Whitfield, H. J., Jr., May, C., Post, R. M., Chrousos, G. P., and Gold, P. W. (1991). Cerebrospinal fluid immunoreactive corticotropin-releasing hormone and adrenocorticotropin secretion in Cushing's disease and major depression: potential clinical implications. *J. Clin. Endocrinol. Metab.* 79:260–71.

Kling, M. A., Smith, M. A., Glowa, J. R., Pluznik, D., Demas, J., DeBellis, M. D., Gold, P. W., and Schulkin, J. (1993). Facilitation of cocaine kindling by glucocorticoids in rats. *Brain Res.* 629:163–6.

Knapp, R., and Moore, M. C. (1995). Hormonal responses to aggression vary in different types of agonistic encounters in male tree lizards, *Urosaurus ornatus. Horm. Behav.* 29:85–105.

Knudsen, E. I., and Brainard, M. S. (1991). Visual instruction of the neural map of auditory space in the developing optic tectum. *Science* 253:85–7.

Knudsen, E. I., Knudsen, P. F., and Masino, T. (1993). Parallel pathways mediating both sound localization and gaze control in the forebrain and midbrain of the barn owl. *J. Neurosci.* 13:2837–52.

Koegler-Muly, S. M., Owens, M. J., Ervin, G. N., Kilts, C. D., and Nemeroff, C. B. (1993). Potential corticotropin-releasing factor pathways in the rat brain as determined by bilateral electrolytic lesions of the central amygdaloid nucleus and the paraventricular nucleus of the hypothalamus. *J. Neuroendocrinol.* 5:95–8.

Kojima, K., Maki, S., Hirata, K., Higuchi, S., Akazawa, K., and Nobutada, T. (1995). Relation of emotional behaviors to urine catecholamines and cortisol. *Physiol. Behav.* 57:445–9.

Kolb, B. (1995). *Brain Plasticity and Behavior.* Mahway, NJ: Lawrence Erlbaum Associates.

Kollack, S. S., and Newman, S. W. (1992). Mating behavior induces selective expression of fos protein within the chemosensory pathways of the male Syrian hamster brain. *Neurosci. Lett.* 143:223–8.

Kollack-Walker, S., Watson, J. J., and Akil, H. (1997). Social stress in hamsters: defeat activates specific neurocircuits within the brain. *J. Neurosci.* 17:8842–55.

Konishi, M., and Akutagawa, E. (1990). Growth and atrophy of neurons labeled at their birth in a song nucleus of the zebra finch. *Proc. Natl. Acad. Sci. U.S.A.* 87:3538–41.

Konishi, M., Emlen, S. T., Ricklefs, R. E., and Wingfield, J. C. (1989). Contributions of bird studies to biology. *Science* 246:465–72.

Konishi, M., and Gurney, M. E. (1982). Sexual differentiation of brain and behaviour. *Trends Neurosci.* 5:20–3.

Koob, G. F., and Bloom, F. E. (1985). Corticotropin-releasing factor and behavior. *Fed. Proc.* 44:259–63.

Koob, G. F., Heinrichs, S. C., Menzaghi, F., Pich, E. M., and Britton, K. T. (1994). Corticotropin releasing factor, stress and behavior. *The Neurosciences* 6:2219–29.

Koob, G. F., Heinrichs, S. C., Pich, E. M., Menzaghi, F., Baldwin, H., Miczek, K., and Britton, K. T. (1993). The role of corticotropin-releasing factor in behavioral responses to stress. In: *Corticotropin-releasing Factor,* ed. K. Chadwick, J. Marsh, and K. Ackrill. New York: Wiley.

Kooslyn, S. M., and Koenig, O. (1992). *Wet Mind.* New York: Free Press.

Korte, S. M., De Boer, S. F., De Kloet, E. R., and Bohus, B. (1994). Anxiolytic effects of selective mineralocorticoid and glucocorticoid antagonists on fear-enhanced behavior in the elevated plus-maze. *Psychoneuroendocrinology* 20:385–94.

Kott, K. S., Moore, B. J., Fournier, L., and Horwitz, B. A. (1989). Hyperprolactinemia prevents short photoperiod-induced changes in brown fat. *Am. J. Physiol.* 256:R174–80.

Kotz, C. M., Grace, M. K., Briggs, J., Levine, A. S., and Billington, C. J. (1995). Effects of opioid antagonists naloxone and naltrexone on neuropeptide Y-induced feeding and brown fat thermogenesis in the rat. Neural site of action. *J. Clin. Invest.* 96:163–70.

Krebs, J. R., and Davies, N. B. (1991). *Behavioral Ecology.* Oxford: Blackwell. (Originally published 1978.)

Krecek, J. (1972). Sex differences in the taste preference for a salt solution in the rat. *Physiol. Behav.* 8:183–8.

Krecek, J. (1973a). Sex differences in salt taste: the effect of testosterone. *Physiol. Behav.* 10:683–8.

Krecek, J. (1973b). Sex differences in salt taste: the effect of testosterone. Physiology and rats. *Dev. Psychobiol.* 9:181–8.

Krecek, J. (1974). Critical periods of development and the regulation of salt intake. Ontogenesis of the brain. *Neuroontogenti,* pp. 249–58.

Krecek, J. (1975). The pineal gland and the effect of neonatal administration of androgen upon the development of spontaneous salt and water intake in female rats. *Neuroendocrinology* 28:137–43.

Krettek, J. E., and Price, J. L. (1978). Amygdaloid projections to subcortical structures within the basal fore brain and brain stem in the rat and cat. *J. Comp. Neurol.* 178: 225–54.

Krieckhaus, E. E. (1970). Innate recognition aids rats in sodium regulation. *J. Comp. Physiol. Psychol.* 73:117–22.

Krieckhaus, E. E., and Wolf, G. (1968). Acquisition of sodium by rats: interaction of innate mechanisms and latent learning. *J. Comp. Physiol. Psychol.* 2:197–201.

Kucharczyk, J. (1984). Neuroendocrine mechanisms mediating fluid intake during the estrous cycle. *Brain Res. Bull.* 12:175–80.

Kuhn, C. M., Pauk, J., and Schanberg, S. M. (1990). Endocrine responses to mother-infant separation in developing rats. *Dev. Psychobiol.* 23:395–410.

Kuhn, C. M., Schanberg, S. M., Field, T., Symanski, R., Zimmerman, E., Scafidi, F., and Roberts, J. (1991). Tactile-kinesthetic stimulation effects on sympathetic and adrenocortical function in preterm infants. *J. Pediatr.* 119:434–40.

Kumar, B. A., Papamichael, M., and Leibowitz, S. F. (1988). Feeding and macronutrient selection patterns in rats: adrenalectomy and chronic corticosterone replacement. *Physiol. Behav.* 42:581–9.

Kvetnansky, R., Fukuhara, K., Pacak, K., Cizza, G., Goldstein, D. S., and Kopin, I. J. (1993). Endogenous glucocorticoids restrain catecholamine synthesis and release at rest and during immobilization stress in rats. *Endocrinology* 133:1411–19.

Kyrkouli, S. E., Stanley, B. G., Seirafi, R. D., and Leibowitz, S. F. (1990). Stimulation of feeding by galanin: anatomical localization and behavioral specificity of this peptide's effects in the brain. *Peptides* 11:995–1001.

Ladd, C. O., Owens, M. J., and Nemeroff, C. B. (1996). Persistent changes in corticotropin releasing factor neuronal systems induced by maternal deprivation. *Endocrinology* 137:1212–18.

Lambert, P. D., Phillips, P. J., Wilding, J. P., Bloom, S. R., and Herbert, J. (1995). c-fos expression in the paraventricular nucleus of the hypothalamus following intracerebroventricular infusions of neuropeptide Y. *Brain Res.* 670:59–65.

Lambert, P. D., Wilding, J. P., al-Dokhayel, A. A., Bohuon, C., Comoy, E., Gilbey, S. G., and Bloom, S. R. (1993). A role for neuropeptide-Y, dynorphin, and noradrenaline in the central control of food intake after food deprivation. *Endocrinology* 133:29–32.

Lambert, P. D., Wilding, J. P., Turton, M. D., Ghatei, M. A., and Bloom, S. R. (1994). Effect of food deprivation and streptozotocin-induced diabetes on hypothalamic neuropeptide Y release as measured by a radioimmunoassay-linked microdialysis procedure. *Brain Res.* 656:135–40.

Lance, V. A., and Elsey, R. M. (1986). Stress-induced suppression of testosterone secretion in male alligators. *J. Exp. Zool.* 239:241–6.

Landsberg, J.-W., and Weiss, J. (1975). Stress and increase of the corticosterone level prevent imprinting in ducklings. *Behaviour* 57:3–4.

Lang, P. J. (1995). The emotion probe, studies of motivation and attention. *Am. Psychol.* 50:372–85.

Lansdowne, A. T. G., and Provost, S. C. (1998). Vitamin D_3 enhances mood in healthy subjects during winter. *Psychopharmacology* 135:319–23.

Larsen, P. J., Jessop, D. S., Chowdrey, H. S., Lightman, S. L., and Mikkelsen, J. D. (1994). Chronic administration of glucocorticoids directly upregulates prepro-neuropeptide Y and Y1-receptor mRNA levels in the arcuate nucleus of the rat. *J. Neuroendocrinol.* 6:153–9.

Lashley, K. S. (1938). An experimental analysis of instinctive behavior. *Psychol. Rev.* 45:445–71.

Laudenslager, M. L., Boccia, M. L., Berger, C. L., Gennaro-Ruggles, M. M., McFerran, B., and Reite, M. L. (1995). Total cortisol, free cortisol, and growth hormone associated with brief social separation experiences in young macaques. *Dev. Psychobiol.* 28:199–211.

LeDoux, J. E. (1987). Emotion. In: *Handbook of Physiology. The Nervous System,* pp. 419–59. Washington, DC: American Physiological Society.

LeDoux, J. E. (1995). Emotion: clues from the brain. *Annu. Rev. Psychol.* 46:209–35.

LeDoux, J. E. (1996). *The Emotional Brain.* New York: Simon & Schuster.

LeDoux, J. E., Cicchettie, P., Xagoraris, A., and Romanski, L. M. (1990). The lateral amygdaloid nucleus: sensory interface of the amygdala in fear conditioning. *J. Neurosci.* 10:1062–9.

LeDoux, J. E., Iwata, J., Cicceti, P., and Reis, D. J. (1988). Different projections of the central amygdaloid nucleus mediate behavioral correlates of conditioned fear. *J. Neurosci.* 8:2517–29.

Lee, E. H. Y., Lee, C. P., Wang, H. L., and Lin, W. R. (1993). Hippocampal CRF, NE and NMDA system interactions in memory processing in the rat. *Synapse* 14:144–53.

Lee, L. Y., and Davis, M. (1997). Role of the hippocampus, the bed nucleus of the stria terminalis, and the amygdala in the excitatory effect of corticotropin-releasing hormone on the acoustic startle reflex. *J. Neurosci.* 17:6434–46.

Lee, L. Y., Schulkin, J., and Davis, M. (1994). Effect of corticosterone on the enhancement of the acoustic startle reflex by corticotropin-releasing hormone. *Brain Res.* 666: 93–8.

Lee, M. C., Mannon, P. J., Grant, J. P., and Pappas, T. N. (1997). Total parenteral nutrition alters NPY/PYY receptor levels in the rat brain. *Physiol. Behav.* 62:1219–23.

Lee, T. M., Pelz, K., Licht, P., and Zucker, I. (1990a). Testosterone influences hibernation in golden-mantled ground squirrels. *Am. J. Physiol.* 259:R760–7.

Lee, T. M., Smale, L., Zucker, I., and Dark, J. (1987). Role of photoperiod during pregnancy and lactation in the meadow vole, *Microtus pennsylvanicus. J. Reprod. Fertil.* 81:343–50.

Lee, W. S., Smith, M. J., and Hoffman, G. C. (1990b). LHRH neurons express c-fos expression during preestrus. *Proc. Natl. Acad. Sci. U.S.A.* 87:5163–7.

Lehman, M. N., Powers, J. B., and Winans, S. S. (1983). Stria terminalis lesions alter the temporal pattern of copulatory behavior in the male golden hamster. *Behav. Brain Res.* 8:109–14.

Lehman, M. N., Winans, S. S., and Powers, J. B. (1980). Medial nucleus of the amygdala mediates chemosensory control of male hamster sexual behavior. *Science* 210:557–60.

Lehrman, D. S. (1955). The physiological basis of parental feeding behavior in the ring dove (*Streptopelia risoria*). *Behaviour* 7:241–86.

Lehrman, D. S. (1958). Induction of broodiness by participation in courtship and nest-building in the ring dove (*Streptopelia risoria*). *J. Comp. Physiol.* 51:32–6.

Lehrman, D. S. (1961). Hormonal regulation of parental behavior in birds and infrahuman mammals. In: *Sex and Internal Secretions,* ed. W. C. Young, pp. 1268–82. Baltimore: Williams & Wilkins.

Lehrman, D. S., and Friedman, M. (1968). Physiological conditions for the stimulation of prolactin secretion by external stimuli in the male ring dove. *Anim. Behav.* 16:233–7.

Leibenluft, E. (1993). Do gonadal steroids regulate circadian rhythms in humans? *J. Affect. Disord.* 29:175–81.

Leibenluft, E., Fiero, P. L., and Rubinow, D. R. (1994). Effects of the menstrual cycle on dependent variables in mood disorder research. *Arch. Gen. Psychiatry* 51:761–81.

Leibowitz, S. F. (1995). Brain peptides and obesity: pharmacological treatment. *Obesity Research* 3:573–89.

Leibowitz, S. F., Diaz, S., and Tempel, D. (1989). Norepinephrine in the paraventricular nucleus stimulates corticosterone release. *Brain Res.* 496:219–27.

Leibowitz, S. F., Roland, C. R., Hor, L., and Squillari, V. (1984). Noradrenergic feeding elicited via the paraventricular nucleus is dependent upon circulating corticosterone. *Physiol. Behav.* 32:857–64.

LeMagnen, J. (1985). *Hunger.* Cambridge University Press.

Lemoine, J., and Kucharczyk, J. (1985). Fluid regulation and reproductive cyclicity in female rats treated neonatally with testosterone and methandrostenolone. *Horm. Res.* 29:291–300.

Lennenberg, E. H. (1967). *Biological Foundations of Language.* New York: Wiley.

Leshem, M., DelCanho, S., and Schulkin, J. (1996). A possible role for vitamin D in calcium appetite. Presented at the Virginia meeting of the Comp. Nutrition Soc.

Leshem, M., and Epstein, A. N. (1989). Ontogeny of renin-induced salt appetite in the rat pup. *Dev. Psychobiol.* 22:437–45.

Leshem, M., Maroun, M., and DelCanho, S., (1995). Sodium depletion and maternal separation in the suckling rat increase its salt intake when adult. *Physiol. Behav.* 58:6–10.

LeVay, S. (1991a). A difference in hypothalamic structure between heterosexual and homosexual men. *Science* 253:1034–6.

LeVay, S. (1991b). *The Sexual Brain.* MIT Press.

Levine, A. S., and Billington, C. J. (in press). Peptides in regulation of energy metabolism and body weight.

Levine, A. S., and Morley, J. E. (1984). Neuropeptide Y: a potent inducer of consummatory behavior in rats. *Peptides* 5:1025–9.

Levine, A. S., Weldon, D. T., Grace, M., Cleary, J. P., and Billington, C. J. (1995). Naloxone blocks that portion of feeding driven by sweet taste in food restricted rats. *Am. J. Physiol.* 268:248–52.

Levine, S. (1993). The influence of social factors on the response to stress. *Psychother. Psychosom.* 60:33–8.

Levine, S. (1994). The ontogeny of the hypothalamic-pituitary-adrenal axis. The influence of maternal factors. *Ann. N.Y. Acad. Sci.* 746:275–87.

Levine, S., Chamoux, M., and Wiener, S. G. (1991). Social modulation of the stress response. In: *Stress and Related Disorders from Adaption to Dysfunction,* ed. A. R. Genazzani et al. New Jersey: Parthenon Publishing Co.

Levine, S., Coe, C., and Wiener, S. (1989). The psychoneuroendocrinology of stress – a psychobiological perspective. In: *Psychoendocrinology,* ed. S. Levine and R. Brusch. New York: Academic Press.

Levine, S., Huchton, D. M., Wiener, S. G., and Rosenfeld, P. (1991). Time course of the effect of maternal deprivation on the hypothalamic-pituitary-adrenal axis in the infant rat. *Dev. Psychobiol.* 24:547–58.

Levine, S., and Mullins, R. F. (1968). Hormones and infancy. In: *Early Experience and Behavior,* ed. G. Newton and S. Levine. Springfield, IL: Thomas.

Levy, F., Kendrick, K. M., Keverne, E. B., Piketty, V., and Poindron, P. (1992). Intracerebral oxytocin is important for the onset of maternal behavior in inexperienced ewes delivered under peridural anesthesia. *Behav. Neurosci.* 106:427–32.

Lewy, A. J., Wehr, T. A., Goodwin, F. K., Newsome, D. A., and Markey, S. P. (1980). Light suppresses melatonin secretion in humans. *Science* 210:1267–8.

Liang, K. C., Melia, R. K., Campeau, S., Falls, W. A., Miserendino, J. D., and Davis, M. (1992). Lesions of the central nucleus of the amygdala, but not the paraventricular nucleus of the hypothalamus, block the excitatory effects of corticotropin-releasing factor on the acoustic startle reflex. *J. Neurosci.* 19:2313–20.

Licht, P., Zucker, I., Hubbard, G., and Boshes, M. (1982). Circannual rhythms of plasma testosterone and luteinizing hormone levels in golden-mantled ground squirrels (*Spermophilus lateralis*). *Biol. Reprod.* 27:411–18.

Lincoln, G. A. (1979). Light-induced rhythms of prolactin secretion in the ram and the effect of cranial sympathectomy. *Acta Endocrinol.* 91:421–7.

Lind, R. W. (1988). Sites of action of angiotensin in the brain. In: *Angiotensin and Blood Pressure Regulation,* ed. J. Harding, H. Wright, R. C. Speth, and N. Barnes. New York: Academic Press.

Lind, R. W., Swanson, L. W., and Ganten, D. (1985). Organization of angiotensin II immunoreactive cells and fibers in the rat central nervous system. *Neuroendocrinology* 40:9–24.

Lindzey, J., and Crews, D. (1992). Interactions between progesterone and androgens in the stimulation of sex behaviors in male little striped whiptail lizards, *Cnemidophorus inornatus. Gen. Comp. Endocrinol.* 86:52–8.

Lindzey, J., and Crews, D. (1993). Effects of progesterone and dihydrotestosterone on stimulation of androgen-dependent sex behavior, accessory sex structures, and in vitro binding characteristics of cytosolic androgen receptors in male whiptail lizards (*Cnemidophorus inornatus*). *Horm. Behav.* 27:269–81.

Linkowski, P., Van Cauter, E., L'Hermite-Baleriaux, M., Kerkhofs, M., Hubain, P., L'Hermite, M., and Mendlewicz, J. (1989). The 24-hour profile of plasma prolactin in men with major endogenous depressive illness. *Arch. Gen. Psychiatry* 46:813–19.

Lisciotto, C. A., and Morrell, J. I. (1993). Circulating gonadal steroid hormones regulate estrogen receptor mRNA in the male rat forebrain. *Mol. Brain Res.* 20:79–90.

Liu, D., Diorio, J., Tannenbaum, B., Caldji, C., Francis, D., Freedman, A., Sharma, S., Pearson, D., Plotsky, P. M., and Meaney, M. J. (1997a). Maternal care, hippocampal glucocorticoid receptors, and hypothalamic-pituitary-adrenal responses to stress. *Science* 277:1659–62.

Liu, J.-W., and Ben-Jonathan, N. (1994). Prolactin-releasing activity of neurohypophysial hormones: structure–function relationship. *Endocrinology* 134:114–18.

Liu, Y. C., Salamone, J. D., and Sachs, B. D. (1997b). Lesions of the preoptic area and bed nucleus of the stria terminalis; differential effects on copulatory behavior and noncontact erection in male rats. *J. Neurosci.* 17:5245–53.

Lo, D. C. (1995). Neurotrophic factors and synaptic plasticity. *Neuron* 15:979–81.

Logan, C. A., and Wingfield, J. C. (1995). Hormonal correlates of breeding status, nest construction, and parental care in multiple-brooded northern mockingbirds, *Mimus polyglottos. Horm. Behav.* 29:17–30.

Lois, C., and Alvarez-Buylla, A. (1994). Long-distance neuronal migration in the adult mammalian brain. *Science* 264:1145–8.

Lonstein, J. L., and Stern, J. M. (1997). Role of the midbrain periaqueductal gray in maternal nurturance and aggression: c-fos and electrolytic lesion studies in lactating rats. *J. Neurosci.* 17:3364–78.

Lorenz, D. N. (1994). Effects of CCK-8 on ingestive behaviors of suckling and weanling rats. *Dev. Psychobiol.* 27:39–52.

Lorenz, D. N. and Goldman, S. A. (1982). Vagal mediation of the cholecystokinin satiety effect in rats. *Physiol. Behav.* 29:599–604.

Lorenz, D. N., Kreielsheimer, G., and Smith, G. P. (1979). Effect of cholecystokinin, gastrin, secretin and GIP on sham feeding in the rat. *Physiol. Behav.* 23:1065–72.

Lorenz, K. L. (1981). *The Foundations of Ethology.* Berlin: Springer-Verlag.

Louch, C. D., and Higginbotham, M. (1967). The relation between social rank and plasma corticosterone levels in mice. *Gen. Comp. Endocrinol.* 8:441–4.

Lovenberg, T. W., Liaw, C. W., Grigoriadis, D. E., Clevenger, W., Chalmers, D. T., De-Souza, E. B., and Oltersdorf, T. (1995). Cloning and characterization of a functionally distinct corticotropin-releasing factor receptor subtype from rat brain. *Neurobiology* 92:836–40.

Lozoff, B., Felt, B. T., Nelson, E. C., Wolf, A. W., Meltzer, H. W., and Jimenez, E. (1995). Serum prolactin levels and behavior in infants. *Biol. Psychiatry* 37:4–12.

Lu, J., and Cassone, V. M. (1993). Daily melatonin administration synchronizes circadian patterns of brain metabolism and behavior in pinealectomized house sparrows. *J. Comp. Physiol.* 173:775–82.

Lupien, S., Lecours, A. R., Lussier, I., Schwartz, G., Nair, N. P. V., and Meaney, M. J. (1994). Basal cortisol levels and cognitive deficits in human aging. *J. Neurosci.* 14: 2893–903.

Lynch, K. R., Hawelu-Johnson, C. L., and Guyenet, P. G. (1987). Localization of brain angiotensinogen mRNA by hybridization histochemistry. *Brain Res.* 388:149–58.

Lyons, D. M., and Levine, S. (1994). Social regulatory effects on squirrel monkey pituitary-adrenal activity: a longitudinal analysis of cortisol and ACTH. *Psychoneuroendocrinology* 19:983–91.

Ma, L. Y., Itharat, P., Fluharty, S. J., and Sakai, R. R. (1997). Intracerebroventricular administration of mineralocorticoid receptor antisense oligonucleotides attenuates salt appetite in the rat. *Stress* 2:37–50.

Ma, L. Y., McEwen, B. S., Sakai, R. R., and Schulkin, J. (1993). Glucocorticoids facilitate mineralocorticoid-induced sodium intake in the rat. *Horm. Behav.* 27:240–50.

Ma, L. Y., Polidori, C., Schulkin, J., Epstein, A. N., Stellar, E., McEwen, B. S., and Sakai, R. R. (1992). Effect of centrally administered mineralocorticoid or glucocorticoid receptor antagonist on aldosterone-induced sodium intake in rats. *Society of Neuroscience Abstracts.*

McCance, R. A. (1936). Medical problems in mineral metabolism. III: Experimental human salt deficiency. *Lancet* 230:823–30.

McCance, R. A. (1938). The effect of salt deficiency in man on the volume of the extra-cellular fluids and on the composition of sweat, saliva, gastric juice and cerebrospinal fluid. *J. Physiol. (Lond.)* 92:208–18.

McCann, S. M., and Antunes-Rodrigues, J. S. (1996). Atrial natriuretic peptide: water, electrolyte dynamics, and behavior. In: *Hormonal Modulation of Brain and Behavior,* ed. U. Halbreich and S. J. Wamback. Berlin: Springer-Verlag.

McCann, S. M., Franci, C. R., and Antunes-Rodrigues, J. A. (1989). Hormonal control of water and electrolyte intake and output. *Acta Physiol. Scand. [Suppl. 1]* 583:97–104.

Maccari, S., Piazza, P. V., Deminiere, J. M., Lemaire, V., Mormede, P., Simon, H., An-gelucci, L., and Le Moal, M. (1991). Life events-induced decrease of corticosteroid type I receptors is associated with reduced corticosterone feedback and enhanced vulnerability to amphetamine self-administration. *Brain Res.* 547:7–12.

McCarthy, M. M., Kleopoulos, S. P., Mobbs, C. V., and Pfaff, D. W. (1993a). Infusion of antisense oligodeoxynucleotides to the oxytocin receptor in the ventromedial hypo-thalamus reduces estrogen-induced sexual receptivity and oxytocin receptor binding in the female rat. *Neuroendocrinology* 59:432–40.

McCarthy, M. M., Schlenker, E. H., and Pfaff, D. W. (1993b). Enduring consequences of neonatal treatment with antisense oligodeoxynucleotides to estrogen receptor mes-senger ribonucleic acid on sexual differentiation of rat brain. *Endocrinology* 133:433–9.

McCaughey, A. S., and Scott, T. R. (in press). The taste of sodium. *Neurosci. Biobehav. Rev.*

MacDougald, O. A., Hwang, C. S., Fan, H., and Lane, M. D. (1995). Regulated expression of the obese gene product (leptin) in white adipose tissue and 3T3-L1 adipocytes. *Proc. Natl. Acad. Sci. U.S.A.* 92:9034–7.

McEwen, B. S. (1991). Non-genomic and genomic effects of steroids on neural activity. *TIPS Reviews* 12:141–7.

McEwen, B. S. (1992). Re-examination of the glucocorticoid hypothesis of stress and aging. *Prog. Brain Res.* 93:365–85.

McEwen, B. S. (1995). Neuroendocrine interactions. In: *Psychopharmacology: The Fourth Generation of Progress,* ed. F. E. Bloom and D. J. Kupfer. New York: Raven Press.

McEwen, B. S. (1998). Protective and damaging effects of stress mediators. *N. Engl. J. Med.* 338:171–9.

McEwen, B. S., Jones, K. J., and Pfaff, D. W. (1987). Hormonal control of sexual behavior in the female rat: molecular, cellular and neurochemical studies. *Biol. Reprod.* 36:37–45.

McEwen, B. S., Lambdin, L. T., Rainbow, T. C., and DeNicola, A. F. (1986). Aldosterone effects on salt appetite in adrenalectomized rats. *Neuroendocrinology* 43:38–43.

McEwen, B. S., and Sapolsky, R. (1995). Stress and cognitive function. *Neurobiology* 5:205–16.

McEwen, B. S., and Schmeck, H. M., Jr. (1994). *The Hostage Brain.* New York: Rocke-feller University Press.

McEwen, B. S., and Stellar, E. (1993). Stress and the individual: mechanism leading to disease. *Arch. Intern. Med.* 153:2093–101.

McEwen, G. S., Lieberburg, I., Chaptal, C., and Krey, L. C. (1977). Aromatization: im-portant for sexual differentiation of the neonatal rat brain. *Horm. Behav.* 9:249–63.

McKinley, M. J., Congiu, M., Oldfield, B. J., and Pennington, G. (1988). Cerebral regu-lation of vasopressin secretion. *Prog. Endocrinol.* 90:1189–94.

Mackintosh, N. J. (1975). A theory of attention: variations in the associability of stimulus and reinforcement. *Psychol. Rev.* 82:276–98.

MacLean, P. D. (1949). Psychosomatic disease and the "visceral brain." *Psychosom. Med.* 11:338–53.

Madden, J., Akil, H., Patrick, R. L., and Barchas, J. D. (1977). Stress-induced parallel changes in central opioid levels and pain responsiveness in the rat. *Nature* 265:358–60.

Madeira, M. D., and Lieberman, A. R. (1995). Sexual dimorphism in the mammalian limbic system. *Prog. Neurobiol.* 45:275–333.

Maffei, M., Halaas, J., Ravussin, E., Pratley, R. E., Lee, G. H., Zhang, Y., Fei, H., Kim, S., Lallone, R., Ranganathan, S., Kern, P. A., and Friedman, J. M. (1995). Leptin levels in human and rodent: measurement of plasma leptin and ob RNA in obese and weight-reduced subjects. *Nature Medicine* 1:1155–61.

Magarinos, A. M., Coirini, H., DeNicola, A. F., and McEwen, B. S. (1986). Mineralocorticoid regulation of salt intake is preserved in hippocampectomized rat. *Neuroendocrinology* 44:494–7.

Mah, S. J., Ades, A. M., Mir, R., Siemens, I. R., Williamson, J. R., and Fluharty, S. J. (1992). Association of solubilized angiotensin II-receptors with phospholipase c-α in murine neuroblastoma NIE-115 cells. *Mol. Pharmacol.* 42:428–37.

Mahler, M. S., Pine, F., and Bergman, A. (1975). *The Psychological Birth of the Human Infant.* New York: Basic Books.

Mahon, J. M., Allen, M., Herbert, J., and Fitzsimons, J. T. (1995). The association of thirst, sodium appetite and vasopressin release with c-fos expression in the forebrain of the rat after intracerebroventricular injection of angiotensin II or carbachol. *Neuroscience* 69:199–208.

Mai, F. M., Shaw, B. F., Jenner, M. R., Wielgosz, G., and Iles, D. (1985). Nocturnal prolactin secretion in depression. *Br. J. Psychiatry* 147:314–17.

Maier, S. F., Grahn, R. E., Kalman, B. A., Sutton, L. C., Wiertelak, E. P., and Watkins, L. R. (1993). The role of the amygdala and dorsal raphe nucleus in mediating the behavioral consequences of inescapable shock. *Behav. Neurosci.* 107:377–88.

Maier, S. F., and Ryan, S. M. (1986). Stressor controllability and the pituitary-adrenal system. *Behav. Neurosci.* 100:669–74.

Majewska, M. D., Bisserbem, J. C., and Eskay, R. L. (1985). Glucocorticoids are modulators of GABA receptors in brain. *Brain Res.* 339:178–85.

Maki, R., He, P., Zhang, D. M., Williamson, J. R., and Fluharty, S. J. (1992). Corticosteroid regulation of PLC-α in rat brain and cultured neuronal cells. *Society for Neuroscience Abstracts* 18:1163.

Makino, S., Gold, P. W., and Schulkin, J. (1994a). Corticosterone effects on CRH mRNA in the central nucleus of the amygdala and the paraventricular nucleus of the hypothalamus. *Brain Res.* 640:105–12.

Makino, S., Gold, P. W., and Schulkin, J. (1994b). Effects of corticosterone on CRH mRNA and content in the bed nucleus of the stria terminalis; comparison with the effects in the central nucleus of the amygdala and the paraventricular nucleus of the hypothalamus. *Brain Res.* 657:141–9.

Makino, S., Schulkin, J., Smith, M. A., Pacak, K., Palkovits, M., and Gold, P. W. (1995). Regulation of corticotropin-releasing hormone receptor mRNA in the rat brain and pituitary by glucocorticoids and stress. *Endocrinology* 136:4517–25.

Maney, D. L., Goode, C. T., and Wingfield, J. C. (1997). Intraventricular infusions of arginine vasotocin induces vocal behavior in a female songbird. *J. Neuroendocrinol.* 9:487–91.

Mangurian, L. P., Walsh, R. J., and Posner, B. I. (1992). Prolactin enhancement of its own uptake at the choroid plexus. *Endocrinology* 131:698–702.

Mani, S. K., Allen, J. M. C., Clark, J. H., Blaustein, J. D., and O'Malley, B. W. (1994). Convergent pathways for steroid hormone- and neurotransmitter-induced rat sexual behavior. *Science* 265:1246–7.

Mani, S. K., Blaustein, J. D., and O'Malley, B. W. (1997). Progesterone receptor function from a behavioral perspective. *Horm. Behav.* 31:244–56.

Manogue, K. R., Leshner, A. I., and Candland, D. K. (1975). Dominance status and adrenocortical reactivity to stress in squirrel monkeys (*Saimiri sciureus*). *Primates* 16: 457–63.

Marchant, E. G., Watson, N. V., and Mistlberger, R. E. (1997). Both neuropeptide Y and serotonin are necessay for entrainment of circadian rhythms in mice by daily treadmill running schedules. *J. Neurosci.* 17:7974–87.

Marin, M. L., Tobias, M. L., and Kelley, D. B. (1990). Hormone-sensitive stages in the sexual differentiation of laryngeal muscle fiber number in *Xenopus laevis. Development* 110:703–12.

Marinelli, M., Piazza, P. V., Deroche, V., Maccari, S., LeMoal, M., and Simon, H. (1994). Corticosterone circadian secretion differentially facilitates dopamine-mediated psychomotor effect of cocaine and morphine. *J. Neurosci.* 14:2724–31.

Mariscal, G. G., Melo, A. I., Jimenez, P., Beyer, C., and Rosenblatt, J. S. (1996). Estradiol, progesterone and prolactin regulate maternal nest-building in rabbits. *J. Neuroendocrinol.* 8:901–7.

Marler, C. A., Chu, J., and Wilczynski, W. (1995). Arginine vasotocin injection increases probability of calling in cricket frogs, but causes call changes characteristic of less aggressive males. *Horm. Behav.* 29:554–70.

Marler, P. (1961). The logical analysis of animal communication. *J. Theor. Biol.* 1:295–317.

Marler, P. (1992). Functions of arousal and emotion in primate communication: a semiotic approach. In: *Topics in Primatology,* ed. T. Nishida, W. C. McGrew, P. Marler, M. Pickford, and F. B. M. De Waal. University of Tokyo Press.

Marler, P., and Hamilton, W. J., III (1966). *Mechanisms of Animal Behavior.* New York: Wiley.

Marler, P., Peters, S., Ball, G. F., Dufty, A. M., Jr., and Wingfield, J. C. (1988). The role of sex steroids in the acquisition and production of birdsong. *Nature* 336:770–2.

Marr, D. (1982). *Vision.* San Francisco: Freeman.

Martinet, L., Bonnefond, C., Peytevin, J., Monnerie, R., and Marcilloux, J. C. (1995). Vasoactive intestinal polypeptide in the suprachiasmatic nucleus of the mink (*Mustela vison*) could play a key role in photic induction. *J. Neuroendocrinol.* 7:69–79.

Masler, E. P., Kelly, T. J., and Menn, J. J. (1993). Insect neuropeptides: discovery and application in insect management. *Arch. Insect Biochem. Physiol.* 22:87–111.

Mason, J. R., and Reidinger, R. F. (1981). Effects of social facilitation and observational learning on feeding behavior of the red-winged blackbird (*Agelaius phoeniceus*). *Auk* 98:778–84.

Mason, J. R., and Reidinger, R. F. (1982). Observational learning of food aversions in red-winged blackbirds (*Agelaius phoeniceus*). *Auk* 99:548–54.

Mason, J. W. (1975). Emotions as reflected as patterns of endocrine integration. In: *Emotions: Their Parameters and Measurements,* ed. L. Levi. New York: Raven Press.

Mason, J. W., Brady, J., Polish, E., Bauer, J., Robinson, J., Rose, R., and Taylor, E. (1961). Patterns of corticosteroid and pepsinogen change related to emotional stress in the monkey. *Science* 133:1596–601.

Mason, J. W., Brady, J. V., and Sidman, M. (1957). Plasma 17-hydroxycorticosteroid levels and conditioned behavior in the rhesus monkey. *Endocrinology* 60:741–52.

Masotto, C., and Negro-Vilar, A. (1985). Inhibition of spontaneous or angiotensin II-stimulated water intake by atrial natriuretic factor. *Brain Res. Bull.* 15:523–6.

Massi, M., and Epstein, A. N. (1990). Angiotensin/aldosterone synergy governs the salt appetite of the pigeon. *Appetite* 14:181–92.

Massi, M., Gentili, L., Perfumi, M., de Caro, G., and Schulkin, J. (1990). Inhibition of salt appetite in the rat following injection of tachykinins into the medial amygdala. *Brain Res.* 513:1–7.

Massi, M., Polidori, C., Perfumi, M., Gentili, L., and de Caro, G. (1991). Tachykinin receptor subtypes involved in the central effects of tachykinins on water and salt intake. *Brain Res. Bull.* 26:155–60.

Mauro, L. J., Elde, R. P., Youngren, O. M., Phillips, R. E., and El Halawani, M. E. (1989). Alterations in hypothalamic vasoactive intestinal peptide-like immunoreactivity are associated with reproduction and prolactin release in the female turkey. *Endocrinology* 125:1795–804.

Mead, M. (1974). *Coming of Age in Samoa.* New York: William Morris & Co. (Originally published 1928.)

Meaney, M. J., Diorio, J., Francis, D., LaRocque, S., O'Donnell, D., Smythe, J. W., Sharma, S., and Tannenbaum, B. (1994). Environmental regulation of the development of glucocorticoid receptor systems in the rat forebrain. The role of serotonin. *Ann. N.Y. Acad. Sci.* 746:260–73.

Meaney, M. J., Mitchell, J. B., Aitken, D. H., Bhatnagar, S., Bondoff, S. R., Iny, L. J., and Sarrieau, A. (1991). The effects of neonatal handling on the development of the adrenocortical response to stress: implications for neuropathology and cognitive deficits in later life. *Psychoneuroendocrinology* 16:85–103.

Meddle, S. L., King, U. M., Follett, B. K., Wingfield, J. C., Ramenofsky, M., Foidart, A., and Balthazart, J. (1997). Copulation activates fos-like immunoreactivity in the male quail forebrain. *Behav. Brain Res.* 85:143–59.

Meisel, R. L., and Sachs, B. D. (1994). The physiology of male sexual behavior. In: *The Physiology of Reproduction,* 2nd ed., ed. E. Knobil and J. D. Neill. New York: Raven Press.

Menani, J. V., Thunhorst, R. L., and Johnson, A. K. (in press). Lateral parabrachial nucleus and serotonergic mechanisms in the control of salt appetite in rats. *Am. J. Physiol.*

Mendelsohn, F. A. O., Allen, A. M., Clevers, J., Denton, D. A., Tarjan, E., and McKinley, M. J. (1988). Localization of angiotensin II receptor binding in rabbit brain by in vitro autoradiography. *J. Comp. Neurol.* 270:372–84.

Mendlewicz, J., Van Cauter, E., Linkowski, P., L'Hermite, M., and Robyn, C. (1980). Current concepts: the 24-hour profile of prolactin in depression. *Life Sci.* 27:2015–24.

Menendez-Pelaez, A., and Reiter, R. J. (1993). Distribution of melatonin in mammalian tissues: the relative importance of nuclear versus cytosolic localization. *J. Pineal Res.* 15:59–69.

Menzaghi, F. E., Heinrichs, S. C., Pich, E. M., Tilders, F. J., and Koob, G. F. (1993).

Functional impairment of hypothalamic corticotropin-releasing neurons with immunotargeted toxins enhances food intake induced by neuropeptide Y. *Brain Res.* 618:76–82.

Mercer, J. G., Lawrence, C. B., and Atkinson, T. (1996). Hypothalamic NPY and CRF gene expression in the food-deprived Syrian hamster. *Physiol. Behav.* 60:121–7.

Mermelstein, P. G., and Becker, J. B. (1995). Increased extracellular dopamine in the nucleus accumbens and striatum of the female rat during paced copulatory behavior. *Behav. Neurosci.* 109:354–65.

Meyer-Bahlburg, H. F. L., Ehrhardt, A. A., Rosen, L. R., and Gruen, R. S. (1995). Prenatal estrogens and the development of homosexual orientation. *Dev. Psychol.* 31: 12–21.

Micevych, P. E., and Hammer, R. P., Jr. (1995). *Neurobiological Effects of Sex Steroid Hormones.* Cambridge University Press.

Michell, A. R. (1976). Relationships between individual differences in salt appetite of sheep and their plasma electrolyte status. *Physiol. Behav.* 17:215–19.

Midkiff, E. E., and Bernstein, I. L. (1983). The influence of age and experience on salt preference of the rat. *Dev. Psychobiol.* 16:385–94.

Millan, M. A., Jacobowitz, D. M., Aguilera, G., and Catt, K. J. (1991). Differential distribution of AT1 and AT2 angiotensin II receptor subtypes in the rat brain during development. *Proc. Natl. Acad. Sci. U.S.A.* 88:11440–4.

Miller, F. R., and Sherrington, C. S. (1915). Some observations on the buccopharyngeal reflex deglutition in the cat. *Q. J. Exp. Physiol.* 9:147–86.

Miller, N. E. (1959–71) *Selected Papers.* Chicago: Aldine.

Miller, W. L. (1988). Molecular biology of steroid hormone synthesis. *Endocr. Rev.* 9: 295–312.

Mills, D. E., and Robertshaw, D. (1981). Response of plasma prolactin to changes in ambient temperature and humidity in man. *J. Clin. Endocrinol. Metab.* 52:279–83.

Minami, M., Kuraishi, Y., Yamaguchi, T., Nakai, S., Hirai, Y., and Satoh, M. (1991). Immobilization stress induces interleukin-1B mRNA in the rat hypothalamus. *Neurosci. Lett.* 123:254–6.

Miselis, R. R. (1981). The efferent projections of the subfornical organ of the rat: a circumventricular organ within a neural network subserving water balance. *Brain Res.* 230:1–23.

Mistlberger, R. E. (1994). Circadian food-anticipatory activity: formal models and physiological mechanisms. *Neurosci. Biobehav. Rev.* 18:171–95.

Mobbs, C. V., Fink, G., and Pfaff, D. W. (1990). HIP-70: a protein induced by estrogen in the brain and LH-RH in the pituitary. *Science* 247:1477–9.

Moga, M. M., Saper, C. B., and Gray, T. S. (1989). The bed nucleus of the stria terminalis. *J. Comp. Neurol.* 283:315–32.

Mogenson, G. J. (1987). Limbic-motor integration. In: *Progress in Psychobiology and Physiological Psychology,* ed. A. N. Epstein and J. Sprague. New York: Academic Press.

Monaghan, E. P., Arjomand, J., and Breedlove, S. M. (1993). Brain lesions affect penile reflexes. *Horm. Behav.* 27:122–31.

Money, J. (1961). Components of eroticism in man. I. The hormones in relation to sexual morphology and desire. *J. Nerv. Ment. Dis.* 132:239–48.

Mook, D. (1963). Oral and postingestional determinants of the intake of various solutions in rats with esophageal fistulas. *J. Comp. Physiol. Psychol.* 56:645–59.

Mook, D. G. (1987). *Motivation. The Organization of Action.* New York: Norton.

Moore, B. J., Gerardo-Gettens, T., Horwitz, B. A., and Stern, J. S. (1986). Hyperprolactinemia stimulates food intake in the female rat. *Brain Res. Bull.* 17:563–9.

Moore, F. L. (1992). Evolutionary precedents for behavioral actions of oxytocin and vasopressin. *Ann. N.Y. Acad. Sci.* 652:156–65.

Moore, F. L., Lowry, C. A., and Rose, J. D. (1994). Steroid-neuropeptide interactions that control reproductive behaviors in an amphibian. *Psychoneuroendocrinology* 19: 381–92.

Moore, F. L., and Orchinik, M. (1991). Multiple molecular actions for steroids in the regulation of reproductive behaviors. *Sem. Neurosci.* 3:489–96.

Moore, F. L., Wood, R. E., and Boyd, S. K. (1992). Sex steroids and vasotocin interact in a female amphibian to elicit female-like egg laying behavior of male-like courtship. *Horm. Behav.* 26:156–66.

Moore, M. C. (1983). Effect of female sexual displays on the endocrine physiology and behaviour of male white-crowned sparrows, *Zonotrichia leucophrys. J. Zool. (Lond.)* 199:137–48.

Moore, M. C. (1986). Elevated testosterone levels during nonbreeding-season territoriality in a fall-breeding lizard, *Sceloporus jarrovi. J. Comp. Physiol. A* 158:159–63.

Moore, M. C. (1987). Circulating steroid hormones during rapid aggressive responses of territorial male mountain spiny lizards, *Sceloporus jarrovi. Horm. Behav.* 21:511–22.

Moore, M. C. (1988). Testosterone control of territorial behavior: tonic-release implants fully restore seasonal and short-term aggressive responses in free-living castrated lizards. *Gen. Comp. Endocrinol.* 70:450–9.

Moore, M. C. (1991). Application of organization–activation theory to alternative male reproductive strategies: a review. *Horm. Behav.* 25:154–79.

Moore, M. C., Thompson, C. W., and Marleer, C. A. (1991). Reciprocal changes in corticosterone and testosterone levels following acute and chronic handling stress in the tree lizard, *Urosaurus ornatus. Gen. Comp. Endocrinol.* 81:217–26.

Moore, R. Y. (1993). GABA is the principal neurotransmitter of the circadian system. *Neurosci. Lett.* 150:112–16.

Moore, R. Y., and Card, J. P. (1994). Neuropeptide Y in the circadian timing system. *Ann. N.Y. Acad. Sci.* 739:247–8.

Moore, R. Y., and Eichler, V. B. (1972). Loss of a circadian adrenal corticosterone rhythm following suprachiasmatic lesions in the rat. *Brain Res.* 42:201–6.

Moore-Ede, M. C. (1986). Physiology of the circadian timing system: predictive versus reactive homeostasis. *Am. J. Physiol.* 250:737–52.

Moore-Ede, M. C., Sulzman, F. M., and Fuller, C. A. (1982). *The Clocks that Time Us.* Harvard University Press.

Moran, T. H., Norgren, R., Crosby, R. J., and McHugh, P. R. (1990). Central and peripheral vagal transport of cholecystokinin binding sites occurs in afferent fibers. *Brain Res.* 526:95–102.

Moran, T. H., Robinson, P. H., Goldrich, M. S., and McHugh, P. R. (1986). Two brain cholecystokinin receptors: implications for behavioral actions. *Brain Res.* 362:175–9.

Moran, T. H., Smith, G. P., Hostetler, A. M., and McHugh, P. R. (1987). Transport of cholecystokinin (CCK) binding sites in subdiaphragmatic vagal branches. *Brain Res.* 415:149–52.

Morel, G., Ouhtit, A., and Kelly, P. A. (1994). Prolactin receptor immunoreactivity in rat anterior pituitary. *Neuroendocrinology* 59:78–84.

Morgan, C., and Stellar, J. (1950). *Physiological Psychology* 2nd ed. New York: McGraw-Hill.

Morgan, M. A., and LeDoux, J. E. (1995). Differential contribution of dorsal and ventral medial prefrontal cortex to the acquisition and extinction of conditioned fear in rats. *Behav. Neurosci.* 109:681–8.

Moriguchi, A., Ferrario, C. M., Brosnihan, K. B., Ganten, D., and Morris, M. (1994). Differential regulation of central vasopressin in transgenic rats harboring the mouse *Ren-2* gene. *Am. J. Physiol.* 267:R786–91.

Morin, L. P. (1980). Effect of ovarian hormones on synchrony of hamster circadian rhythms. *Physiol. Behav.* 24:741–9.

Morin, L. P., and Cummings, L. A. (1982). Splitting of wheel running rhythms by castrated or steroid treated male and female hamsters. *Physiol. Behav.* 29:665–75.

Morin, L. P., and Dark, J. (1992). Hormones and biological rhythms. In: *Behavioral Endocrinology,* ed. J. B. Becker, M. C. Breedlove, and D. Crews. MIT Press.

Morin, L. P., Fitzgerald, K. M., and Zucker, I. (1977). Estradiol shortens the period of hamster circadian rhythms. *Science* 196:305–7.

Morley, J. E., Levine, A. S., Gossnel, B. A., Kneip, J., and Grace, M. (1987). Effect of neuropeptide Y on ingestive behavior in the rat. *Am. J. Physiol.* 252:R599–609.

Morris, J. S., Frith, C. D., Perrett, D. I., Rowland, D., Young, A. W., Calder, A. J., and Dolan, R. J. (1996). A differential neural response in the human amygdala to fearful and happy expressions. *Nature* 383:812–15.

Morris, M., and Lucion, A. B. (1995). Antisense oligonucleotides in the study of neuroendocrine systems. *J. Neuroendocrinol.* 7:493–500.

Moyer, J. A., O'Donohue, T. O., Herrenkohl, L. R., Gala, R. R., and Jacobowitz, D. M. (1979). Effects of suckling on serum prolactin levels and catecholamine concentrations and turnover in discrete brain regions. *Brain Res.* 176:125–33.

Mrosovsky, N. (1995). A non-photic gateway to the circadian clock of hamsters. In: *Circadian Clocks and Their Adjustment.* Ciba Foundation Symposium 183.

Munck, A., Guyre, P. M., and Holbrook, N. J. (1984). Physiological regulation of glucocorticoids in stress and their regulation to pharmacological actions. *Endocr. Rev.* 5:25–44.

Munck, A., and Naray-Fejes-Togh, A. (1992). The ups and downs of glucocorticoid physiology, permissive and suppressive effects revisited. *Mol. Cell. Endocrinol.* 90: C1–4.

Myers, J. E., Buysse, D. J., Thase, M. E., Perel, J., Miewald, J. M., Cooper, T. B., Kupfer, D. J., and Mann, J. J. (1993). The effects of fenfluramine on sleep and prolactin in depressed inpatients: a comparison of potential indices of brain serotonergic responsivity. *Biol. Psychiatry* 34:753–8.

Nachman, M. (1963). Learned aversion to the taste of lithium chloride and generalization to other salts. *J. Comp. Physiol. Psychol.* 56:343–9.

Nachman, M., and Ashe, J. H. (1974). Effects of basolateral amygdala lesions on neophobia learned taste aversions, and sodium appetite in rats. *J. Comp. Physiol. Psychol.* 87:622–43.

Nachmias, M., Gunnar, M., Mangelsdorf, S., Parritz, R. H., and Buss, K. (1996). Behavioral inhibition and stress reactivity: the moderating role of attachment security. *Child Dev.* 67:508–22.

Nakamura, K., and Norgren, R. (1995). Sodium-deficient diet reduces gustatory activity in the nucleus of the solitary tract of behaving rats. *Am. J. Physiol.* 269:R647–61.

Nauta, J. H. (1961). Fibre degeneration following lesions of the amygdaloid complex in the monkey. *J. Anat.* 95:515–31.

Neal, C. R., Jr., Swann, J. M., and Newman, S. W. (1989). The colocalization of substance P and prodynorphin immunoreactivity in neurons of the medial preoptic area, bed nucleus of the stria terminalis and medial nucleus of the amygdala of the Syrian hamster. *Brain Res.* 496:1–13.

Needleman, P., and Greenwals, J. E. (1986). Atriopeptin: a cardiac hormone intimately involved in fluid, electrolyte, and blood-pressure homeostasis. Mechanisms of disease. *N. Engl. J. Med.* 314:828–9.

Nelson, E., and Panksepp, J. (1996). Oxytocin mediates acquisition of maternally associated odor prefences in preweanling rat pups. *Behav. Neurosci.* 110:583–92.

Nelson, R. J. (1995). *An Introduction to Behavioral Endocrinology.* Sunderland, MA: Sinauer Associates.

Nelson, R. J. (1997). The use of genetic knockout mice in behavioral endocrinology research. *Horm. Behav.* 31:188–96.

Nelson, R. J., Gubernick, D. J., and Blom, J. M. (1995). Influence of photoperiod, food, and water availability on reproduction in male California mice (*Peromyscus californicus*). *Physiol. Behav.* 57:1175–80.

Nemeroff, C. B. (ed.) (1991). *Neuroendocrinology.* London: CRF Press.

Nemeroff, C. B., Owens, M. J., Bissette, G., Andorn, A. C., and Stanley, M. (1988). Reduced corticotropin releasing factor binding sites in the frontal cortex of suicide victims. *Arch. Gcn. Psychiatry* 45:577–79.

Nemeroff, C. B., Widerlov, E., Bissette, G., Walleus, H., Karlsson, I., Eklund, K., Kilts, C. D., Loosen, P. T., and Vale, W. (1984). Elevated concentrations of CSF corticotropin-releasing factor-like immunoreactivity in depressed patients. *Science* 226:1342–3.

Nespor, A. A., Lukazewicz, M. J., Dooling, R. J., and Ball, G. F. (1996). Testosterone induction of male like vocalizations in female budgerigars *(Melopsittacus undulatus).* *Horm. Behav.* 30:162–9.

Nice, M. M. (1941). Studies in the life history of the song sparrow. *Trans. Linnaean Soc.*

Nice, M. M. (1947). *Studies in the Life History of the Song Sparrow.* New York: Dover.

Nichols, N. R., and Finch, C. E. (1994). Gene products of corticosteroid action in hippocampus. *Ann. N.Y. Acad. Sci.* 746:145–54.

Nicolaidis, S., Galaverna, O., and Metzler, C. H. (1990). Extracellular dehydration during pregnancy increases salt appetite of offspring. *Am. J. Physiol.* 258:263.

Nicolaidis, S., Ishibashi, S., Gueguen, B., Thornton, S. N., and De Beaurepaire, R. (1983). Iontophoretic investigation of identified SFO angiotensin responsive neurons firing in relation to blood pressure changes. *Brain Res. Bull.* 10:357–63.

Nissen, E., Lilja, G., Widstrom, A.-M., and Uvnas-Moberg, K. (1995). Elevation of oxytocin levels in early post partum women. *Acta Obstet. Gynaecol. Scand.* 74:530–3.

Nitabach, M., Schulkin, J., and Epstein, A. N. (1989). The medial amygdala is part of a mineralocorticoid-sensitive circuit controlling NaCl intake in the rat. *Behav. Brain Res.* 35:197–204.

Norgren, R. (1976). Taste pathways to hypothalamus and amygdala. *J. Comp. Neurol.* 166:17–30.

Norgren, R. (1984). Central neural mechanisms of taste. In: *Handbook of Physiology and the Nervous System. Vol. 3: Sensory Processes*, pt. 2.1, ed. J. M. Brookhart and V. B. Mountcastle, pp. 1087–128. Bethesda, MD: American Physiological Society.

Norgren, R. (1995). Gustatory system. In: *The Rat Nervous System*. Orlando: Academic Press.

Norgren, R., and Smith, G. P. (1988). Central distribution of subdiaphragmatic vagal branches in the rat. *J. Comp. Neurol.* 273:207–23.

Norgren, R., and Wolf, G. (1975). Projections of thalamic gustatory and lingual areas in the rat. *Brain Res.* 92:123–9.

Nose, H., Mach, G. W., Shi, X. R, and Nadel, E. R. (1988). Involvement of sodium retention hormones during rehydration in humans. *J. Appl. Physiol.* 65:332–6.

Nottebohm, F. (1993). The search for neural mechanisms that define the sensitive period for song learning in birds. *Neth. J. Zool.* 43:193–234.

Nowlis, G. H. (1977). From reflex to representation: taste-elicited tongue movements in the human newborn. In: *Taste and Development*, ed. J. M. Weiffenbach. Washington, DC: U.S. Department of Health, Education, and Welfare.

Numan, M. (1994). Maternal behavior. In: *The Physiology of Reproduction*, 2nd ed. ed. E. Knobil and J. D. Neill. New York: Raven Press.

Numan, M., and Numan, M. J. K. (1994). Expression of fos-like immunoreactivity in the preoptic area of maternally behaving virgin and postpartum rats. *Behav. Neurosci.* 108:379–94.

Numan, M., and Numan, M. J. (1995). Importance of pup-related sensory inputs and maternal performance for the expression of fos-like immunoreactivity in the preoptic area and ventral bed nucleus of the stria terminalis of postpartum rats. *Behav. Neurosci.* 109:135–49.

Obal, F., Jr., Kacsoh, B., Alfoldi, P., Payne, L., Markovic, O., Grosvenor, C., and Krueger, J. M. (1992). Antiserum to prolactin decreases rapid eye movement sleep (REM sleep) in the male rat. *Physiol. Behav.* 52:1063–8.

Obal, F., Jr., Payne, L., Kacsoh, B., Opp, M., Kapas, L., Grosvenor, C. E., and Krueger, J. M. (1994). Involvement of prolactin in the REM sleep-promoting activity of systemic vasoactive intestinal peptide (VIP). *Brain Res.* 645:143–9.

Oftedal, O. T. (1991). The nutritional consequences of foraging in primates: the relationship of nutrient intakes to nutrient requirements. *Phil. Trans. R. Soc. Lond. Biol.* 334:161–70.

Oftedal, O. T., Chen, T. C., and Schulkin, J. (1997). Preliminary observations on the relationship of calcium ingestion to vitamin D in the green iguana *(Iguana iguana.)* *Zoo. Biol.* 16:201–7.

Ohman, L. E., and Johnson, A. K. (1986). Lesions of the lateral parabrachial nucleus enhance drinking to angiotensin II and isoproterenol. *Am. J. Physiol.* 251:R504–9.

Olschowka, J. A., O'Donohue, T. L., Mueller, G. P., and Jacobowitz, D. M. (1982). Hypothalamic and extrahypothalamic distribution of CRF-like immunoreactive neurons in the rat brain. *Neuroendocrinology* 35:305–8.

Olson, B. R., Drutarosky, M. D., Stricker, E. M., and Verbalis, J. G. (1991). Brain oxytocin receptors mediate corticotropin-releasing hormone-induced anorexia. *Am. J. Physiol.* 260:R448–52.

Olson, B. R., Freilino, M., Hoffman, G. E., Stricker, E. M., Sved, A. F., and Verbalis, J. G. (1993). C-fos expression in rat brain and brainstem nuclei in response to treatments that alter food intake and gastric motility. *Mol. Cell. Neurosci.* 4:93–106.

Orchinik, M., and McEwen, B. S. (1994). Rapid steroid actions in the brain: a critique of genomic and nongenomic mechanisms. In: *Genomic and Non-genomic Effects of Aldosterone,* ed. M. Wehling, Boca Raton: CRC Press.

Orchinik, M., Murray, T. F., and Moore, F. L. (1991). A corticosteroid receptor in neuronal membranes. *Science* 252:1848–51.

Orchinik, M., Murray, T. F., and Moore, F. L. (1994). Steroid modulation of GABA receptors in an amphibian brain. *Brain Res.* 646:258–66.

Oren, D. A., Brainard, G. C., Johnston, S. H., Joseph-Vanderpool, J. R., Sorek, E., and Rosenthal, N. E. (1991). Treatment of seasonal affective disorder with green light and red light. *Am. J. Psychiatry* 148:509–11.

Oren, D. A., Schulkin, J., and Rosenthal, N. E. (1994). 1,25 $(OH)_2$ vitamin D_3 levels in seasonal affective disorder: effects of light. *Psychopharmacology* 116:515–16.

Osborne, T. B., and Mendel, L. B. (1918). The choice between adequate and inadequate diets as made by rats. *J. Biol. Chem.* 35:19–27.

Owens, M. J., Barolome, J., Schanberg, S. M., and Nemeroff, C. B. (1990). Corticotropin-releasing factor concentrations exhibit an apparent diurnal rhythm in hypothalamic and extrahypothalamic brain regions: differential sensitivity to corticosterone. *Neuroendocrinology* 52:626–31.

Owens, M. J., and Nemeroff, C. B. (1991). Physiology and pharmacology of CRF. *Pharmacol. Rev.* 43:425–73.

Pacak, K., Palkovits, M., Kvetnasky, R., Fukuhara, K., Armando, I., Kopin, I. J., and Goldstein, D. (1993). Effects of single or repeated immobilization on release of norepinephrine and its metabolites in the central nucleus of the amygdala. *Neuroendocrinology* 57:626–33.

Panksepp, J. (1992). Oxytocin effects on emotional processes: separation distress, social bonding and relationships to psychiatric disorders. *Ann. N.Y. Acad. Sci.* 652:243–52.

Panksepp, J., Normansell, L., Herman, B., Bishop, P., and Crepeau, L. (1988). Neural and neurochemical control of the separation distress call. In: *The Physiological Control of Mammalian Vocalization,* ed. J. D. Newman, pp. 263–99. New York: Plenum Press.

Pannzica, G. C., Pannzica, C. V., and Balthazart, J. (1996). The sexually dimorphic medial preoptic nucleus of quail: a key brain area mediating steroid action on male sexual behavior. *Front. Neuroendocrinol.* 17:51–125.

Papez, J. W. (1937). A proposed mechanism of emotion. *Arch. Neurol. Psychiatry* 38:725–44.

Parrott, W. G., and Schulkin, J. (1993). Neuropsychology and the cognitive nature of the emotions. *Cognition and Emotion* 1:43–59.

Parsons, B., Rainbow, T. C., MacLusky, N. J., and McEwen, B. S. (1982). Progestin receptor levels in rat hypothalamic and limbic nuclei. *J. Neurosci.* 2:1445–52.

Pauk, J., Kuhn, C. M., Field, T. M., and Schanberg, S. M. (1986). Positive effects of tactile versus kinesthetic or vestibular stimulation on neuroendocrine and ODC activity in maternally-deprived rat pups. *Life Sci.* 39:2081–7.

Paul, S. M., and Purdy, R. H. (1992). Neuroactive steroids. *FASEB J.* 6:2311–22.

Paul-Pagano, L., Roky, R., Valat, J.-L., Kitahama, K., and Jouvet, M. (1993). Anatomical distribution of prolactin-like immunoreactivity in the rat brain. *Neuroendocrinology* 58:682–95.

Pavcovich, L. A., and Valentino, R. J. (1997). Regulation of a putative neurotransmitter effect of corticotrophin-releasing factor: effects of adrenalectomy. *J. Neurosci.* 17: 401–8.

Pavlov, I. P. (1927). *Conditioned Reflexes.* New York: Dover.

Peach, M. J., and Chiu, A. T. (1974). Stimulation and inhibition of aldosterone biosynthesis in vitro by angiotensin II and analogs. *Circ. Res.* 34:1–14.

Pearce, D., and Yamamoto, K. R. (1993). Mineralocorticoid and glucocorticoid receptor activities distinguished by nonreceptor factors at a composite response element. *Science* 259:1161–4.

Pedersen, C. A., Ascher, J. A., Monroe, Y. L., and Prange, A. J., Jr. (1982). Oxytocin induces maternal behavior in virgin female rats. *Science* 216:648–50.

Pedersen, C. A., Caldwell, J. D., McGuire, M., and Evans, D. L. (1991). Corticotropin-releasing hormone inhibits maternal behavior and induces pup-killing. *Life Sci.* 48: 1537–46.

Pedersen, C. A., Caldwell, J. D., Walker, C., Ayers, G., and Mason, G. A. (1994). Oxytocin activates the postpartum onset of rat maternal behavior in the ventral tegmental and medial preoptic areas. *Behav. Neurosci.* 108:1163–71.

Pedersen, C. A., and Prange, A. J., Jr. (1979). Induction of maternal behavior in virgin rats after intracerebroventricular administration of oxytocin. *Proc. Natl. Acad. Sci. U.S.A.* 76:6661–5.

Pedersen, J. M., Glickman, S. E., Frank, L. G., and Beach, F. A. (1992). Sex differences in the play behavior of immature spotted hyenas, *Crocuta crocuta. Horm Behav.* 24: 403–20.

Peirce, C. S. (1992). *Reasoning and the Logic of Things.* Harvard University Press. (Originally published 1898.)

Pepin, M.-C., Pothier, F., and Barden, N. (1992). Impaired type II glucocorticoid-receptor function in mice bearing antisense RNA transgene. *Nature* 355:725–8.

Perrin, M. H., Donaldson, C. J., Chen, R., Lewis, K. A., and Vale, W. W. (1993). Cloning and functional expression of a rat brain corticotropin releasing factor (CRF) receptor. *Endocrinology* 6:3058–61.

Peterson, G., Mason, G. A., Barakat, A. S., and Pedersen, C. A. (1991). Oxytocin selectively increases holding and licking of neonates in preweanling but not postweanling juvenile rats. *Behav. Neurosci.* 105:470–7.

Pfaff, D. W. (1980). *Estrogens and Brain Function: Neural Analysis of a Hormone-controlled Mammalian Reproductive Behavior.* Berlin: Springer-Verlag.

Pfaff, D. W. (1982). *The Physiological Basis of Motivation.* Berlin: Springer-Verlag.

Pfaff, D., Dellovade, T., and Zhu, Y.-S. (1997). Interactions among genes for transcription factors in hypothalamic neurons: implications for reproductive behaviors. *Mol. Psychiatry* 2:448–50.

Pfaff, D. W., Kow, L. M., Zhu, U.S., Scott, R. E., Wu-Peng, S. X., and Delbaude, T. (1996). Hypothalamic cellular and molecular mechanisms helping to satisfy axiomatic requirements for reproduction. *J. Neuroendocrinol.* 8:325–36.

Pfaff, D. W., Schwartz-Giblin, S., McCarthy, M. M., and Kow, L.-M. (1994). Cellular and molecular mechanisms of female reproductive behaviors. In: *The Physiology of Reproduction,* 2nd ed., ed. E. Knobil and J. D. Neill. New York: Raven Press.

Pfaffmann, C. (1960). The pleasures of sensation. *Psychol. Rev.* 67:253–68.

Pfaffmann, C., Norgren, R., and Grill, H. J. (1977). Sensory affect and motivation. In: *Tonic Function of Sensory Systems,* ed. B. M. Wenzel and H. P. Ziegler. New York: New York Academy of Sciences.

Pfeiffer, C. A. (1936). Sexual differences of the hypophyses and their determination by the gonads. *Am. J. Anat.* 58:195–994.

Phifer, C. B., and Hall, W. G. (1988). Ingestive behavior in preweanling rats: emergence of postgastric controls. *Am. J. Physiol.* 255:R191–9.

Philips, P. A. (1978). Angiotensin in the brain. *Neuroendocrinology* 25:354–77.

Phillips, M. I., and Gyurko, R. (1997). Antisense oligonucleotides: new tools for physiology. *News Physiol. Sci.* 12:99–103.

Phillips, R. G., and LeDoux, J. E. (1994). Overlapping and divergent projections of CA1 and the ventral subiculum to the amygdala. *Society of Neuroscience Abstracts.*

Phoenix, C. H., Goy, R. W., Gerall, A. A., and Young, W. C. (1959). Organizing action of prenatally administered testosterone propionate on the tissues mediating mating behavior in the female guinea pig. *Endocrinology* 65:369–89.

Piaget, J. (1974). *The Child and Reality.* New York: Viking Press.

Piazza, P. V., Maccari, S., Deminere, J. M., LeMoal, M., Mormede, P., and Simon, H. (1991). Corticosterone levels determine individual vulnerability to amphetamine self-administration. *Proc. Natl. Acad. Sci. U.S.A.* 88:2088–92.

Pich, E. M., Heinrichs, S. C., Rivier, C., Miczek, K. A., Fisher, D. A., and Koob, G. F. (1993a). Blockade of pituitary-adrenal axis activation induced by peripheral immunoneutralization of corticotropin-releasing factor does not affect the behavioral response to social defeat stress in rats. *Psychoneuroendocrinology* 18:495–507.

Pich, E. M., Koob, G. F., Heilig, M., Menzaghi, F., Vale, W., and Weiss, F. (1993b). Corticotropin-releasing factor release from the mediobasal hypothalamus of the rat as measured by microdialysis. *Neuroscience* 55:695–707.

Pickard, G. E., Turek, F. W., and Sollars, P. J. (1993). Light intensity and splitting in the golden hamster. *Physiol. Behav.* 54:1–5.

Pieper, D. R., Ali, H. Y., Benson, L. L., Shows, M. D., Lobocki, C. A., and Subramanian, M. G. (1995). Voluntary exercise increases gonadotropin secretion in male golden hamsters. *Am. J. Physiol.* 269:R179–85.

Pieper, D. R., Borer, K. T., Lobocki, C. A., and Samuel, D. (1988). Exercise inhibits reproductive quiescence induced by exogenous melatonin in hamsters. *Am. J. Physiol.* 255:R718–23.

Pike, R. L., and Yao, C. (1971). Increased sodium chloride appetite during pregnancy in the rat. *J. Nutr.* 101:169–76.

Pinker, S. (1994). *The Language Instinct.* New York: Morrow.

Pitman, D. L., Ottenweller, J. E., Pritzel, T., Natelson, B. H., McCarty, R., and Tapp, W. N. (1995). Effects of exposure to stressors of varying predictability on adrenal function in rats. *Behav. Neurosci.* 109:767–76.

Pittendrigh, C. S., and Daan, S. (1976). A functional analysis of circadian pacemakers in nocturnal rodents. V. Pacemaker structure: a clock for all seasons. *J. Comp. Physiol.* 106:333–55.

Plapinger, L., Landau, I. T., McEwen, B. S., and Feder, H. H. (1977). Characteristics of estradiol-binding macromolecules in fetal and adult guinea pig brain cytosols. *Biol. Reprod.* 16:586–99.

Plotsky, P. M., and Meaney, M. J. (1993). Early postnatal experience alters hypothalamic corticotropin-releasing factor (CRF) mRNA, median eminence CRF content and stress-induced release in adult rats. *Brain Res. Mol. Brain Res.* 18:195–200.

Plunkett, L. M., Shigematsu, K., Kurihara, M., and Saavedra, J. M. (1987). Localization of angiotensin II receptors along the anteroventral third ventricle area of the rat brain. *Brain Res.* 4:205–19.

Pompei, P. L., Tayebaty, S. J., DeCaro, G., Schulkin, J., and Massi, M. (1992). Bed nucleus of the stria terminalis: site for the anti-natriorexegenic action of tachykinins in the rat. *Pharmacol. Biochem. Behav.* 40:977–81.

Popov, V. I., and Bocharova, L. S. (1992). Hibernation-induced structural changes in synaptic contacts between mossy fibres and hippocampal pyramidal neurons. *Neuroscience* 48:53–62.

Popov, V. I., Bocharova, L. S., and Bragin, A. G. (1992). Repeated changes of dendritic morphology in the hippocampus of ground squirrels in the course of hibernation. *Neuroscience* 48:45–51.

Porges, S. W. (1996). Cardiac vagal tone: a physiological index of stress. *Neurosci. Biobehav. Rev.* 19:225–33.

Potter, E., Sutton, S., Donaldson, C., Chen, R., Perrin, M., Lewis, K., Sawchenko, P. E., and Vale, W. (1994). Distribution of corticotropin-releasing factor receptor mRNA expression in the rat brain and pituitary. *Proc. Natl. Acad. Sci. U.S.A.* 91:8777–81.

Power, M. I., Tardiff, S. D., Layne, D. G., and Schulkin, J. (in press). Common marmosets have an appetite for calcium. *J. Primatol.*

Powers, J. B., Jetton, A. E., Mangels, R. A., and Bittman, E. L. (1997). Effects of photoperiod duration and melatonin signal characteristics on the reproductive system of male Syrian hamsters. *J. Neuroendocrinol.* 9:451–66.

Powers, J. B., Newman, S. W., and Bergondy, M. L. (1987). MPOA and BNST lesions in male Syrian hamsters: differential effects on copulatory and chemo-investigatory behaviors. *Behav. Brain Res.* 23:181–95.

Powley, T. E. (1977). The ventromedial hypothalamic syndrome, satiety and cephalic phase. *Psychol. Rev.* 84:89–126.

Powley, T. E, and Berthoud, H.-R. (1985). Diet and cephalic phase insulin responses. *Am. J. Clin. Nutr.* 42:991–1002.

Pratt, N. C., Alberts, A. C., Fulton-Medler, K. G., and Phillips, J. A. (1992). Behavioral, physiological, and morphological components of dominance and mate attraction in male green iguanas. *Zoo. Biol.* 11:153–63.

Premack, D. A., and Premack, A. (1983). *The Mind of an Ape.* New York: Norton.

Propper, C. R., Hillyard, S. D., and Johnson, W. E. (1995). Central angiotensin II induces thirst-related responses in an amphibian. *Horm. Behav.* 29:74–84.

Propper, C. R., and Moore, F. L. (1991). Effects of courtship on brain gonadotrophin hormone releasing hormone and plasma steroid concentration in a female amphibian. *Gen. Comp. Endocrinol.* 81:304–12.

Proudman, J. A. (1991). Daily rhythm of prolactin and corticosterone in unrestrained, incubating turkey hens. *Domest. Anim. Endocrinol.* 8:265–70.

Proudman, J. A., and Opel, H. (1981). Turkey prolactin: validation of a radioimmunoassay and measurement of changes associated with broodiness. *Biol. Reprod.* 25: 573–80.

Pugh, C. R., Tremblay, D., Fleshner, M., and Rudy, J. W. (1997). A selective role for corticosterone in contextual-fear conditioning. *Behav. Neurosci.* 111:503–14.

Purdie, D. W., Empson, J. A. C., Crichton, C., and MacDonald, L. (1995). Hormone replacement therapy, sleep quality and psychological well-being. *Br. J. Obstet. Gynaecol.* 102:735–9.

Quadri, S. K., and Spies, H. G. (1976). Cyclic and diurnal patterns of serum prolactin in the rhesus monkey. *Biol. Reprod.* 14:495–501.

Quartermain, D., and Wolf, G. (1967). Drive properties of mineralocorticoid-induced sodium appetite. *Physiol. Behav.* 2:261–3.

Raadsheer, F. C. (1994). Increased numbers of corticotropin-releasing hormone expressing neurons in the hypothalamic paraventricular nucleus of depressed patients. *Neuroendocrinology* 60:436–44.

Raadsheer, F. C., Van Heerikhuize, J. J., Lucassen, P. J., Hoogendijk, W. J., Tilders, F. J., and Swaab, D. F. (1995). Corticotropin-releasing hormone mRNA levels in the paraventricular nucleus of patients with Alzheimer's disease and depression. *Am. J. Psychiatry* 152:1371–6.

Raber, J., and Bloom, F. E. (1994). IL-2 induces vasopressin release from the hypothalamus and the amygdala: role of nitric oxide–mediated signaling. *J. Neurosci.* 14: 6187–95.

Raber, J., Pich, E. M., Koob, G. F., and Bloom, F. E. (1994). IL-1B potentiates the acetylcholine-induced release of vasopressin from the hypothalamus in vitro, but not from the amygdala. *Neuroendocrinology* 59:208–17.

Raisman, G., and Field, M. (1973). Sexual dimorphism in the neuropil of the preoptic area of the rat and its dependence on neonatal androgen. *Brain Res.* 54:1–29.

Ramos, C., and Silver, R. (1992). Gonadal hormones determine sex differences in timing of incubation by doves. *Horm. Behav.* 26:586–601.

Ramsay, D. J. (1979). The brain renin angiotensin system: a re-evaluation. *Neuroscience* 4:313–21.

Ramsay, D. J., Thrasher, T. N., and Keil, L. C. (1983). The organum vasculosum laminae terminalis: a critical area for osmoreception. The neurohypophysis: structure, function and control. *Prog. Brain Res.* 60:91–8.

Rand, M. N., and Breedlove, S. M. (1992). Androgen locally regulates rat bulbocavernosus. *J. Neurobiol.* 23:17–30.

Rassnick, S., Heinrichs, S. C., Britton, K. T., and Koob, G. F. (1993). Microinjection of a corticotropin-releasing factor antagonist into the central nucleus of the amygdala reverses anxiogenic-like effects of ethanol withdrawal. *Brain Res.* 605:25–32.

Ravault, J. P., Reinberg, A., and Mechkouri, M. (1987). Circadian and circahemidian rhythms in plasma prolactin of the ram: seasonal changes. *Chronobiol. Int.* 4:209–17.

Reagan, L. P., Flanagan-Cato, L. M., Yee, D. K., Ma, L. Y., Sakai, R. R., and Fluharty, S. J. (1994). Immunohistochemical mapping of angiotensin type 2 (AT2) receptors in rat brain. *Brain Res.* 662:45–59.

Reagan, L. P., Ye, X., Maretzki, C. H., and Fluharty, S. J. (1993). Down-regulation of angiotensin II receptor subtypes and desensitization of cyclic GMP production in neuroblastoma NIE-115 cells. *J. Neurochem.* 60:24–31.

Reilly, J. J., Maki, R., Nardozzi, J., and Schulkin. J. (1994). The effects of lesions of the bed nucleus of the stria terminalis on sodium appetite. *Acta Neurobiol. Exp. (Warsz.)* 54:253–7.

Reilly, J. J., Mamadi, D. B., McEwen, B. S., Schulkin, J., and Sakai, R. R. (in press). Rapid arousal of sodium intake by adrenal steroid implants in the amygdala of the rat brain. *Brain Res.*

Reilly, J. J., Mamadi, D. B., Schulkin, J., McEwen, B. S., and Sakai, R. R. (1993). The effect of amygdala adrenal steroid implants in the rat brain on sodium intake. Presented at the 11th International Conference on the Physiology of Food and Fluid Intake, Oxford, England.

Reilly, J. J., Nardozzi, J., and Schulkin, J. (1995). The ingestion of calcium in multiparous and virgin female rats. *Brain Res. Bull.* 37:301–3.

Reilly, J. J., and Schulkin, J. (1993). Hormonal control of calcium ingestion: the effects of neonatal manipulations of the gonadal steroids both during and after critical stages in development on the calcium ingestion of male and female rats. *Psychobiology* 21:50–4.

Reiter, R. J. (1993). The melatonin rhythm: both a clock and a calendar. *Experientia* 49:654–64.

Reppert, S. M., Perlow, M. J., Ungerleider, L. G., Mishkin, M., Tamar, L., Orloff, D. G., Hoffman, H. J., and Klein, D. C. (1981). Effects of damage to the suprachiasmatic area of the anterior hypothalamus on the daily melatonin and cortisol rhythms in the rhesus monkey. *J. Neurosci.* 1:1414–25.

Reppert, S. M., and Weaver, D. R. (1995). Melatonin madness. *Cell* 83:1059–62.

Reppert, S. M., Weaver, D. R., Cassone, V. M., Godson, C., and Kolakowski, L. F., Jr. (1995). Melatonin receptors are for the birds: molecular analysis of two receptor subtypes differentially expressed in chick brain. *Neuron* 15:1003–15.

Rescorla, R. A., and Wagner, A. R. (1972). A theory of Pavlovian conditioning: variations in the effectiveness of reinforcement and nonreinforcement. In: *Classical Conditioning. Vol. 2: Current Research and Theory,* ed. A. Black and W. Prokasy, pp. 64–99. New York: Appleton-Century-Crofts.

Reul, J. M. H. M., van den Bosch, F. R., and De Kloet, E. R. (1987). Differential response of type I and type II corticosteroid receptors to changes in plasma steroid level and circadian rhythmicity. *Neuroendocrinology* 45:405–12.

Revusky, S. H., and Bedarf, E. W. (1967). Association of illness with prior ingestion of novel foods. *Science* 155:219–20.

Rey, G. (1997). *Contemporary Theories of Mind.* Oxford: Blackwell.

Rice, K. K., and Richter, C. P. (1943). Increased sodium chloride and water intake of normal rats treated with desoxycorticosterone acetate. *Endocrinology* 33:106–15.

Richardson, R. D., Boswell, T., Weatherford, S. C., Wingfield, J. C., and Woods, S. C. (1993). Cholecystokinin octapeptide decreases food intake in white-crowned sparrows. *Am. J. Physiol.* 264:R852–6.

Richardson, R. D., Ramsay, D. S., Lernmark, A., Scheurink, A. J., Baskin, D. G., and Woods, S. C. (1994). Weight loss in rats following intraventricular transplants of pancreatic islets. *Am. J. Physiol.* 266:R59–64.

Richter, C. P. (1922). A behavioristic study of the activity of the rat. *Comp. Psychol. Monograph* 1:1–55.

Richter, C. P. (1936). Increased salt appetite in adrenalectomized rats. *Am. J. Physiol.* 115:155–61.

Richter, C. P. (1941). Sodium chloride and dextrose appetite of untreated and treated adrenalectomized rats. *Endocrinology* 29:115–25.

Richter, C. P. (1943). Total self-regulatory functions in animals and human beings. *Harvey Lect.* 38:63–103.

Richter, C. P. (1953). Experimentally produced reactions to food poisoning in wild and domesticated rats. *Ann. N.Y. Acad. Sci.* 56:225–39.

Richter, C. P. (1956). Salt appetite of mammals: its dependence on instinct and metabolism. In: *L'instinct dans le comportement des animaux et de l'homme,* pp. 577–629. Paris: Masson.

Richter, C. P. (1965). *Biological Clocks in Medicine and Psychiatry.* Springfield, IL: Thomas.

Richter, C. P. (1976). *Psychobiology of Curt Richter.* Baltimore: York Press.

Richter, C. P., and Barelare, B., Jr. (1938) Nutritional requirements of pregnant and lactating rats studied by the self-selection method. *Endocrinology* 23:15.

Richter, C. P., and Birmingham, J. R. (1941). Calcium appetite of parathyroidectomized rats used to bioassay substances which affect blood calcium. *Endocrinology* 29:645–60.

Richter, C. P., and Eckert, J. F. (1937). Increased calcium appetite of parathyroidectomized rats. *Endocrinology* 21:50–4.

Richter, C. P., Holt, L. E., Jr., and Barelare, B., Jr. (1938). Nutritional requirement for normal growth and reproduction in rats studied by the self-selection method. *Am. J. Physiol.* 122:734–44.

Rickman, M. D., and Davidson, R. J. (1994). Personality and behavior in parents of temperamentally inhibited and uninhibited children. *Dev. Psychobiol.* 30:346–54.

Riddle, O., Lahr, E. L., and Bates, R. W. (1935a). Effectiveness and specificity of prolactin in the induction of the maternal instinct in virgin rats. *Am. J. Physiol.* 113:109.

Riddle, O., Lahr, E. L., and Bates, R. W. (1935b). Maternal behavior induced in virgin rats by prolactin. *Proc. Soc. Exp. Biol. Med.* 32:49–56.

Riftina, F., Angulo, J., Pompei, P., and McEwen, B. (1995). Regulation of angiotensinogen gene expression in the rat forebrain by adrenal steroids and relation to salt appetite. *Mol. Brain Res.* 33:201–8.

Rinaman, L., Verbalis, J. G., Stricker, E. M., and Hoffman, G. E. (1993). Distribution and neurochemical phenotypes of caudal medullary neurons activated to express c-fos following peripheral administration of cholecystokinin. *J. Comp. Neurol.* 338:475–90.

Rivier, C., Brownstein, M., Spiess, J., Rivier, J., and Vale, W. (1981). In vivo corticotropin releasing factor-induced secretion of adrenocorticotropin, β-endorphin and corticosterone. *Endocrinology* 110:272–8.

Rivist, S., and Rivier, C. (1994). Stress and interleukin-1 β-induced activation of c-fos, NGFI-B and CRF gene expression in the hypothalamic PVN: comparison between Sprague-Dawley, Fisher-344 and Lewis rats. *J. Neuroendocrinol.* 6:101–17.

Robertson, J. C., Watson, J. T., and Kelley, D. B. (1994). Androgen directs sexual differentiation of laryngeal innervation in developing *Xenopus laevis. J. Neurobiol.* 25:1625–36.

Robinson, B. G., Arbiser, J. L., Emanuel, R. L., and Majzoub, J. A. (1989). Species-specific placental corticotropin releasing hormone messenger RNA and peptide expression. *Mol. Cell. Endocrinol.* 62:337–41.

Robinson, S. M., Fox, T. O., Dikkes, P., and Pearlstein, R. A. (1986). Sex differences in the shape of the sexually dimorphic nucleus of the preoptic area and suprachiasmatic nucleus of the rat: 3-D computer constructions and morphometrics. *Brain Res.* 371:380–4.

Roca, A. L., Godson, C., Weaver, D. R., and Reppert, S. M. (1996). Structure, characterization and expression of the gene encoding the mouse melatonin receptor (subtype). *Endocrinology* 137:3469–77.

Roca, A. L., Weaver, D. L., and Reppert, S. M. (1993). Serotonin receptor gene expression in the rat suprachiasmatic nuclei. *Brain Res.* 608:159–65.

Rodd, Z. A., Dubek, B. C., Phelan, L. L., Heinrichs, K. K., and McCutcheon, B. N. (in press). Sodium appetite in the mouse: regulation by the mineralocorticoid and glucocorticoid systems. *Behav. Neurosci.*

Rodgers, W. L. (1967). Specificity of specific hungers. *J. Comp. Physiol. Psychol.* 64:49–58.

Rodgers, W. L., and Rozin, P. (1966). Novel food preferences in thiamine deficient rats. *J. Comp. Physiol. Psychol.* 61:194.

Roky, R., Valatx, J.-L., and Jouvet, M. (1993). Effect of prolactin on the sleep–wake cycle in the rat. *Neurosci. Lett.* 156:117–20.

Roky, R., Valatx, J. L., Paut-Pagano, L., and Jouvet, M. (1994). Hypothalamic injection of prolactin or its antibody alters the rat sleep–wake cycle. *Physiol. Behav.* 55:1015–19.

Rolls, B. J. (1986). Changing hedonic responses to foods during and after a meal. In: *Interaction of the Chemical Senses and Ingestion,* ed. M. A. Kare and J. G. Brand. New York: Academic Press.

Rolls, B. J., and Rolls, E. (1982). *Thirst.* Cambridge University Press.

Roozendaal, B., Koolhaas, J. M., and Bohus, B. (1992). Central amygdaloid involvement in neuroendocrine correlates of conditioned stress responses. *J. Endocrinol.* 4:46–52.

Rosen, J. B., and Davis, M. (1988). Enhancement of acoustic startle by electrical stimulation of the amygdala. *Behav. Neurosci.* 102:195–202.

Rosen, J. B., Hamerman, E., Sitcoske, M., Glowa, J. R., and Schulkin, J. (1996). Hyperexcitability: exaggerated fear-potentiated startle produced by partial amygdala kindling. *Behav. Neurosci.* 110:43–50.

Rosen, J. B., Hitchcock, J. M., Sananes, C. B., Miserendino, M. J. D., and Davis, M. (1991). A direct projection from the central nucleus of the amygdala to the acoustic startle pathway: anterograde and retrograde tracing studies. *Behav. Neurosci.* 105: 817–25.

Rosen, J. B., Pishevar, S., Weiss, S. B., Smith, M. A., Kling, M., Gold, P., and Schulkin, J. (1994). Glucocorticoid potentiation of CRH-induced seizures. *Neurosci. Lett.* 174: 113–16.

Rosen, J. R., and Schulkin, J. (1998). From normal fear to pathological anxiety. *Psychol. Rev.* 105:325–50.

Rosenbaum, J. F., Biederman, J., Gersten, M., Hirshfeld, D. R., Meminger, S. R., Herman, J. B., Kagan, J., Reznick, J. S., and Snidman, N. (1988). Behavioral inhibition in children of parents with panic disorder and agoraphobia. *Arch. Gen. Psychiatry* 45: 463–70.

Rosenberg, J., and Hurwitz, S. (1987). Concentration of adrenocortical hormones in relation to cation homeostasis in birds. *Am. J. Physiol.* 253:20–4.

Rosenblatt, J. S. (1970). Views on the onset and maintenance of maternal behavior in the rat. In: *Development and Evolution of Behavior,* ed. L. R. Aronson, E. Tobach, D. S. Lehrman, and J. S. Rosenblatt. San Francisco: Freeman.

Rosenblatt, J. S. (1994a) Psychobiology of maternal behavior: contribution to the clinical understanding of maternal behavior among humans. *Acta Paediatr.* [*Suppl.*] 397: 3–8.

Rosenblatt, J. S. (1994b). Hormonal priming and triggering of maternal behavior in the rat with special reference to the relations between estrogen receptor binding in specific brain regions. *Psychoneuroendocrinology* 19:543–52.

Rosenblatt, J. S., Hazelwood, S., and Poole, J. (1996). Maternal behavior in male rats: effects of medial preoptic area lesions and presence of maternal aggression. *Horm. Behav.* 30:201–15.

Rosenfeld, P., Suchecki, D., and Levine, S. (1992). Multifactorial regulation of the hypothalamic-pituitary-adrenal axis during development. *Neurosci. Biobehav. Rev.* 16:553–68.

Rosenfeld, P., Sutanto, W., Levine, S., and de Kloet, R. (1990). Ontogeny of mineralocorticoid (type I) receptors in brain and pituitary: an in vivo autoradiographic study. *Dev. Brain Res.* 52:57–62.

Rosenthal, N. E., Genhart, M. J., Caballero, B., Jacobsen, F. M., Skwerer, R. G., Coursey, R. D., Rogers, S., and Spring, B. J. (1989). Psychobiological effects of carbohydrate- and protein-rich meals in patients with seasonal affective disorder and normal controls. *Biol. Psychiatry* 25:1029–40.

Rosenthal, N. E., Sack, D. A., Gillin, C., Lewy, A. J., Goodwin, F. K., Mueller, P. S., Newsome, D. A., and Wehr, T. A. (1984). Seasonal affective disorder, a description of the syndrome and preliminary findings with light therapy. *Arch. Gen. Psychiatry* 41:72–80.

Rosenwasser, A. M., and Adler, A. N. (1986). Structure and function in circadian timing systems. *Neurosci. Biobehav. Rev.* 10:431–48.

Rosenwasser, A. M., Schulkin, J., and Adler, A. N. (1988). The behavior of salt hungry rats to limited periods of salt. *Anim. Behav.* 16:324–9.

Rouge-Pont, F., Marinelli, M., LeMoal, M., Simon, H., and Piazza, P. V. (1995). Stress-induced sensitization. II: Sensitization of the increase in extracellular dopamine induced by cocaine depends on stress-induced corticosterone secretion. *J. Neurosci.* 15:7189–95.

Rowe, B. P., Saylor, D. L., and Speth, R. C. (1992). Analysis of angiotensin II receptor subtypes in individual rat brain. *Neuroendocrinology* 55:563–73.

Rowland, N. E., Bellush, L. L., and Fregly, M. J. (1985). Nyctohemeral rhythms and sodium chloride appetite in rats. *Am. J. Physiol.* 249:375–8.

Rowland, N. E., and Fregly, M. J. (1988). Sodium appetite: species and strain differences and role of renin-angiotensin-aldosterone system. *Appetite* 11:143–78.

Rowland, N. E., and Fregly, M. J. (1992). Role of gonadal hormones in hypertension in the Dahl salt-sensitive rat. *Clin. Exp. Hypertens.* A14:367–75.

Rowland, N. E., Li, B.-H., Rozelle, A. K., and Smith, G. C. (1994). Comparison of fos-like immunoreactivity induced in rat brain by central injection of angiotensin II and carbachol. *Am. J. Physiol.* 267:R792–8.

Rowland, N. E., Rozelle, A., Riley, P. J., and Fregly, M. J. (1992). Effect of nonpeptide angiotensin receptor antagonists on water intake and salt appetite in rats. *Brain Res. Bull.* 29:389–93.

Roy, A., Linnoila, M., Karoum, F., and Pickar, D. (1988). Urinary-free cortisol in depressed patients and controls: relationship to urinary indices of noradrenergic function. *Psychol. Med.* 18:93–8.

Rozin, P. (1967a). Specific aversions as a component of specific hungers. *J. Comp. Physiol. Psychol.* 64:237–42.

Rozin, P. (1967b). Thiamine specific hunger. In: *Handbook of Physiology,* ed. C. F. Code, pp. 414–30. Washington, DC: American Physiological Society.

Rozin, P. (1969). Adaptive food sampling patterns in vitamin deficient rats. *J. Comp. Physiol. Psychol.* 69:126–32.

Rozin, P. (1976a). The evolution of intelligence: access to the cognitive unconscious. In: *Progress in Physiological Psychology,* ed. J. Sprague and A. N. Epstein. New York: Academic Press.

Rozin, P. (1976b). *Curt Richter: The Compleat Psychobiologist,* ed. E. M. Blass. Baltimore: York Press.

Rozin, P. (1996). Towards a psychology of food and eating: from motivation to module to model to marker, morality, meaning and metaphor. *Curr. Direct. Psychol. Sci.* 5:18–24.

Rozin, P., and Fallon, A. E. (1987). A perspective on disgust. *Psychol. Rev.* 94:23–46.

Rozin, P., and Schulkin, J. (1990). Food selection. In: *Handbook of Behavioral Neurobiology,* ed. E. M. Stricker, pp. 297–328. New York: Plenum.

Rubinow, D. R., Hoban, C., Grover, G. N., Galloway, D. S., Byrne, P. R., Anderson, R., and Merriam, G. R. (1998), Changes in plasma hormones across the menstrual cycle in patients with menstrually related mood disorder and in control subjects. *Am. J. Obstet. Gynecol.* 158:5–11.

Rubinow, D. R., and Schmidt, P. J. (1987). Mood disorders and the menstrual cycle. *J. Reprod. Med.* 32:389–94.

Ruby, N. F., Nelson, R. J., Licht, P., and Zucker, I. (1993). Prolactin and testosterone inhibit torpor in Siberian hamsters. *Am. J. Physiol.* 264:R123–8.

Ruby, N. F., and Zucker, I. (1992). Daily torpor in the absence of the suprachiasmatic nucleus in Siberian hamsters. *Am. J. Physiol.* 263:353–62.

Rusak, B., and Zucker, I. (1979). Neural regulation of circadian rhythms. *Physiol. Rev.* 59:449–513.

Saad, A. H., and Ridi, R. E. (1988). Endogenous corticosteroids mediate seasonal cyclic changes in immunity of lizards. *Immunobiology* 177:390–403.

Sachar, E. J., Hellman, I., Fukushima, D. K., and Gallagher, T. F. (1970). Cortisol production in depressive illness: a clinical and biochemical clarification. *Arch. Gen. Psychiatry* 23:289–98.

Sakai, R. R. (1986). The hormones of renal sodium conservation act synergistically to arouse a sodium appetite in the rat. In: *The Physiology of Thirst and Sodium Appetite,* ed. G. de Caro, A. N. Epstein, and M. Massi. New York: Plenum.

Sakai, R. R., and Epstein, A. N. (1990). The dependence of adrenalectomy-induced sodium appetite on the action of angiotensin II in the brain of the rat. *Behav. Neurosci.* 104:167–76.

Sakai, R. R., He, P. F., Yank, X. D., Ma, L. Y., Guo, Y. F., Reilly, J. J., Moga, C. N., and Fluharty, S. J. (1994). Intracerebroventricular administration of AT I receptor antisense oligonucleotides inhibits the behavioral actions of angiotensin II. *J. Neurochem.* 62:2053–9.

Sakai, R. R., Nicolaidis, S., and Epstein, A. N. (1989). Salt appetite is suppressed by interference with angiotensin II and aldosterone. *Am. J. Physiol.* 251:R762–8.

Saldanha, C. J., Leak, R. K., and Silver, R. (1994). Detection and transduction of daylength in birds. *Psychoneuroendocrinology* 19:641–56.

Salmon, U. J., and Geist, S. H. (1943). Effect of androgens upon libido in women. *J. Clin. Endocrinol.* 3:235–8.

Saltzman, W., Schultz-Darken, N. J., Scheffler, G., Wegner, F. H., and Abbott, D. H. (1994). Social and reproductive influences on plasma cortisol in female marmoset monkeys. *Physiol. Behav.* 56:801–10.

Samson, W. K., Lumpkin, M. D., and McCann, S. M. (1986). Evidence for a physiological role for oxytocin in the control of prolactin secretion. *Endocrinology* 119:554.

Sanford, L. D., Mann, G. L., Schulkin, J., Wehr, T. A., Ross, R. J., and Morrison, A. R. (1997). Prolactin microinjections into the amygdaloid central nucleus modulate behavioral state. *Sleep* 1.

Sanford, L. D., Tejani-Butt, S. M., Ross, R. J., and Morrison, A. R. (1995). Amygdaloid control of alerting and behavioral arousal in rats: involvement of serotonergic mechanisms. *Arch. Ital. Biol.* 134:81–99.

Santana, P., Akana, S. F., Hanson, E. S., Strack, A. M., Sebastian, R. J., and Dallman, M. F. (1995). Aldosterone and dexamethasone both stimulate energy acquisition whereas only the glucocorticoid alters energy storage. *Endocrinology* 136: 2214–21.

Saper, C. B. (1995). Of hearts and minds: natriuretic peptides in the brain. *J. Comp. Neurol.* 356:166–17.

Saper, C. B., and Levisohn, D. (1983). Afferent connections of the median preoptic nucleus in the rat: anatomical evidence for a cardiovascular interactive mechanism in the anteroventral third ventricular region. *Brain Res.* 288:21–31.

Saphier, D. (1989). Neurophysiological and endocrine consequences of immune activity. *Psychoneuroendocrinology* 14:63–87.

Sapolsky, R. M. (1990). Stress in the wild. *Sci. Am.* 262:116–23.

Sapolsky, R. M. (1992). *Stress, the Aging Brain, and the Mechanisms of Neuron Death.* MIT Press.

Sapolsky, R. M., Krey, L. W., and McEwen, B. S. (1986). The neuroendocrinology of stress and aging: the glucocorticoid cascade hypothesis. *Endocr. Rev.* 7:284–312.

Sapolsky, R. M., and Meaney, M. J. (1986). Maturation of the adrenocortical stress response: neuroendocrine control mechanisms and the stress hyporesponsive period. *Brain Res. Rev.* 11:65–76.

Sassoon, D. A., Gray, G. E., and Kelley, D. B. (1987). Androgen regulation of muscle fiber type in the sexually dimorphic larynx of *Xenopus laevis. J. Neurosci.* 7:3198–206.

Sassoon, D. A., and Kelley, D. B. (1986). The sexually dimorphic larynx of *Xenopus laevis:* development and androgen regulation. *Am. J. Anat.* 177:457–72.

Sassoon, D. A., Segil, N., and Kelley, D. (1986). Androgen-induced myogenesis and chondrogenesis in the larynx of *Xenopus laevis. Exp. Biol.* 113:135–40.

Sawchenko, P. (1993). The functional neuroanatomy of corticotropin-releasing factor. In: *Corticotropin-releasing Factor,* ed. D. J. Chadwick, J. Marsh, and K. Ackrill. New York: Wiley.

Scalera, G., Spector, A. C., and Norgren, R. (1995). Excitotoxic lesions of the parabrachial nuclei prevent conditioned taste aversion and sodium appetite in rat. *Behav. Neurosci.* 109:997–1008.

Schaller, G. B. (1963). *The Mountain Gorilla.* University of Chicago Press.

Schanberg, S. M., and Field, T. M. (1988). Stress and coping across development. In: *Maternal Deprivation and Supplemental Stimulation,* ed. T. M. Field, P. M. McCabe, and N. Schneiderman, pp. 3–25. Hillsdale, NJ: Erlbaum.

Schanberg, S. M., Field, T., Kuhn, C. M., and Bartolame, J. (1993). Touch: a biological regulator of growth and development in the neonate. *Verhaltenstherapie* 3:15–20.

Scheidler, M. G., Verbalis, J. G., and Stricker, E. M. (1994). Inhibitory effects of estrogen on stimulated salt appetite in rats. *Behav. Neurosci.* 108:141–50.

Schlinger, B. A., and Arnold, A. P. (1992). Circulating estrogens in a male songbird originate in the brain. *Proc. Natl. Acad. Sci. U.S.A.* 89:7650–3.

Schlinger, B. A., and Arnold, A. P. (1993). Estrogen synthesis in vivo in the adult zebra finch: additional evidence that circulating estrogens can originate in brain. *Endocrinology* 133:2610.

Schmidt, L. A., Fox, N. A., Rubin, K. H., Sternberg, E. M., Gold, P. W., Smith, C. C., and Schulkin, J. (1997). Behavioral and neuroendocrine responses in shy children. *Dev. Psychobiol.* 30:127–40.

Schmidt, L. A., Fox, N. A., Rubin, K. H., Sternberg, E. M., Gold, P. W., Smith, C. C., and Schulkin, J. (in press). Salivary cortisol responses in shy children. *Individual Differences and Personality.*

Schneider, J. E., and Wade, G. N. (1989). Availability of metabolic fuels controls estrous cyclicity of Syrian hamsters. *Science* 244:1326–8.

Schneider, J. E., and Wade, G. N. (1990). Decreased availability of metabolic fuels induces anestrus in golden hamsters. *Am. J. Physiol.* 258:750–5.

Schneider, M. L. (1992a). The effect of mild stress during pregnancy on birthweight and neuromotor maturation in rhesus monkey infants (*Macaca mulatta*). *Infant Behavior and Development* 15:389–403.

Schneider, M. L. (1992b). Prenatal stress exposure alters postnatal behavioral expression under conditions of novelty challenge in rhesus monkey infants. *Dev. Psychobiol.* 25:529–40.

Schnierla, T. C. (1959). An evolutionary and developmental biphasic process underlying approach and withdrawal. In: *Selected Writings of T. C. Schnierla,* ed. L. R. Aronson et al. San Francisco: Freeman.

Schobitz, B., De Kloet, E. R., and Holsboer, F. (1994a). Gene expression and function of interleukin 1, interleukin 6 and tumor necrosis factor in the brain. *Prog. Neurobiol.* 44:397–432.

Schobitz, B., Holsboer, F., Sutanto, W., Gross, G., Schonbaum, E., and De Kloet, E. R. (1994b). Corticosterone modulates interleukin-evoked fever in the rat. *Neuroendocrinology* 59:387–95.

Schobitz, B., Voorhuis, D. A. M., and De Kloet, E. R. (1992). Localization of interleukin 6 mRNA and interleukin 6 receptor mRNA in rat brain. *Neurosci. Lett.* 136:189–92.

Schuhr, B. (1987). Social structure and plasma corticosterone level in female albino mice. *Physiol. Behav.* 40:689–93.

Schulkin, J. (1978). Mineralocorticoids, dietary conditions and sodium appetite. *Behav. Biol.* 23:197–205.

Schulkin, J. (1982). Behavior of sodium deficient rats: the search for a salty taste. *J. Comp. Physiol. Psychol.* 96:628–34.

Schulkin, J. (1991a). *Sodium Hunger.* Cambridge University Press.

Schulkin, J. (1991b). The allure of salt. *Psychobiology* 19:116–21.

Schulkin, J. (1991c). The ingestion of calcium in female and male rats. *Psychobiology* 19:262–4.

Schulkin, J. (1993). Melancholic depression and the hormones of adversity. *Current Directions in Psychological Sciences* 5:41–4.

Schulkin, J. (1996). Eliot Stellar: from physiological psychology to behavioral neuroscience. *Proc. Natl. Acad. Sci. U.S.A.* 69:3–10.

Schulkin, J., Arnell, P., and Stellar, E. (1985). Running to the taste of salt in mineralocorticoid treated rats. *Horm. Behav.* 19:413–25.

Schulkin, J., Eng, R., and Miselis, R. R. (1983). The effects of disconnecting the subfornical organ on behavioral and physiological responses to alterations of body sodium. *Brain Res.* 263:351–5.

Schulkin, J., and Fluharty, S. J. (1993). Neuroendocrinology of sodium hunger: angiotensin, corticosteroids, and atrial natriuretic hormone. In: *Hormonally Induced Changes in Mind and Brain,* ed. J. Schulkin. San Diego: Academic Press.

Schulkin, J., Gold, P. S., and McEwen, B. S. (in press). Induction of corticotropin releasing hormone gene expression by glucocorticoids: implication for understanding the states of fear and anxiety and allostatic load. *Psychoneuroendocrinology.*

Schulkin, J., and Grill, H. J. (1980). Compensatory ingestion in the decorticate rat. Presented at an international conference on Food and Fluid Intake, Warsaw, Poland.

Schulkin, J., Liebman, D., Ehrman, R. N., Norton, D. L., and Ternes, J. (1984). Sodium hunger in the rhesus monkey. *Behav. Neurosci.* 98:753–6.

Schulkin, J., McEwen, B. S., and Gold, P. W. (1994). Allostasis, amygdala and anticipatory angst. *Neurosci. Biobehav. Rev.* 18:385–96.

Schulkin, J., Marini, J., and Epstein, A. N. (1989). A role for the medial region or the amygdala in mineralocorticoid-induced salt hunger. *Behav. Neurosci.* 103:178–85.

Schulkin, J., Rozin, P., and Stellar, E. (1994b). Curt Richter: a great inquirer. *Proc. Natl. Acad. Sci. U.S.A.* 65:311–20.

Schulze, K. J., and Rasmussen, K. M. (1993). Nutritional status, suckling behavior, and prolactin release during lactation. *Physiol. Behav.* 54:1015–19.

Schumacher, M., Coirini, H., Frankfurt, M., and McEwen, B. S. (1989). Localized actions of progesterone in hypothalamus involve oxytocin. *Proc. Natl. Acad. Sci. U.S.A.* 86: 6798–801.

Schumacher, M., Coirini, H., Pfaff, D. W., and McEwen, B. S. (1990). Behavioral effects of progesterone associated with rapid modulation of oxytocin receptors. *Science* 250: 691–4.

Schumacher, M., and McEwen, B. S. (1989). Steroid and barbiturate modulation of the GABAa receptor: possible mechanisms. *Mol. Neurobiol.* 3:275–303.

Schwaber, J. D., Kapp, B. S., Higgins, G. A., and Rapp, P. R. (1982). Amygdaloid and basal forebrain direct connections with the nucleus of the solitary tract and the dorsal motor nucleus. *J. Neurosci.* 2:1424–38.

Schwabl, H. (1993). Yolk is a source of maternal testosterone for developing birds. *Proc. Natl. Acad. Sci. U.S.A.* 90:11446–50.

Schwartz, G. J., McHugh, P. R., and Moran, T. H. (1993a). Gastric loads and cholecystokinin synergistically stimulate rat gastric vagal afferents. *Am. J. Physiol.* 265:R872–6.

Schwartz, G. J., and Moran, T. H. (1996). Sub-diaphragmatic vagal afferent integration of meal-related gastrointestinal signals. *Neurosci. Biobehav. Rev.* 20:47–56.

Schwartz, M. W., Dallman, M. F., and Woods, S. C. (1995). Hypothalamic response to starvation: implications for the study of wasting disorders. *Am. J. Physiol.* 269:R949–57.

Schwartz, M. W., Figlewicz, D. P., Woods, S. C., Porte, D., Jr., and Baskin, D. G. (1993b). Insulin, neuropeptide Y, and food intake. *Ann. N.Y. Acad. Sci.* 692:60–71.

Sclafani, A., and Nissenbaum, J. W. (1985). On the role of the mouth and gut in the control of saccharin and sugar intake: a reexamination of the sham-feeding preparation. *Brain Res. Bull.* 14:569–76.

Sclera, G., Spector, A. C., and Norgren, R. (1995). Excitotoxic lesions of the parabrachial nuclei prevent conditioned taste aversions and sodium appetite in rats. *Behav. Neurosci.* 109:997–1008.

Scott, T. R., and Mark, G. P. (1986). Feeding and taste. *Prog. Neurobiol.* 27:293–317.

Searle, J. (1986). *Minds, Brains and Science.* Harvard University Press.

Seeley, R. J., Payne, C. J., and Woods, S. C. (1995). Neuropeptide Y fails to increase intraoral intake in rats. *Am. J. Physiol.* 268:423–7.

Seeley, R. J., van Dijk, G., Campfield, L. A., Smith, F. J., Burn, P., Nelligan, J. A., Bell, S. M., Baskin, D. G., Woods, S. C., and Schwartz, M. W. (1996). Intraventricular leptin reduces food intake and body weight of lean rats but not obese Zucker rats. *Horm. Metab. Res.* 28:1–5.

Seeley, R. J., and Schwartz, M. J. (1997). The regulation of energy balance: peripheral endocrine signals and hypothalamic neuropeptides. *Psychol. Sci.* 6:41–4.

Seligman, M. E. P. (1975). *Helplessness.* San Francisco: Freeman.

Semple, P. F., Nichols, M. G., Tree, M., and Fraser, R. (1978). Angiotensin II in the dog: blood levels and effect on aldosterone. *Endocrinology* 4:1476–82.

Sergent, D., Berbigier, P., and Ravault, J. P. (1988). Effect of prolactin inhibition on thermophysiological parameters, water and feed intake of sun-exposed male creole goats (*Capra hircus*) in Guadeloupe (French West Indies). *J. Therm. Biol.* 13:53–9.

Servatius, R. J., Ottenweller, J. E., Bergen, M. T., Soldan, S., and Natelson, B. H. (1994). Persistent stress-induced sensitization of adrenocortical and startle responses. *Physiol. Behav.* 56:945–54.

Shaikh, M. B., Steinberg, A., and Siegel, A. (1993). Evidence that substance P is utilized in medial amygdaloid facilitation of defensive rage behavior in the cat. *Brain Res.* 625:282–94.

Shapiro, L. E., and Insel, T. R. (1990). Infant's response to social separation reflects adult differences in affiliative behavior: a comparative developmental study in prairie and montane voles. *Dev. Psychobiol.* 93:375–93.

Shapiro, L. E., Leonard, C. W., Sessions, C. E., Dewsbury, D. A., and Insel, T. R. (1991). Comparative neuroanatomy of the sexually dimorphic hypothalamus in monogamous and polygamous voles. *Brain Res.* 541:232–40.

Shapiro, R. E., and Miselis, R. R. (1985). The central organization of the vagus nerve innervating the stomach of the rat. *J. Comp. Neurol.* 238:473–88.

Shaywitz, B. A., Shaywitz, S. E., Pugh, K. R., Constable, R. T., Skudlarski, P., Fulbright, R. K., Bronen, R. A., Fletcher, J. M., Shankweller, D. P., Katz, L., and Gore, J. C. (1995). Sex differences in the functional organization of the brain for language. *Nature* 373: 607–9.

Shen, P., Schlinger, B. A., Campagnoni, A. T., and Arnold, A. P. (1995). An atlas of aromatase mRNA expression in the zebra finch brain. *J. Comp. Neurol.* 360:172–84.

Sherwin, B. B. (1996). Estrogen, the brain and memory. *J. North Am. Menopause Soc.* 2:97–105.

Shettleworth, S. J., and Krebs, J. R. (1982). How marsh tits find their hoards: the role of site preference and spatial memory. *J. Exp. Psychol. Anim. Behav. Process.* 8:354–75.

Shoemaker, W. J., and Kehoe, P. (1995). Effect of isolation conditions on brain regional enkephalin and β-endorphin levels and vocalizations in 10-day-old rat pups. *Behav. Neurosci.* 109:117–22.

Shulkes, A. A., Covelli, M. D., Denton, D. A., and Nelson, J. F. (1972). Hormonal factors influencing salt appetite in lactation. *Exp. Biol. Med. Sci.* 50:819–26.

Siaud, P., Manzoni, O., Balmefrezol, M., Barbanel, G., Assenmacher, I., and Alonso, G. (1989). The organization of prolactin-like immunoreactive neurons in the rat central nervous system. *Cell Tissue Res.* 255:107–15.

Sichel, D. A., Cohen, L. S., Robertson, L. M., Ruttenberg, A., and Rosenbaum, J. F. (1995). Prophylactic estrogen in recurrent postpartum affective disorder. *Biol. Psychiatry* 38:814–18.

Siemens, I. R., Adler, H. J., Addya, K., Mah, S. J., and Fluharty, S. J. (1991). Biochemical analysis of solubilized angiotensin II receptors from murine neuroblastoma N1E-115 cells by covalent cross-linking and affinity purification. *Mol. Pharmacol.* 40: 717–26.

Silver, R. (1984). Prolactin and parenting in the pigeon family. *J. Exp. Zool.* 232:617–25.

Silver, R., and Buntin, J. (1973). Role of adrenal hormones in incubation behavior of male ring doves (*Streptopelia risoria*). *J. Comp. Physiol. Psychol.* 84:453–63.

Silver, R., and Feder, H. H. (1973a). Inhibition of crop-sac growth by dexamethasone in ring doves (*Streptopelia risoria*). *Endocrinology* 92:1568–71.

Silver, R., and Feder, H. H. (1973b). Role of gonadal hormones in incubation behavior of male ring doves (*Streptopelia risoria*). *J. Comp. Physiol. Psychol.* 84:464–71.

Silver, R., and Ramos, C. (1990). Vasoactive intestinal polypeptide in avian reproduction. *Comp. Physiol. (Basel)* 8:191–204.

Silverman, A.-J., Don Carlos, L. L., and Morrell, J. I. (1991). Ultrastructural characteristics of estrogen receptor-containing neurons of the ventrolateral nucleus of the guinea-pig hypothalamus. *J. Neuroendocrinol.* 3:623–34.

Simerly, R. B. (1991). Prodynorphin and proenkephalin gene expression in the anteroventral periventricular nucleus of the rat: sexual differentiation and hormonal regulation. *Mol. Cell. Neurosci.* 2:473–84.

Simerly, R. B. (1995). Hormonal regulation of limbic and hypothalamic pathways. In: *Neurobiological Effects of Sex Steroids,* ed. P. E. Micevych and R. P. Hammer. Cambridge University Press.

Simerly, R. B., Chang, C., Muramatsu, M., and Swanson, L. W. (1990). Distribution of androgen and estrogen receptor mRNA-containing cells in the rat brain: an in situ hybridization study. *J. Comp. Neurol.* 294:76–95.

Simerly, R. B., McCall, L. D., and Watson, S. J. (1988). Distribution of opioid peptides in the preoptic region: immunohistochemical evidence for a steroid-sensitive enkephalin sexual dimorphism. *J. Comp. Neurol.* 276:442–59.

Simerly, R. B., and Swanson, L. W. (1988). Projections of the medial preoptic nucleus: a *Phaseolus vulgaris* leucoagglutinin anterograde tract-tracing study in the rat. *J. Comp. Neurol.* 270:209–42.

Simpkins, J. W., Singh, M., and Bishop, J. (1994). The potential role for estrogen replacement therapy in the treatment of the cognitive decline and neurodegeneration associated with Alzheimer's disease. *Neurobiol. Aging* 15:195–7.

Simpson, J. B., and Routtenberg, A. (1973). Subfornical organ: site of drinking elicitation by angiotensin II. *Science* 181:1172–4.

Sipols, A. J., Gaskin, D. G., and Schwartz, M. W. (1995). Effect of intracerebroventricular insulin infusion on diabetic hyperphagia and hypothalamic neuropeptide gene expression. *Diabetes* 44:147–51.

Skinner, B. F. (1938). *Science and Human Behavior.* New York: Macmillan.

Skofitsch, G., and Jacobowitz, D. M. (1985a). Immunohistochemical mapping of galanin-like neurons in the rat central nervous system. *Peptides* 6:509–46.

Skofitsch, G., and Jacobowitz, D. M. (1985b). Distribution of corticotropin releasing factor-like immunoreactivity in the rat brain by immunohistochemistry and radioimmunoassay: comparison and characterization of ovine and rat/human CRF antisera. *Peptides* 6:319–36.

Skofitsch, G., and Jacobowitz, D. M. (1988). Atrial natriuretic peptide in the central nervous system of the rat. *Cell. Mol. Neurobiol.* 8:339–91.

Slawski, B. A., and Buntin, J. D. (1995). Preoptic area lesions disrupt prolactin-induced parental feeding behavior in ring doves. *Horm. Behav.* 29:248–66.

Smale, L., Cassone, V. M., Moore, R. Y., and Morin, L. P. (1989). Paraventricular nucleus projections mediating pineal melatonin and gonadal responses to photoperiod in the hamster. *Brain Res. Bull.* 22:263–9.

Smale, L., Dark, J., and Zucker, I. (1988). Pineal and photoperiodic influences on fat deposition, pelage, and testicular activity in male meadow voles. *J. Biol. Rhythms* 3: 349–55.

Smale, L., Pelz, K., Zucker, I., and Licht, P. (1986). Neonatal androgenization in ground squirrels: influence on sex differences in body mass and luteinizing hormone levels. *Biol. Reprod.* 34:507–11.

Smedley, S. R., and Eisner, T. (1995). Sodium uptake by puddling in a moth. *Science* 270:1816–18.

Smith, G. P. (1997). *Satiation: From Gut to Brain.* Oxford University Press.

Smith, G. P., and Gibbs, J. (1975). Cholecystokinin: a putative satiety signal. *Pharmacol. Biochem. Behav.* 3:135–8.

Smith, G. P., Jerome, C., Cushin, B. J., Eterno, R., and Simansky, K. J. (1981). Abdominal vagotomy blocks the satiety effect of cholecystokinin in the rat. *Science* 213: 1036–7.

Smith, G. P., Jerome, C., and Norgren, R. (1985). Afferent axons in abdominal vagus mediate satiety effect of cholecystokinin in rats. *Am. J. Physiol.* 24:R638–41.

Smith, G. T., Brenowitz, E. A., and Wingfield, J. C. (1997). Seasonal changes in the size of the avian song control nucleus HVC defined by multiple histological markers. *J. Comp. Neurol.* 381:253–61.

Smith, J. W. (1977). *The Behavior of Communicating.* Harvard University Press.

Smith, M. A., Makino, S., Altemus, M., Michelson, D., Hong, S.-K., Kvetnansky, R., and Post, R. M. (1995a). Stress and antidepressants differentially regulate neurotrophin 3 mRNA expression in the locus coeruleus. *Proc. Natl. Acad. Sci. U.S.A.* 92: 8788–92.

Smith, M. A., Makino, S., Kim, S. Y., and Kvetnansky, R. (1995b). Stress increases brain-derived neurotrophic factor messenger ribonucleic acid in the hypothalamus and pituitary. *Endocrinology* 136:3743–50.

Smith, M. A., Makino, S., Kvetnansky, R., and Post, R. M. (1995c). Stress and glucocorticoids affect the expression of brain-derived neurotrophic factor and neurotrophin-3 mRNAs in the hippocampus. *J. Neurosci.* 15:1768–77.

Smith, M. A., Weiss, S. R., Abedin, T., Kim, H., Post, R. M., and Gold, P. W. (1991). Effects of amygdala kindling and electroconvulsive seizures on the expression of corticotropin releasing hormone in the rat brain. *Mol. Cell. Neurosci.* 2:103–6.

Smith, R. D., Turek, F. W., and Takahashi, J. S. (1992). Two families of phase-response curves characterize the resetting of the hamster circadian clock. *Am. J. Physiol.* 262: R1149–53.

Sohrabji, F., Greene, L. A., Miranda, R. C., and Toran-Allerand, C. D. (1994). Reciprocal regulation of estrogen and NGF receptors by their ligands in PC12 cells. *J. Neurobiol.* 25:974–88.

Somers, W., Ultsch, M., DeVos, A. M., and Kosslakoff, A. A. (1994). The x-ray structure of a growth hormone–prolactin receptor complex. *Nature* 372:478–81.

Sonnenberg, J., Luine, V. N., Krey, L. C., and Christakos, S. (1986). 1,25-dihydroxyvitamin D_3 treatment results in increased choline acotyltransferase activity in specific brain nuclei. *Endocrinology* 118:1433–7.

Sonnenberg, J., Pansini, A. R., and Christakos, S. (1984). Vitamin D-dependent rat renal neural calcium-binding protein: development of a radioimmunoassay, tissue distribution, and immunologic identification. *Endocrinology* 115:640–8.

Spears, N., Meyer, J. S., Whaling, C. S., Wade, G. N., Zucker, I., and Darik, J. (1990). Long day lengths enhance myelination of midbrain and hindbrain regions of developing meadow voles. *Dev. Brain Res.* 55:103–8.

Spector, A. (1995). Gustatory function in parabrachial nuclei: implications from lesion studies. *Rev. Neurosci.* 6:143–75.

Spencer, H. (1901). *The Principles of Psychology*. New York: Dover. (Originally published 1890.)

Spencer, R. L., Miller, A. H., Moday, H., Stein, M., and McEwen, B. S. (1993). Diurnal differences in basal and acute stress levels of type I and type II adrenal steroid receptor activation in neural and immune tissues. *Endocrinology* 133:1941–50.

Spiegel, K., Follenius, M., Simon, C., Saini, J., Ehrhart, J., and Brandenberger, G. (1994). Prolactin secretion and sleep. *Sleep* 17:20–7.

Spies, H. G., Norman, R. L., and Buhl, A. E. (1979). Twenty-four-hour patterns in serum prolactin and cortisol after partial and complete isolation of the hypothalamic-pituitary unit in rhesus monkeys. *Endocrinology* 105:1361–8.

Spina, M., Merlo-Pich, E., Chan, R. K., Basso, A. M., Rivier, J., Vale, W., and Koob, G. F. (1996). Appetite-suppressing effects of urocortin, a CRF-related neuropeptide. *Science* 273:1561–4.

Squire, L. R. (1987). *Memory and Brain*. Oxford University Press.

Standaert, D. G., Saper, C. B., and Needleman, P. (1985). Atriopeptin: potent hormone and potential neuromediator. *Trends Neurosci.* 8:509–11.

Standifer, K. M., Chien, C. C., Wahlestedt, C., Brown, G. P., and Pasternak, G. W. (1994). Selective loss of opioid analgesia and binding by antisense oligodeoxynucleotides to an opioid receptor. *Neuron* 12:805–10.

Stanley, B. G., Lanthier, D., Chin, A. S., and Leibowitz, S. F. (1989). Suppression of neuropeptide Y elicited eating by adrenalectomy or hypophysectomy: reversal with corticosterone. *Brain Res.* 501:32–6.

Stanley, B. G., and Leibowitz, S. F. (1984). Neuropeptide Y: stimulation of feeding and drinking by injection into the paraventricular nucleus. *Life Sci.* 35:2635–42.

Stein, L. J., Dorsa, D. M., Baskin, D. G., Figlewicz, D. P., Ikeda, H., Frankmann, S. P., Greenwood, M. R. C., Porte, D. J., and Woods, S. C. (1983). Immunoreactive insulin levels are elevated in the cerebrospinal fluid of genetically obese Zucker rats. *Endocrinology* 113:2299.

Stein, M., Miller, A. H., and Trestman, R. L. (1991). Depression, the immune system, and health and illness. *Arch. Gen. Psychiatry* 48:171–7.

Steiner, J. C. (1977). Facial expressions of the neonate infant indicating the hedonics of food-related chemical stimuli. In: *Taste and Development: The Genesis of Sweet Preference*, ed. J. M. Weiffenbach. Bethesda, MD: NIH.

Stellar, E. (1954). The physiology of motivation. *Psychol. Rev.* 61:5–22.

Stellar, J. R., and Stellar, E. (1985). *The Neurobiology of Reward*. Berlin: Springer-Verlag.

Stenzel-Poore, M. P., Heinrichs, S. C., Rivest, S., Koob, G. F., and Vale, W. W. (1994). Overproduction of corticotropin-releasing factor in transgenic mice: a genetic model of anxiogenic behavior. *J. Neurosci.* 14:2579–84.

Stephens, T. W., Basinsci, M., Bristow, P. K., Bue-Valleskey, J. M., Burgett, S. G., Craft, L., Hale, J., Hoffman, J., Hsiung, H. M., and Kriancinnes, A. (1995). The role of neuropeptide Y in the antiobesity action of the obese gene product. *Nature* 377:530–2.

Stephenson, G., Hammet, M., Hadaway, G., and Funder, J. W. (1984). Ontogeny of renal mineralocorticoid receptors and urinary electrolyte responses in the rat. *Am. J. Physiol.* 247:F665–71.

Stewart, P. M., Wallace, A. M., Valentino, R., Burt, D., Shackleton, C. H., and Edwards, C. R. (1987). Mineralocorticoid activity of liquorice: 11-beta-hydroxysteroid dehydrogenase deficiency comes of age. *Lancet* 2:821–4.

Stopa, E. G., Johnson, J. K., Friedman, D. I., Ryer, H. I., Reidy, J., Kuo-LeBlanc, V., and Albers, H. E. (1995). Neuropeptide Y receptor distribution and regulation in the suprachiasmatic nucleus of the Syrian hamster (*Mesocricetus auratus*). *Pep. Res.* 8:95–100.

Strack, A. M., Sebastian, R. J., Schwartz, M. W., and Dallman, M. F. (1995). Glucocorticoids and insulin: reciprocal signals for energy balance. *Am. J. Physiol.* 268:R142–9.

Strauss, K. I., Jacobowitz, D. M., and Schulkin, J. (1994). Dietary calcium deficiency causes a reduction in calretinin mRNA in the substantia nigra compacta–ventral tegmental area of rat brain. *Mol. Brain Res.* 25:140–2.

Strauss, K. I., Schulkin, J., and Jacobowitz, D. M. (1995). Corticosterone effects on rat calretinin mRNA in discrete brain nuclei and the testes. *Mol. Brain Res.* 28:81–6.

Stricker, E. M. (1966). Extracellular fluid volume and thirst. *Am. J. Physiol.* 211:232–8.

Stricker, E. M. (1983). Thirst and sodium appetite after colloid treatment in rats; role of the renin-angiotensin-aldosterone system in rats. *Behav. Neurosci.* 97:725–37.

Stricker, E. M., and Jalowiec, J. E. (1970). Restoration of intravascular fluid volume following acute hypovolemia in rats. *Am. J. Physiol.* 218:191–6.

Stricker, E. M., and Sterritt, G. M. (1966). Osmoregulation in the newly hatched domestic chick. *Physiol. Behav.* 2:117–19.

Stricker, E. M., Thiels, E., and Verbalis, J. G. (1991). Sodium appetite in rats after prolonged dietary sodium deprivation: a sexually dimorphic phenomenon. *Am. J. Physiol.* 260:1082–8.

Stricker, E. M., Vagnucci, A. H., McDonald, R. H., Jr., and Leenen, F. H. (1979). Renin and aldosterone secretions during hypovolemia in rats: relation to NaCl intake. *Am. J. Physiol.: Regulatory Integrative Comp. Physiol.* 6:R45–51.

Stricker, E. M., and Verbalis, J. G. (1987). Central inhibitory control of sodium appetite in rats: correlation with pituitary oxytocin secretion. *Behav. Neurosci.* 4:560–7.

Stricker, E. M., and Verbalis, J. G. (1990). Sodium appetite. In: *Handbook of Behavioral Neurobiology. Vol. 10: Neurobiology of Food and Fluid Intake,* ed. E. M. Stricker. New York: Plenum.

Stricker, E. M., and Verbalis, J. G. (1991). Caloric and noncaloric controls of food intake. *Brain Res. Bull.* 27:299–303.

Stricker, E. M., and Wolf, G. (1969). Behavioral control of intravascular fluid volume: thirst and sodium appetite. *Ann. N.Y. Acad. Sci.* 157:553–68.

Stricker, E. M., and Zigmond, M. J. (1974). Effects on homeostasis of intraventricular injections of 6-hydroxydopamine in rats. *J. Comp. Physiol. Psychol.* 86:973–94.

Stumpf, W. E., and O'Brien, L. P. (1987). 1,25(OH)vitamin D sites of action in the brain. *Histochemistry* 87:393–406.

Stumpf, W. E., and Privette, T. H. (1989). Light, vitamin D and psychiatry. Role of vitamin D in etiology and therapy of seasonal affective disorder and other mental processes. *Psychopharmacology* 97:285–94.

Sucheckl, D., Nelson, D. Y., Oers, H. V., and Levine, S. (1995). Activation and inhibition of the hypothalamic-pituitary-adrenal axis of the neonatal rat: effects of maternal deprivation. *Psychoendocrinology* 2:169–82.

Sullivan, R. M., Wilson, D. A., and Leon, M. (1989). Norepinephrine and learning-induced plasticity in infant rat olfactory system. *J. Neurosci.* 9:3998–4006.

Summers, C., Gault, T. R., and Fregly, M. J. (1991a). Potentiation of angiotensin II-induced drinking by glucocorticoids is a specific glucocorticoid Type II receptor (GR)-mediated event. *Brain Res.* 552:283–90.

Sumners, C., and Myers, L. M. (1991). Angiotensin II decreases cGMP levels in neural cultures from rat brain. *Am. J. Physiol.* 2260:C79–87.

Sumners, C., Myers, L. M., Kalberg, C. J., and Raizada, M. K. (1990). Physiological and pharmacological comparison of angiotensin II receptors in neuronal and astrocyte glial cultures. *Prog. Neurobiol.* 34:355–85.

Sumners, C., Tang, W., Zelezna, B., and Raizada, M. K. (1991b). Angiotensin II receptor subtypes are coupled with distinct signal transduction mechanisms in neurons and astrocytes from rat brain. *Proc. Natl. Acad. Sci. U.S.A.* 88:7567–71.

Swaab, D. F., and Hofman, M. A. (1995). Sexual differentiation of the human hypothalamus in relation to gender and sexual orientation. *Trends Neurosci.* 18:264–70.

Swaab, D. F., Raadsheer, F. C., Endert, E., Hofman, M. A., Kamphrost, W., and Ravid, R. (1994). Increased cortisol levels in aging and Alzheimer's disease in postmortem cerebrospinal fluid. *J. Neuroendocrinol.* 6:681–7.

Swann, J. M., and Newman, S. W. (1992). Testosterone regulates substance P within neurons of the medial nucleus of the amygdala, the bed nucleus of the stria terminalis and the medial preoptic area of the male golden hamster. *Brain Res.* 590:18–28.

Swann, J. M., and Turek, F. W. (1985). Multiple circadian oscillators regulate the timing of behavioral and endocrine rhythms in female golden hamsters. *Science* 228:898–900.

Swann, J. M., and Turek, F. A. (1988). Transfer from long to short days reduces the frequency of pulsatile luteinizing hormone release in intact but not in castrated male golden hamsters. *Neuroendocrinology* 47:343–9.

Swanson, L. W. (1991). Biochemical switching in hypothalamic circuits mediating responses to stress. *Prog. Brain Res.* 87:181–200.

Swanson, L. W., Kucharczyk, J., and Mogenson, G. J. (1978). Audiographic evidence for pathways from the medial preoptic area to the midbrain involved in the drinking response to angiotensin II. *J. Comp. Neurol.* 178:645–60.

Swanson, L. W., Sawchenko, P. E., Rivier, J., and Vale, W. W. (1983). Organization of ovine corticotropin-releasing factor immunoreactive cells and fibers in the rat brain: an immunohistochemical study. *Neuroendocrinology* 36:165–86.

Swanson, L. W., and Sharpe, L. G. (1973). Centrally induced drinking: comparison of angiotensin II and carbachol-sensitive sites in rats. *Am. J. Physiol.* 225:566–73.

Swanson, L. W., and Simmons, D. M. (1989). Differential steroid hormone and neural influences on peptide mRNA levels in CRH cells of the paraventricular nucleus: a hybridization histochemical study in the rat. *J. Comp. Neurol.* 285:413–35.

Swiergiel, A. H., Takahashi, L. K., and Kalin, N. H. (1993). Attenuation of stress-induced behavior by antagonism of corticotropin-releasing factor receptors in the central amygdala in the rat. *Brain Res.* 623:229–34.

Takahashi, J. S. (1995). Molecular neurobiology and genetics of circadian rhythms in mammals. *Annu. Rev. Neurosci.* 18:531–53.

Takahashi, L. K. (1994). Organizing action of corticosterone on the development of behavioral inhibition in the preweanling rat. *Dev. Brain Res.* 81:121–7.

Takahashi, L. K. (1995). Glucocorticoids, the hippocampus, and behavioral inhibition in the preweanling rat. *J. Neurosci.* 15:6023–4.

Takahashi, L. K., Kalin, N. H., Vanden Burgt, J. A., and Sherman, J. E. (1989). Corticotropin-releasing factor modulates defensive-withdrawal and exploratory behavior in rats. *Behav. Neurosci.* 103:648–54.

Takahashi, L. K., and Kim, H. (1994). Intracranial action of corticosterone facilitates the development of behavioral inhibition in the adrenalectomized preweanling rat. *Neurosci. Lett.* 176:272–6.

Takahashi, L. K., and Rubin, W. W. (1993). Corticosteroid induction of threat-induced behavioral inhibition in preweanling rats. *Behav. Neurosci.* 107:860–6.

Takei, Y., Hirano, T., and Kobayashi, H. (1979). Angiotensin and water intake in the Japanese eel (*Anguilla japonica*). *Gen. Comp. Endocrinol.* 38:466–75.

Tarjan, E., and Denton, D. A. (1991). Sodium/water intake of rabbits following administration of hormones of stress. *Brain Res. Bull.* 26:133–16.

Tarjan, E., Denton, D. A., and Weisinger, R. S. (1988). Atrial natriuretic peptide inhibits water and sodium intake in rabbits. *Regul. Pept.* 23:63–75.

Tartaglia, L. A., Dembski, M., Weng, X., et al. (1995). Identification and expression cloning of a leptin receptor, OB-R. *Cell* 83:1263–71.

Tarttelin, M. D., and Gorski, R. A. (1971). Variations in food and water intake in the normal and acyclic female rat. *Physiol. Behav.* 7:847–52.

Tempel, D. L., and Leibowitz, S. F. (1989). PVN steroid implants: effect on feeding patterns and macronutrient selection. *Brain Res. Bull.* 23:553–60.

Tempel, D. L., and Leibowitz, S. F. (1993). Glucocorticoid receptors in PVN: interactions with NE, NPY, and Gal in relation to feeding. *Am. J. Physiol.* 265:E794–800.

Tempel, D. L., and Leibowitz, S. F. (1994). Adrenal steroid receptors:interactions with brain neuropeptide systems in relation to nutrient intake and metabolism. *J. Neuroendocrinol.* 6:479–501.

Tennes, K., and Kreye, M. (1985). Children's adrenocortical responses to classroom activities and tests in elementary school. *Psychosom. Med.* 47:451–60.

Thiels, E., Verbalis, J. G., and Stricker, E. M. (1990). Sodium appetite in lactating rats. *Behav. Neurosci.* 104:742–50.

Thomas, E. M., Jewett, M. E., and Zucker, I. (1993). Torpor shortens the period of Siberian hamster circadian rhythms. *Am. J. Physiol.* 265:R951–6.

Thomas, T. L., Devenport, L. D., and Stith, R. D. (1994). Relative contribution of type I and II corticosterone receptors in VMH lesion-induced obesity and hyperinsulinemia. *Am. J. Physiol.* 266:R1623–9.

Thompson, C. I., and Epstein, A. N. (1991). Salt appetite in rat pups: ontogeny of angiotensin II–aldosterone synergy. *Am. J. Physiol.* 260:R421–9.

Thorpe, W. H. (1963). *Learning and Instinct in Animals*. London: Methuen. (Originally published 1936.)

Thunhorst, R. L. (1996). Role of peripheral angiotensin in salt appetite of the sodium-deplete rat. *Neurosci. Biobehav. Rev.* 20:101–6.

Thunhorst, R. L., Ehrlich, K. J., and Simpson, J. B. (1990). Subfornical organ participates in salt appetite. *Behav. Neurosci.* 4:637–42.

Thunhorst, R. L., and Johnson, A. K. (1994). Renin-angiotensin, arterial blood pressure, and salt appetite in rats. *Am. J. Physiol.* 266:R458–65.

Tinbergen, N. (1969). *The Study of Instinct*. Oxford University Press. (Originally published 1951.)

Tobias, M. L., Marin, M. L., and Kelley, D. B. (1991). Development of functional sex differences in the larynx of *Xenopus laevis*. *Dev. Biol.* 147:251–9.

Tobias, M. L., Marin, M. L., and Kelley, D. B. (1993). The roles of sex, innervation, and androgen in laryngeal muscle of *Xenopus laevis*. *J. Neurosci.* 13:324–33.

Tokarz, R. R. (1987). Effects of corticosterone treatment on male aggressive behavior in a lizard (*Anolis sagrei.*) *Horm. Behav.* 21:358–70.

Toran-Allerand, C. D. (1991). Organotypic culture of the developing cerebral cortex and hypothalamus: relevance to sexual differentiation. *Psychoneuroendocrinology* 16:7–24.

Toran-Allerand, C. D., Miranda, R. C., Bentham, W. D. L., Schrabji, F., Brown, T. J., Hochberg, R. B., and MacLusky, N. J. (1992). Estrogen receptors colocalize with low-affinity nerve growth factor receptors in cholinergic neurons of the basal forebrain. *Neurobiology* 89:4668–72.

Tordoff, M. G. (1994). Corticosterone is a physiological regulator of salt intake in the rat (abstract). *FASEB J.* 8:A552.

Tordoff, M. G. (1996). Adrenalectomy decreases NaCl intake of rats fed low-calcium diets. *Am. J. Physiol.* 270:R11–21.

Tordoff, M. G., Hughes, R. L., and Pilchak, D. M. (1993). Independence of salt intake from the hormones regulating calcium homeostasis. *Am. J. Physiol.* 264:500–12.

Tordoff, M. G., Schulkin, J., and Friedman, M. K. (1986). Hepatic contribution to satiation of salt appetite in rats. *Am. J. Physiol.* 251:R1095–102.

Tordoff, M. G., Ulrich, P. M., and Schulkin, J. (1990). Calcium deprivation increases salt intake. *Am. J. Physiol. (Regulatory Integrative Comp. Physiol.* 28) 259:R411–19.

Toth, E., Stelfox, J., and Kaufman, S. (1987). Cardiac control of salt appetite. *Am. J. Physiol.* 252:R925–9.

Toyoda, F., Ito, M., Tanaka, S., and Kikuyama, S. (1993). Hormonal induction of male courtship behavior in the Japanese newt, *Cynops pyrrhogaster. Horm. Behav.* 27: 511–22.

Truman, J. W. (1992). Hormonal regulation of behavior: insights from invertebrate systems. Unpublished manuscript.

Tubbiola, M. L., and Bittman, E. L. (1995). Short days increase sensitivity to methadone inhibition of male copulatory behavior. *Physiol. Behav.* 58:647–51.

Turek, F. W., Desjardins, C., and Menaker, M. (1975). Melatonin: antigonadal and pro-gonadal effects in male golden hamsters. *Science* 190:280–2.

Turek, F. W., Losee-Olson, S., Swann, J. M., Horwath, K., Van Cauter, E., and Milette, J. J. (1987). Circadian and seasonal control of neuroendocrine-gonadal activity. *J. Steroid Biochem.* 27:573–9.

Turek, F. W., Pinto, L. H., Vitaterna, M. H., Peneve, P. D., Zee, P. C., and Takahashi, J. S. (1995). Pharmacological and genetic approaches for the study of circadian rhythms in mammals. *Front. Neuroendocrinol.* 16:191–223.

Turek, F. W., and Van Cauter, E. (1994). Rhythms in reproduction. In: *The Physiology of Reproduction,* 2nd ed. New York: Raven Press.

Ulibarri, C., and Yahr, P. (1996). Effects of androgens and estrogens on sexual differentiation of sex behavior, scent marking and the sexually dimorphic area of the gerbil hypothalamus. *Horm. Behav.* 30:107–30.

Upadhyaya, A. K., Pennell, I., Cowen, P. J., and Deakin, J. F. W. (1991). Blunted growth hormone and prolactin responses to L-tryptophan in depression: a state-dependent abnormality. *J. Affect. Disord.* 21:213–18.

Uvnas-Moberg, K. (1994). Role of efferent and afferent vagal nerve activity during reproduction: integrating function of oxytocin on metabolism and behaviour. *Psychoneuroendocrinology* 19:687–95.

Uvnas-Moberg, K. (1996). Neuroendocrinology of the mother–infant interaction. *Trends Endocrinol. Metab.* 7:126–31.

Uvnas-Moberg, K. (1997). Physiological and endocrine effects of social contact. *Ann. N.Y. Acad. Sci.* 807:146–63.

Valdizan, J. R., Vergara, J. M., Rodriguez, J. P., Guallar, A., and Garcia, C. (1992). Nocturnal prolactin and growth hormone levels in children with complex partial and generalized tonic-clonic seizures. *Acta Neurol. Scand.* 86:139–41.

Vale, W., Spiess, J., River, C., and Rivier, J. (1981). Characterization of 41 residue ovine hypothalamic peptide that stimulates the secretion of corticotropin releasing hormone and β-endorphin. *Science* 213:1394–7.

Valentino, R. J., and Curtis, A. L. (1991). Pharmacology of locus coeruleus spontaneous and sensory evoked activity. *Prog. Brain Res.* 88:249–56.

Valentino, R. J., Foote, S. L., and Jones, G. A. (1983). Corticotropin-releasing factor activates noradrenergic neurons in the locus coeruleus. *Brain Res.* 270:363–7.

Valentino, R. J., Pavcovich, L. A., and Hirata, H. (1995). Evidence for corticotropin-releasing hormone projections from Barrington's nucleus to the periaqueductal gray and dorsal motor nucleus of the vagus in the rat. *J. Comp. Neurol.* 363:402–22.

Van Cauter, E., Shapiro, E. T., Tillil, H., and Polonsky, K. S. (1992). Circadian modulation of glucose and insulin responses to meals: relationship to cortisol rhythm. *Am. J. Physiol.* 262:E467–75.

Van Dam, A.-M., Brouns, M., Louisse, S., and Berkenbosch, F. (1992). Appearance of interleukin-1 in macrophages and in ramified microglia in the brain of endotoxin-treated rats: a pathway for the induction of non-specific symptoms of sickness? *Brain Res.* 588:291–6.

Van de Kar, L. D., Piechowski, R. A., Rittenhouse, P. A., and Gray, T. S. (1991). Amygdaloid lesions: differential effect on conditioned stress and immobilization-induced increases in corticosterone and renin secretion. *Neuroendocrinology* 54:89–96.

Van Lecuwen, F. W., Caffe, A. R., and DeVries, G. J. (1985). Vasopressin cells in the bed nucleus of the stria terminalis of the rat: sex differences and the influence of androgens. *Brain Res.* 325:391–4.

Verbalis, J. G., Blackburn, R. E., Olson, B. R., and Stricker, E. M. (1993). Central oxytocin inhibition of food and salt ingestion: a mechanism for intake regulation of solute homeostasis. *Regul. Pept.* 45:149–54.

Verbalis, J. G., Hoffman, G. E., and Sherman, T. G. (1995). Use of immediate early genes as markers of oxytocin and vasopressin neuronal activation. *Curr. Opin. Endocrinol. Diabetes* 2:157–68.

Verbalis, J. G., Mangione, M. P., and Stricker, E. M. (1991). Oxytocin produces natriuresis in rats at physiological plasma concentrations. *Endocrinology* 91:1317–22.

Vijande, H. I., Costales, M., Schiaffini, O., and Marin, B. (1978) Angiotensin-induced drinking: sexual differences. *Pharmacol. Biochem. Behav.* 8:753–5.

Vinson, G. P., Whitehouse, B. J., Goddard, C., and Sibley, C. P. (1979). Comparative and evolutionary aspects of aldosterone secretion and zona glomerulosa function. *J. Endocrinol.* 81:5P–24P.

Vitaterna, M. H., King, D. P., Chang, A. M., Kornhauser, J. M., Lowrey, P. L., McDonald, J. D., Dove, W. F., Pinto, L. H., Turek, F. W., and Takahashi, J. S. (1994). Mutagenesis and mapping of a mouse gene, *clock,* essential for circadian behavior. *Science* 264: 719–25.

Von Holst, E. (1973). *The Behavioural Physiology of Animals and Man,* vol. 1. Coral Gables: University of Miami Press. (Originally published 1969.)

Voorhuis, T. A., Dc Kloet, E. R., and De Wied, D. (1991). Effect of a vasotocin analog on singing behavior in the canary. *Horm. Behav.* 25:549–59.

Voorhuis, T. A. M., Kiss, J. Z., De Kloet, E. R., and De Wied, D. (1988). Testosterone sensitive vasotocin-immunoreactivity cells and fiber in the canary brain. *Brain Res.* 442:139–46.

Wade, G. N., and Bartness, T. J. (1984). Seasonal obesity in Syrian hamsters: effects of age, diet, photoperiod, and melatonin. *Am. J. Physiol.* 247:328–34.

Wade, G. N., and Schneider, J. E. (1992). Metabolic fuels and reproduction in female mammals. *Neurosci. Biobehav. Rev.* 16:1–38.

Wade, G. N., and Zucker, I. (1969). Hormonal and developmental influences on rat saccharin preferences. *J. Comp. Physiol. Psychol.* 69:291–300.

Wade, J., and Crews, D. (1991). The effects of intracranial implantation of estrogen on receptivity in sexually and asexually reproducing female whiptail lizards, *Cnemidophorus inornatus* and *Cnemidophorus uniparens*. *Horm. Behav.* 25:342–53.

Wade, J., Huang, J. M, and Crews, D. (1993). Hormonal control of sex differences in the brain, behavior and accessory sex structures of whiptail lizards (*Cnemidophorus* species). *J. Neuroendocrinol.* 3:81–93.

Wahlestedt, C., Pich, E. M., Koob, G. F., Yee, F., and Heilig, M. (1993). Modulation of anxiety and neuropeptide Y-Y1 receptors by antisense oligodeoxynucleotides. *Science* 259:528–31.

Wallen, K. (1990). Desire and ability: hormones and the regulation of female sexual behavior. *Neurosci. Biobehav. Rev.* 14:233–41.

Wallen, K., Maestripieri, D., and Mann, D. R. (1995). Effects of neonatal testicular suppression with a GnRH antagonist on social behavior in group-living juvenile rhesus monkeys. *Horm. Behav.* 29:322–37.

Walsh, R. J., Mangurian, L. P., and Posner, B. I. K. (1990). The distribution of lactogen receptors in the mammalian hypothalamus: an in vitro autoradiographic analysis of the rabbit and rat. *Brain Res.* 530:1–11.

Walsh, R. J., Slaby, F. J., and Posner, B. I. (1987). A receptor-mediated mechanism for the transport of prolactin from blood to cerebrospinal fluid. *Endocrinology* 120:1846–50.

Walton, J. S., Evins, J. D., Fitzgerald, B. P., and Cunningham, F. J. (1980). Abrupt decrease in daylength and short-term changes in the plasma concentrations of FSH, LH and prolactin in anoestrous ewes. *J. Reprod. Fertil.* 59:163–71.

Wang, Z., and DeVries, G. J. (1993). Androgen and estrogen effects on vasopressin messenger RNA expression in the medial amygdaloid nucleus in male and female rats. *J. Neuroendocrinol.* 7:827–31.

Wang, Z., Ferris, C. F., and DeVries, G. J. (1994). Role of septal vasopressin innervation in paternal behavior in prairie voles (*Microtus ochrogaster*). *Proc. Natl. Acad. Sci. U.S.A.* 91:400–4.

Wang, Z., Smith, W., Major, D. E., and DeVries, G. J. (1995). Sex and species differences in the effects of cohabitation on vasopressin messenger RNA expression in the bed nucleus of the stria terminalis in prairie voles (*Microtus ochrogaster*) and meadow voles (*Microtus pennsylvanicus*). *Brain Res.* 650:212–18.

Warner, M. D., Sinha, Y. N., and Peabody, C. A. (1993). Growth hormone and prolactin variants in normal subjects. Relative proportions in morning and afternoon samples. *Horm. Metab. Res.* 25:425–9.

Waterman, G. S., Dahl, R. E., Birmaher, B., Ambrosini, P., Rabinovich, H., Williamson, D., Novacenko, H., Nelson, B., Puig-Antich, J., and Ryan, N. D. (1994). The 24-hour pattern of prolactin secretion in depressed and normal adolescents. *Biol. Psychiatry* 35:440–5.

Watson, J. T., and Kelley, D. B. (1992). Testicular masculinization of vocal behavior in juvenile female *Xenopus laevis* reveals sensitive periods for song duration, rate, and frequency spectra. *J. Comp. Physiol.* A171:343–50.

Watson, S. J., Khachaturian, H., Akil, H., Coy, D. H., and Goldstein, A. (1982). Comparison of the distribution of dynorphin systems and enkephalin systems in brain. *Science* 218:1134–6.

Watts, A. G. (1996). The impact of physiological stimuli on the expression of corticotrophin-releasing hormone (CRH) and other neuropetide genes. *Front. Neuroendocrinol.* 17:1–48.

Watts, A. G., and Sanchez-Watts, G. (1995a). A cell-specific role for the adrenal gland in regulating CRH mRNA levels in rat hypothalamic neurosecretory neurons after cellular dehydration. *Brain Res.* 484:63–70.

Watts, A. G., and Sanchez-Watts, G. (1995b). Region-specific regulation of neuropeptide mRNAs in rat limbic forebrain neurones by aldosterone and corticosterone. *J. Physiol. (Lond.)* 484:721–36.

Watts, A. G., Sheward, W. J., Whale, D., and Fink, G. (1989). The effects of knife cuts in the sub-paraventricular zone of the female rat hypothalamus on oestrogen-induced diurnal surges of plasma prolactin and LH, and circadian wheel-running activity. *J. Endocrinol.* 122:593–604.

Watts, A. G., and Swanson, L. W. (1989). Diurnal variations in the content of preprocorticotropin releasing hormone messenger ribonucleic acids in the hypothalamic paraventricular nucleus of rats of both sexes as measured by in situ hybridization. *Endocrinology* 125:1734–8.

Weaver, D. R., Keohan, J. T., and Reppert, S. M. (1987). Definition of a prenatal sensitive period for maternal–fetal communication of day length. *Am. J. Physiol.* 16:E701–4.

Weaver, D. R., Stehle, J. H., Stopa, E. G., and Reppert, S. M. (1993). Melatonin receptors in human hypothalamus and pituitary: implications for circadian and reproductive responses to melatonin, *J. Clin. Endocrinol. Metab.* 76:295–301.

Weeks, H. P., Jr., and Kirkpatrick, C. M. (1976). Adaptations of white-tailed deer to naturally occurring sodium deficiencies. *J. Wildl. Manage.* 40:610–25.

Wehling, M., Eisen, C., and Christ, M. (1992). Aldosterone-specific membrane receptors and rapid non-genomic actions of mineralocorticoids. *Mol. Cell. Endocrinol.* 90:C5–9.

Wehr, T. A. (1989). Sleep loss: a preventable cause of mania and other excited states. *J. Clin. Psychiatry* 50:8–16.

Wehr, T. A. (1991). The durations of human melatonin secretion and sleep respond to changes in daylength (photoperiod). *J. Clin. Endocrinol. Metab.* 73.1976–80.

Wehr, T. A., Giesen, H. A., Moul, D. E., Turner, E. H., and Schwartz, P. J. (1995). Suppression of human responses to seasonal changes in day-length by modern artificial lighting. *Am. J. Physiol.* 269:R173–8.

Wehr, T. A., Moul, D. E., Barbato, G., Giesen, H. A., Seidel, J. A., Barker, C., and Bender, C. (1993). Conservation of photoperiod-responsive mechanisms in humans. *Am. J. Physiol.* 265:846–57.

Weingarten, H. P. (1996). Cytokines and food intake: the relevance of the immune system to the student of ingestive behavior. *Neurosci. Biobehav. Rev.* 20:163–70.

Weisinger, R. S., Blair-West, J. R., Denton, D. A., McBurnie, M., Ong, F., Tarjan, E., and Williams, R. M. (1990a). Effect of angiotensin-converting enzyme inhibitor on salt appetite and thirst of BALB/c mice. *Am. J. Physiol.* 259:R736–40.

Weisinger, R. S., Blair-West, J. R., Denton, D. A., and Tarjan, E. (1997). Role of brain angiotensin II in thirst and sodium appetite of sheep. *Am. J. Physiol.* 42:R187–96.

Weisinger, R. S., Denton, D. A., Di Nicolantonio, R., Hards, D. K., McKinley, M. J., Old-field, B., and Osborne, P. G. (1990b). Subfornical organ lesion decreases sodium appetite in the sodium depleted rat. *Brain Res.* 526:23–30.

Weisinger, R. S., Denton, D. A., McKinley, M. J., and Nelson, J. F. (1978). ACTH induced sodium appetite in the rat. *Pharmacol. Biochem. Behav.* 8:339–43.

Weisinger, R. S., Denton, D. A., Di Nicolantonio, R., McKinley, M. J., Muller, A. F., and Tarjan, E. (1987). Role of angiotensin in sodium appetite of sodium-depleted sheep. *Am. J. Physiol.* 253:482–8.

Weiss, J. M., Goodman, P. A., Losito, B. G., Corrigan, S., Charry, J. M., and Bailey, W. H. (1981). Behavioral depression produced by an uncontrollable stressor: relationship to norepinephrine, dopamine, and serotonin levels in various regions of rat brain. *Brain Res. Rev.* 3:167–205.

Weiss, J. M., Stout, J. C., Aaron, M. F., Quan, N., Owens, M. J., Butler, P. D., and Nemeroff, C. B. (1994). Depression and anxiety: role of the locus coeruleus and corticotropin-releasing factor. *Brain Res. Bull.* 35:561–72.

Weiss, M. L., Moe, K. E., and Epstein, A. N. (1986). Interference with central actions of angiotensin II suppresses sodium appetite. *Am. J. Physiol.* 250:R250–9.

Weiss, S. R. B., Nirenberg, J., Lewis, R., and Post, R. M. (1991). Corticotropin-releasing hormone: potentiation of cocaine-kindled seizures and lethality. *Epilepsia* 33:248–54.

Welch, C. C., Grace, M. K., Billington, C. J., and Levine, A. S. (1994). Preference and diet type affect macronutrient selection after morphine, NPY, norepinephrine and deprivation. *Am. J. Physiol.* 35:R426–33.

Weller, A., Smith, G. P., and Gibbs, J. (1990). Endogenous cholecystokinin reduces feeding in young rats. *Science* 247:1589–91.

Wentworth, B. C., Proudman, J. A., Opel, H., Wineland, M. J., Zimmermann, N. G., and Lapp, A. (1983). Endocrine changes in the incubating and brooding turkey hen. *Biol. Reprod.* 29:87–92.

Westlind-Danielsson, A., Gould, E., and McEwen, B. S. (1991). Thyroid hormone causes sexually distinct neurochemical and morphological alterations in rat septal–diagonal band neurons. *J. Neurochem.* 56:119–28.

Wetzel, D. M., Haerter, U. L., and Kelley, D. B. (1985). A proposed neural pathway for vocalization in South African clawed frogs, *Xenopus laevis. J. Comp. Physiol.* A157:749–61.

Whaling, C. S., Zucker, I., Wade, G. N., and Dark, J. (1990). Sexual dimorphism in brain weight of meadow voles: role of gonadal hormones. *Dev. Brain Res.* 53:270–5.

Whitman, D. C., and Albers, H. E. (1995). Role of oxytocin in the hypothalamic regulation of sexual receptivity in hamsters. *Brain Res.* 680:73–9.

Whitnall, M. H. (1993). Regulation of the hypothalamic corticotropin-releasing hormone neurosecretory system. *Prog. Neurobiol.* 40:573–629.

Wibbels, T., Bull, J. J., and Crews, D. (1991). Synergism between temperature and estradiol: a common pathway in turtle sex determination? *J. Exp. Zool.* 260:130–14.

Wibbels, T., Bull, J. J., and Crews, D. (1992). Steroid hormone-induced male sex determination in an amniotic vertebrate. *J. Exp. Zool.* 262:454–7.

Wiersma, A., Baauw, A. D., Bohus, B., and Koolhaas, J. M. (1995). Behavioural activation produced by CRH but not a-helical CRH (CRH-receptor antagonist) when microinfused into the central nucleus of the amygdala under stress-free conditions. *Psychoneuroendocrinology* 20:423–32.

Wilding, J. P., Gilbey, S. G., Gailey, C. J., Batt, R. A., Williams, G., Ghatei, M. A., and

Bloom, S. R. (1993a). Increased neuropeptide-Y messenger ribonucleic acid (mRNA) and decreased neurotensin mRNA in the hypothalamus of the obese (ob/ob) mouse. *Endocrinology* 132:1939–44.

Wilding, J. P., Gilbey, S. G., Lambert, P. D., Ghatei, M. A., and Bloom, S. R. (1993b). Increases in neuropeptide Y content and gene expression in the hypothalamus of rats treated with dexamethasone are prevented by insulin. *Neuroendocrinology* 57:581–7.

Wilkins, L., and Richter, C. P. (1940). A great craving for salt by a child with cortico-adrenal insufficiency. *JAMA* 114:866–8.

Wilkinson, M. F., Mathieson, W. B., and Pittman, Q. J. (1993). Interleukin-1B has excitatory effects on neurons of the bed nucleus of the stria terminalis. *Brain Res.* 625: 342–6.

Williams, C. L., and Meck, W. H. (1991). The organizational effects of gonadal steroids on sexually dimorphic spatial ability. *Psychoneuroendocrinology* 16:155–76.

Williams, J. R., Insel, T. R., Harbaugh, C. R., and Carter, C. S. (1994). Oxytocin administered centrally facilitates formation of a partner preference in female prairie voles (*Microtus ochrogaster*). *J. Neuroendocrinol.* 6:247–50.

Williams, R. F., and Hodgen, G. D. (1980). The diurnal rhythm of prolactin secretion in postpartum rhesus monkeys. *Biol. Reprod.* 23:276–80.

Wilson, C., Nomikos, G. G., Collu, M., and Fibiger, H. C. (1995). Dopaminergic correlates of motivated behavior: importance of drive. *J. Neurosci.* 15:5169–78.

Wilson, K. M., Sumners, C., Hathaway, S., and Fregly, M. J. (1986). Mineralocorticoids modulate central angiotensin II receptors in rats. *Brain Res.* 382:87–96.

Wingfield, J. C. (1992). Hormonal responses to removal of a breeding male in the cooperatively breeding white-browed sparrow weaver, *Plocepasser mahali*. *Horm. Behav.* 26:145–55.

Wingfield, J. C. (1993). Control of testicular cycles in the song sparrow, *Melospiza melodia melodia:* interaction of photoperiod and an endogenous program? *Gen. Comp. Endocrinol.* 92:388–401.

Wingfield, J. C. (1994). Control of territorial aggression in a changing environment. *Psychoneuroendocrinology* 19:709–21.

Wingfield, J. C., and Monk, D. (1992). Control and context of year-round territorial aggression in the non-migratory song sparrow *Zonotrichia melodia morphna. Ornis Scandinavica* 23:298–303.

Wingfield, J. C., Vleck, C. M., and Moore, M. C. (1992). Seasonal changes of the adrenocortical response to stress in birds of the Sonoran Desert. *J. Exp. Zool.* 264: 419–28.

Wingfield, J. C., Whaing, C. S., and Marler, P. (1994). Communication in vertebrate aggression and reproduction: the role of hormones. In: *Physiology of Reproduction,* 2nd ed., ed. E. Knobil and J. D. Neill. New York: Raven Press.

Winslow, J. T., Hastings, N., Carter, C. S., Harbaugh, C. R., and Insel, T. R. (1993). A role for central vasopressin in pair bonding in monogamous prairie voles. *Nature* 365: 545–8.

Wisor, J. P., and Takahashi, J. S. (1997). Regulation of the vgf gene in the golden hamster suprachiasmatic nucleus by light and by the circadian clock. *J. Comp. Neurol.* 378: 229–38.

Wolf, G. (1964). Sodium appetite elicited by aldosterone. *Psychonomic Sci.* 1:211–12.

Wolf, G. (1965). Effect of deoxycorticosterone on sodium appetite of intact and adrenalectomized rats. *Am. J. Physiol.* 208:1281–5.

Wolf, G. (1968). Projections of thalamic and cortical gustatory areas in the rat. *J. Comp. Neurol.* 132:510–30.

Wolf, G. (1969a). Effects of a mineralocorticoid antagonist on sodium appetite. Presented at the Seventh International Congress of Nutrition, Prague.

Wolf, G. (1969b). Innate mechanisms for regulation of sodium appetite. In: *Olfaction and Taste,* ed. C. Pfaffmann. New York: Rockefeller University Press.

Wolf, G. (1982). Refined salt appetite methodology for rats demonstrated by assessing sex differences. *J. Comp. Physiol. Psychol.* 96:1120–4.

Wolf, G., and Handel, P. J. (1966). Aldosterone induced sodium appetite: dose response and specificity. *Endocrinology* 78:1120–4.

Wolf, G., and Schulkin, J. (1980). Brain lesions and sodium appetite: an approach to the neurological analysis of homeostatic behavior. In: *Biological and Behavioral Aspects of Salt Intake,* ed. M. Kare. New York: Academic Press.

Wolkowitz, O. M. (1994). Prospective controlled studies of the behavioral and biological effects of exogenous corticosteroids. *Psychoneuroendocrinology* 19:237–55.

Wolkowitz, O. M., Rubinow, D., Doran, A. R., Breier, A., Berrettini, W. H., Kling, M. A., and Pickar, D. (1990). Prednisone effects on neurochemistry and behavior. *Arch. Gen. Psychiatry* 47:963–8.

Wollnik, F., and Turek, F. W. (1988). Estrous correlated modulations of circadian and ultradian wheel-running activity rhythms in Lewis/Fisher rats. *Physiol. Behav.* 43:389–96.

Wong, R., and Jones, W. (1978). Effects of aldactazide and DOCA injections on saline preference in gerbils (*Meriones unguiculatus*). *Behav. Biol.* 23:460–8.

Wood, R. I., Brabec, R. K., Swann, J. M., and Newman, S. W. (1992). Androgen and estrogen concentrating neurons in chemosensory pathways of the male Syrian hamster brain. *Brain Res.* 596:89–98.

Wood, R. I., and Newman, S. W. (1993a). Mating activates androgen receptor-containing neurons in chemosensory pathways of the male Syrian hamster brain. *Brain Res.* 614:65–77.

Wood, R. I., and Newman, S. W. (1993b). Intracellular partitioning of androgen receptor immunoreactivity in the brain of the male Syrian hamster: effects of castration and steroid replacement. *J. Neurobiol.* 24:925–38.

Wood, R. I., and Newman, S. W. (1995a). The medial amygdaloid nucleus and medial preoptic area mediate steroidal control of sexual behavior in the male Syrian hamster. *Horm. Behav.* 29:338–53.

Wood, R. I., and Newman, S. F. (1995b). Integration of chemosensory and hormonal cues is essential for mating in the male Syrian hamster. *J. Neurosci.* 11:7261–9.

Wood, R. I., and Newman, S. W. (1995c). Hormonal influence on neurons of the mating behavior pathway in male hamsters. In: *Neurobiological Effects of Sex Steroid Hormones,* eds. P. E. Micevych and R. P. Hammer, Jr. Cambridge University Press.

Woodmansee, K. B., Zabel, C. J., Glickman, S. E., Frank, L. G., and Keppel, G. (1991). Scent marking (pasting) in a colony of immature spotted hyenas (*Crocuta crocuta*): a developmental study. *J. Comp. Psychol.* 105:10–14.

Woods, S. C., Chavez, M., Park, C. R., Riedy, C., Kaiyala, K., Richardson, R. D., Figlewicz, D. P., Schwartz, M. W., Porte, D., Jr., and Seeley, R. J. (1996). The evaluation of insulin as a metabolic signal influencing behavior via the brain. *Neurosci. Biobehav. Rev.* 20:139–44.

Woods, S. C., Hutton, R. A., and Makous, W. (1970). Conditioned insulin secretion in the albino rat. *Proc. Soc. Exp. Biol. Med.* 133:965–8.

Woods, S. C., Lotter, E. C., McKay, L. D., and Porte, D, Jr. (1979). Chronic intracerebroventricular infusion of insulin reduces food intake and body weight of baboons. *Nature* 282:503–5.

Woods, S. C., Taborsky, G. J., Jr., and Porte, D., Jr. (1986). Central nervous system control of nutrient homeostasis. In: *Handbook of Physiology:* Sect. 1, *The Nervous System. Vol. 4: Intrinsic Regulatory Systems of the Brain,* ed. F. Plum. Bethesda, MD: American Physiological Society.

Woods, S. C., Vasselli, J. R., Kaestner, E., Szakmary, G. A., Milburn, P., and Vitiello, M. V. (1977). Conditioned insulin secretion and meal feeding in rats. *J. Comp. Physiol. Psychol.* 91:128–33.

Woodside, B., Leon, M., Attard, M., Feder, H. H., Siegel, H. I., and Fischette, C. (1981). Prolactin-steroid influences on the thermal basis for mother–young contact in Norway rats. *J. Comp. Physiol. Psychol.* 95:771–80.

Woodside, B., and Millelire, L. (1987). Self-selection of calcium during pregnancy and lactation in rats. *Physiol. Behav.* 39:291–5.

Woolley, C. S., Gould, E., Frankfurt, M., and McEwen, B. S. (1990). Naturally occurring fluctuation in dendritic spine density on adult hippocampal pyramidal neurons. *J. Neurosci.* 10:4035–9.

Woolley, C. S., and McEwen, B. S. (1993). Roles of estradiol and progesterone in regulation of hippocampal dendritic spine density during the estrous cycle in the rat. *J. Comp. Neurol.* 336:293–306.

Woolley, C. S., Weiland, N. G., McEwen, B. S., and Schwartzkroin, P. A. (1997). Estradiol increases the sensitivity of hippocampal ca1 pyramidal cells to NMDA receptor-mediated synaptic input: correlation with dendritic spine density. *J. Neurosci.* 17: 1848–59.

Wurtman, J. J. (1993). Depression and weight gain: the serotonin connection. *J. Affect. Disord.* 29:183–92.

Wurtman, R. J., and Ahdanova, I. (1995). Improvement of sleep quality by melatonin. *Lancet* 346:541–4.

Xu, Z., and Herbert, J. (1994). Regional suppression by water intake of c-fos expression induced by intraventricular infusions of angiotensin II. *Brain Res.* 659:157–68.

Yahr, P. (1995). Neural circuitry for the hormonal control of male sexual behavior. In: *Neurobiological Effects of Sex Steroid Hormones,* ed. P. E. Micevych, and R. P. Hammer, Jr. Cambridge University Press.

Yahr, P., and Jacobsen, C. H. (1994). Hypothalamic knife cuts that disrupt mating in male gerbils sever efferents and forebrain afferents of the sexually dimorphic area. *Behav. Neurosci.* 108:735–42.

Yalcinkaya, T. M., Silteri, P. K., Vigne, J.-L., Licht, P., Pavgi, S., Frank, L. G., and Glickman, S. E. (1993). A mechanism for virilization of female spotted hyenas in utero. *Science* 260:1029 31.

Yehuda, R. (1997). Sensitization of the hypothalamic-pituitary-adrenal axis in posttraumatic stress disorder. In: *Psychobiology of Posttraumatic Stress Disorder. Ann. N.Y. Acad. Sci.* 821.

Yehuda, R., Levengood, R. A., Schmeidler, J., Wilson, S., Guo, L. S., and Gerber, D. (1994). Increased pituitary activation following metyrapone administration in posttraumatic stress disorder. *Psychoneuroendocrinology* 21:1–16.

Yellon, S. M., and Goldman, B. D. (1984). Photoperiod control of reproductive development in the male Djungarian hamster (*Phodopus sungorus*). *Endocrinology* 114: 664–70.

Yellon, S. M., and Goldman, B. D. (1987). Influence of short days on diurnal patterns of serum gonadotrophins and prolactin concentrations in the male Djungarian hamster, *Phodopus sungorus. J. Reprod. Fertil.* 80:167–74.

Young, E. Z., and Akil, H. (1985). Corticotropin-releasing factor stimulation of adrenocorticotropin and β-endorphin release: effects of acute and chronic stress. *Endocrinology* 117:23–30.

Young, E. Z., and Akil, H. (1988). Paradoxical effect of corticosteroids on pituitary ACTH/β-endorphin release in stressed animals. *Psychoneuroendocrinology* 13:317–23.

Young, L. J., Greenberg, N., and Crews, D. (1991). The effects of progesterone on sexual behavior in male green anole lizards (*Anolis carolinensis*). *Horm. Behav.* 25:477–88.

Young, L. J., Lopreato, G. F., Horan, K., and Crews, D. (1994). Cloning and in situ hybridization analysis of estrogen receptor, progesterone receptor, and androgen receptor expression in the brain of whiptail lizards (*Cnemidophorus uniparens* and *C. inornatus). J. Comp. Neurol.* 346:1–13.

Young, L. J., Nag, P. K., and Crews, D. (1995). Regulation of estrogen receptor and progesterone receptor messenger ribonucleic acid by estrogen in the brain of the whiptail lizard (*Cnemidophorus uniparens*). *J. Neuroendocrinol.* 7:119–25.

Young, L. J., Winslow, J. T., Wang, Z., Gingrich, B., Guo, Q., Matzuk, M. M., and Insel, T. R. (1997). Gene targeting approaches to neuroendocrinology: oxytocin, maternal behavior, and affiliation. *Horm. Behav.* 31:221–31.

Young, P. T. (1949). Palatability versus appetite as determinants of the critical concentrations of sucrose and sodium chloride. *Comp. Psychol. Monogr.* 19:102.

Young, W. C., Dempsey, E. W., Hagquist, C. W., and Boling, J. H. (1939). Sexual behavior and sexual receptivity in the female guinea pig. *J. Comp. Psychol.* 27:49–68.

Youngren, O. M., El Halawani, M., Phillips, R. E., and Silsby, J. L. (1989). Effects of preoptic and hypothalamic lesions in female turkeys during a photoinduced reproductive cycle. *Biol. Reprod.* 41:610–17.

Youngren, O. M., El Halawani, M. E., Silsby, J. L., and Phillips, R. E. (1991). Intracranial prolactin perfusion induces incubation behavior in turkey hens. *Biol. Reprod.* 44:425–31.

Youngren, O. M., El Halawani, M. E., Silsby, J. L., and Phillips, R. E. (1993). Effect of reproductive condition on luteinizing hormone and prolactin release induced by electrical stimulation of the turkey hypothalamus. *Gen. Comp. Endocrinol.* 89:220–8.

Zamir, N., Skoifitsch, G., Eskay, R. L., and Jacobowitz, D. M. (1986). Distribution of immunoreactive atrial natriuretic peptides in the central nervous system of the rat. *Brain Res.* 365:105–11.

Zardetto-Smith, A. M., Beltz, T. G., and Johnson, A. K. (1994). Role of the central nucleus of the amygdala and bed nucleus of the stria terminalis in experimentally-induced salt appetite. *Brain Res.* 645:123–34.

Zardetto-Smith, A. M., and Gray, T. S. (1995). Catecholamine and NPY efferents from the ventrolateral medulla to the amygdala in the rat. *Brain Res. Bull.* 38:253–60.

Zarrow, M. X., Farooq, A., and Denenberg, V. H. (1963). Maternal behavior in the rabbit: endocrine control of maternal behavior nest-building. *J. Reprod. Fertil.* 6:375–83.

Zhang, D. M., Epstein, A. N., and Schulkin, J. (1993). Medial region of the amygdala involvement in adrenal-steroid induced sodium appetite. *Brain Res.* 600:20–6.

Zhang, Y., Proenca, R., Mattei, M., Barone, M., Leopold, L., and Friedman, J. M. (1994). Positional cloning of the mouse obese gene and its human homologue. *Nature* 372:425–32.

Zhang, D.-M., Stellar, E., and Epstein, A. N. (1984). Together intracranial angiotensin and systemic mineralocorticoid produce avidity for salt in the rat. *Physiol. Behav.* 32: 677–81.

Zhou, L., Blaustein, J. D., and DeVries, G. J. (1994). Distribution of androgen receptor immunoreactivity in vasopressin and oxytocin immunoreactive neurons in the male rat. *Endocrinology* 134:2622–7.

Zhuang, X., Silverman, A.-J., and Silver, R. (1993). Reproductive behavior, endocrine state, and the distribution of GnRH-like immunoreactive mast cells in dove brain. *Horm. Behav.* 27:283–95.

Ziegler, T. E., Wegner, F. H., and Snowdon, C. T. (1996). Hormonal responses to parental and nonparental conditions in male cotton-top tamarins, *Saguinus oedipus,* a New World primate. *Horm. Behav.* 30:287–97.

Zoli, M., Ferraguti, F., Gustafsson, J. A., Toffano, G., Fuxe, K., and Agnati, L. F. (1991). Selective reduction of glucocorticoid receptor immunoreactivity in the hippocampal formation and central amygdaloid nucleus of the aged rat. *Brain Res.* 545:199–207.

Zucker, I. (1985). Pineal gland influences period of circannual rhythms of ground squirrels. *Am. J. Physiol.* 249:R111–14.

Zucker, I. (1988). Neuroendocrine substrates of circannual rhythms. In: *Biological Rhythms and Mental Disorders,* ed. D. J. Kupfer, T. H. Monk, and J. D. Barchas. New York: Guilford Press.

Zucker, I., Boshes, M., and Dark, J. (1983). Suprachiasmatic nuclei influence circannual and circadian rhythms of ground squirrels. *Am. J. Physiol.* 244:R472–80.

Zucker, I., Fitzgerald, K. M., and Morin, L. P. (1980). Sex differentiation of the circadian system in the golden hamster. *Am. J. Physiol.* 7:R97–101.

Zucker, I., and Licht, P. (1983). Seasonal variations in plasma luteinizing hormone levels of gonadectomized male ground squirrels (*Spermophilus lateralis*) *Biol. Reprod.* 29:278–83.

Name Index

Subject Index

CHESTER COLLEGE LIBRARY